THE ENGINEERING DESIGN OF SYSTEMS

THE ENGINEERING DESIGN OF SYSTEMS

MODELS AND METHODS

Fourth Edition

DENNIS M. BUEDE

ITA International LLC
Reston, Virginia

WILLIAM D. MILLER

ITA International LLC
Berkeley Heights, New Jersey

Library of Congress Cataloging-in-Publication Data

Names: Buede, Dennis M., author. | Miller, William D., 1949- author.
Title: The engineering design of systems : models and methods / Dennis M.
    Buede, ITA International LLC, William D. Miller, ITA International LLC.
Description: Fourth edition. | Hoboken, New Jersey : John Wiley & Sons,
    Inc., [2024] | Includes index.
Identifiers: LCCN 2023057866 (print) | LCCN 2023057867 (ebook) | ISBN
    9781119984016 (hardback) | ISBN 9781119984023 (adobe pdf) | ISBN
    9781119984030 (epub)
Subjects: LCSH: Systems engineering. | Engineering design. | System design.
Classification: LCC TA168 .B83 2024 (print) | LCC TA168 (ebook) | DDC
    620.001/171–dc23/eng/20240102
LC record available at https://lccn.loc.gov/2023057866
LC ebook record available at https://lccn.loc.gov/2023057867

Cover Design: Wiley
Cover Image: © William D. Miller. Ralf Hiemisch/Getty Images

Set in 10/12pt TimesLTStd by Straive, Chennai, India

SKY10068942_030524

Dennis: *In Memory of my Mother and Father*

Bill: *To John Lewis, Mal Buchner, Jake Schaefer,*
and Les Kleinberg who inspired and
nurtured this nascent systems engineer

Contents

Preface

This book is meant to be a basic text for courses in the engineering design of systems at both the upper-division undergraduate and beginning graduate levels. The book is the product of many years of consulting on numerous portions of the system development process, research into the use of systems engineering in industry, and six years of developing a course on the engineering design of systems. During the development of this book, I found that many engineers did not understand systems engineering. Even those who do may not have a good perspective on a complete and unified process for engineering a system. The desire to suppress the number of decisions being made during design is quite strong in most engineers. While engineers have learned modeling throughout their academic lives, and most have developed models during the practice of engineering, very few engineers working on systems are knowledgeable of the modeling techniques required in systems engineering. In addition, most engineers are not aware of methods for using models during the systems engineering process. As a result, I adopted the following themes in formulating this book:

1. Defining the design problem in systems engineering is one of several keys to success and can be approached systematically using engineering techniques.
2. The design problem in systems engineering is defined in terms of requirements. These requirements evolve from a high-level set of mission and stakeholders' requirements to detailed sets of derived requirements.
3. The design process will fail if the requirements are defined too narrowly, leaving little if any room for design decisions and raising the possibility that no feasible solution exists. The design problem should be well defined and decision rich.
4. For the design problem to be well defined, the evolving sets of requirements must be complete (none missing), consistent (no contradictions), correct (valid for an acceptable solution), and attainable (an acceptable solution exists). While it is not possible at this time to state requirements mathematically and prove these properties, it is possible to develop mathematical and heuristic representations of the design problem to assist in evaluating the presence of these properties.
5. These characteristics of the requirements will not be achieved if scenarios defining how the system will be used are not elaborated in detail, the interactions among the system and other

systems are not defined, and the stakeholders' objectives are not understood. Each of these requires a different kind of modeling to be successful.

6. The design problem is not likely to be well defined if the requirements do not address every relevant phase of the system's life cycle.

7. The design problem is not likely to be well defined if the requirements do not contain stakeholder preferences for comparing feasible designs against each other.

8. The keys to understanding many of the modeling techniques for developing requirements, defining architectures, and deriving requirements are found in discrete mathematics: set theory, relations and functions, and graph theory.

9. Integration requires a well-defined design, including a design of the qualification process for verification, validation, and acceptance. A systematic process of design provides all the necessary inputs for defining the qualification process.

10. Early validation of the evolution of the definition of the design problem needs to be pursued vigorously to ensure that the definition of the design problem does not change as the problem is defined in greater detail.

11. Qualification of the system is the key issue in integration. Qualification includes verification and validation of both the requirements and the system design, followed by the stakeholders' acceptance. There are many methods for qualifying the system; these methods must be chosen judiciously.

The successful qualification also requires that decisions about what should be tested be made in a systematic way that balances the two conflicting objectives of not wasting resources and obtaining stakeholder acceptance.

The above themes for the methods and models in this book are fundamental to the engineering of systems that have been validated in use beginning in the twentieth century CE and are independent of and realizable using the several systems modeling standards introduced in the twenty-first century CE including the Object Management Group (OMG) Systems Modeling Language (SysML®), ISO/PAS 19450 Object-Process Methodology (OPM), and the Lifecycle Modeling Organization (LMO) Lifecycle Modeling Language (LML).

The major changes for the fourth edition reflect the emphasis on SysML to visualize system models while still keeping legacy IDEF0 diagrams to visualize systems engineering process modeling. Chapter 1 was rewritten to address the updates that were made throughout the original chapters. As SysML2 becomes available as a standard and is implemented in software tools, the SysML diagrams in the book will be revised as SysML 2.0 diagrams and will be available on the Wiley companion website. The chapter on graphical modeling techniques was removed from the book but will still be available on the Wiley companion website for this book.

The book is divided into three major parts: (1) Introduction, Overview, and Basic Knowledge, (2) Design and Integration Topics, and (3) Supplemental Topics. The first part introduces the issues associated with the engineering of a system. Next, an overview of the engineering process is provided so that readers will have a context for the more detailed material. Finally, the basic knowledge needed for the core material is presented. Homework problems are provided at the end of each chapter.

Chapter 1 defines a system, systems engineering, the life cycle of a system and then introduces systems engineering processes. This material sets the stage for the details that follow.

Chapter 2 provides an overview of the details that are to come by presenting a number of basic concepts; these concepts include an operational concept, objectives, requirements, functions, items, components, interfaces verification, validation, and acceptance. The relations among these concepts are also addressed.

Chapter 3 provides an overview of modeling and the types of modeling needed in engineering systems. Modeling methods associated with SysML are then introduced and described. While IDEF0 is not part of SysML, this topic has been kept in Chapter 3 as an important part of the modeling concepts described in this book.

Chapter 4 presents basic discrete mathematics. The purpose of discrete mathematics is to demonstrate the mathematical rigor for which systems engineering must strive and to provide a language with which we can discuss key issues. Examples of such important concepts are the distinction between a relation and a function and why this is critical for engineering a system, a partition of the elements of a set that can be applied to many systems engineering concepts (e.g., requirements), and partial orders of functional execution.

Chapter 5 extends the discussion of discrete mathematics to graph theory, so that the graphical communication structures commonly used in the engineering of systems can be seen to have substantial problems as rigorous mathematical representations. On the other hand, the difficult concepts in Chapter 4 can be effectively represented with graphs for analysis and communication.

Part 2 covers the critical material required to understand the major elements needed in the engineering design of any system: requirements, architectures (functional, physical, and allocated), interfaces, and qualification.

Requirements development is approached as a systematic process in Chapter 6. This systematic process involves the definition of an operational concept of the system (including usage scenarios), a description of the involvement of the system with other systems, and an objectives hierarchy of the stakeholders across all phases of the system's life cycle. A partition of requirements is employed to discuss the systematic approach to defining requirements.

Definitions of the functional, physical, and allocated architectures are provided as well as the detailed methods for developing these architectures in Chapters 7–9. Chapter 7 begins with several definitions that are needed to enable a meaningful discussion of the topic. The notion of a functional architecture is defined. An emphasis is placed on process modeling in Chapter 7. However, additional material is presented in Chapter 3 and the graphical modeling techniques on the companion website on data and behavioral modeling methods, as well as other approaches for process modeling. (This material can be used while discussing Chapters 7–9.) Modeling approaches for partitioning a function into segments are discussed. Key topics are feedback and control within the functional decomposition and evaluating the architecture for shortfalls and overlaps. Chapter 7 also addresses the functionality needed for error detection and recovery as well as tracing the input/output requirements to functions and items.

Chapter 8 introduces the distinction between generic and instantiated physical architectures. The morphological box is used to demonstrate the generation of multiple instantiated physical architectures. The graphical representation of the physical architecture is discussed along with notions of centralized, decentralized, and distributed architectures. Finally, fault-tolerant architectures are described.

Chapter 9 defines the allocated architecture and discusses the allocation of functions to components, the tracing and derivation of requirements, the analysis of activation and control structures, and the conduct of various analyses (risk, performance, and trade-off).

Chapter 10 characterizes interfaces, discusses the functions associated with interfaces in several contexts (communications systems and software design), describes interface architectures, and discusses interface design as it impacts system performance as part of the design process.

Finally, qualification of the system (Chapter 11) during integration requires an understanding of the stakeholders' needs and the qualification methods that are typically used. Deciding what to test and how to test it is critical in this phase of the development process.

Chapter 12 provides a comprehensive review of Chapters 6–11, which is a very useful way to end a quarter or semester by tying everything together with a simple system – an automated soda machine.

All these topics in Chapters 6–12 are addressed in a rigorous and systematic manner, consistent with the general practical application of systems engineering in industry.

Homework exercises are provided on each of these topics from Part 2 for several real but simple systems that are familiar to all students: an automatic teller machine, an airbag, and the OnStar system of Cadillac. A case study is available on the web to give the students a sample of the solutions to the homework. Readers are encouraged to access and apply commercial system engineering software products to enhance the conduct of these homework exercises and the educational mission of this book.

Finally, three additional key topics are introduced in the third part: decision analysis, system science and analytics, and the value of systems engineering. Chapter 13 presents ideas on the value of systems engineering. This chapter identifies six key value propositions that should interest most stakeholders: problem and solution discovery, communication, identification and solution of show stoppers as early as possible, error reduction in both the product and design systems, risk reduction in both the product and design systems and continuous improvement of the design system.

Chapter 14 presents the key topics of decision analysis as an integrative way of supporting the many decisions that are part of the design and integration of a system. These decision analytic topics include the development and quantification of values (objectives, value functions, and trade-offs), and the modeling of uncertainty regarding facts.

Chapter 15 addresses the bedrock of the science of systems in general systems theory, systems science, natural systems, cybernetics, and systems thinking. Engineering successful systems accounts for the pervasiveness of uncertainty throughout the system lifecycles. Successful engineering is critically dependent on system analytics: quantitative characterization of systems, system dynamics, constraint theory, and approximate methods. Illustrative examples are aircraft performance, flight control system stability, elevator system performance, soda machine performance, and Fermi's classic problem to estimate the number of piano tuners in Chicago. The chapter also introduces the challenges of artificial (or augmented) intelligence and machine learning (AI/ML) embedded in systems and approaches to ensure desired behaviors and mitigate undesired behaviors.

The homework problems and the case study of the elevator are defined with the express purpose of having the student demonstrate the level of understanding necessary to perform the engineering activities described in the book. In developing these homework exercises, I have taken the position that demonstrating an ability to discuss how to do systems engineering is a necessary but not a sufficient level of understanding. The GENESYS software (that is appropriate for use with this book and is available as a downloadable package via the web: http://www.vitechcorp.com) takes the tedium out of performing these systems engineering activities as well as reinforcing the basic concepts behind the activities. The case material related to an elevator system can be downloaded from the companion website.

Several colleagues have provided many useful comments and suggestions. We wish to thank Kathryn Laskey, Bob Kenley, Mike Pennotti, and Tony Barrese.

Several students and teaching assistants have contributed to sections of these notes. Cathy Brown provided a substantial extension of the requirements for the elevator case study. John Van Ormer extended the physical architecture of the elevator. Jahan Araghi extended my initial case study on the automatic teller machine (ATM) as part of his master's project. Tong Zhang and Parham Pasha provided some examples of sets, relations, and graphs. Christine Salter provided extensive support in addressing topics that needed revision, developed solutions for homework problems, and provided solution material for the OnStar and ATM problems. Several student groups provided material on which the airbag case is based. Meg Giordana and Barry Liner provided extensive comments on the qualification material. Tim Parker developed two case studies for use in Chapters 8 and 9: the FBI Fingerprint Identification System and the Wide-Area Augmentation System of the Federal Aviation Administration. Steve Charbonneau provided interesting insights about state charts as part of his MS thesis. The SYST 520 class at George Mason University during the spring of 1998 provided many extensive and useful comments on an early draft of the first edition.

We wish to thank all these individuals, as well as many others with whom we have conversed on these topics, for stimulating me to complete this effort.

One of the most difficult aspects of writing this book has been to decide which material to include and which to leave out. There is still a great deal more to be said on the topics covered in this book and on some additional topics that were not included. More importantly, there is still a great deal more to discover, at least on our part. The following Wiley Companion Website contains some material that we have dropped from the third edition and some material that will be posted as the second version of SysML is released.

DENNIS M. BUEDE
Reston, Virginia

WILLIAM D. MILLER
Berkeley Heights, New Jersey
August 2023

About the Companion Website

This book is accompanied by a companion website

www.wiley.com/go/engineeringdesignofsystems4e

Once registered at the website, the student will receive access to:

- Elevator Case Study
- Power Point Slides
- Related Links
- Solutions Manual for Instructors Only

Part **1**

Introduction, Overview, and Basic Knowledge

Chapter 1

Introduction to Systems Engineering

1.1 INTRODUCTION

A *system* is commonly defined to be "a collection of hardware, software, people, facilities, and procedures organized to accomplish some common objectives." The stakeholders in the system hold these objectives. Never forget that the system being addressed by one group of engineers is the subsystem of another group and the supersystem of yet a third group. The objective of the engineers for a system is to provide a system that accomplishes the primary objectives set by the stakeholders, including those objectives associated with the creation, production, and disposal of the system. To accomplish this engineering task, the engineers must identify the system's stakeholders throughout the system's life cycle and define the objectives of all of these stakeholders. These objectives typically address the triad of cost, schedule, and performance – cheaper, faster, and better.

A *system of systems* is a "set of systems or system elements that interact to provide a unique capability that none of the constituent systems can accomplish on its own" [ISO/IEC/IEEE 21841, 2019; Maier, 1998; Dahmann, 2015; DoD, 2010]. A system of systems (SoS) is characterized by the managerial and operational independence of its constituent systems. A SoS exhibits emergent capabilities beyond the mere aggregation of the constituent systems, just as a system has its own emergent capabilities. SoS categories are acknowledged, collaborative, directed, and virtual. Acknowledged SoS fall between directed and collaborative SoS and have recognized objectives, a designated manager, and resources; changes in the constituent systems are based on collaboration between the SoS and the system, for example, an aircraft carrier battle group of several ships, an embarked air wing, and other assets. Collaborative SoS have component systems interacting more or less voluntarily to fulfill agreed-upon central purposes, for example, the Internet, which originated as a directed SoS but evolved. The Internet Engineering Task Force (IETF) works out standards but has no power to enforce them. Agreements among the central players on service provision an rejection provide what enforcement mechanism there is to maintain standards. Directed SoS are created and managed to fulfill specific purposes, and the constituent systems are subordinated to the SoS; for example, US ballistic missile defense managed by the Missile Defense Agency (MDA) directly reporting

The Engineering Design of Systems: Models and Methods, Fourth Edition. Dennis M. Buede and William D. Miller
© 2024 John Wiley & Sons, Inc. Published 2024 by John Wiley & Sons, Inc.
Companion website: www.wiley.com/go/engineeringdesignofsystems4e

to the US Department of Defense (DoD). Virtual SoS lack a central management authority and a centrally agreed-upon purpose for the SoS, for example, national economies.

A major characteristic of the engineering of systems is the attention devoted to the entire life cycle of the system. This life cycle has been characterized as "birth to death" and "lust to dust." That is, the *life cycle* begins with the gleam in the eyes of the users or stakeholders, is followed by the definition of the stakeholders' needs by the systems engineers, includes developmental design and integration, goes through production and operational use, usually involves refinement, and finishes with the retirement and disposal of the system. Ignoring any part of this life cycle while engineering the system can lead to sufficiently negative consequences, including failure at the extreme. In particular, developing a system that has not adequately addressed the stakeholders' needs leads to failures such as the "highway to nowhere" near San Francisco, which was stopped by political pressure brought to bear by homeowners on the surrounding hills overlooking the bay. The view of the bay that these homeowners enjoyed and thought was an associated right of the property they owned would have been blocked by the highway. Similar commercial failures that did not consider the needs of the stakeholders in sufficient detail include the personal computers IBM PC Jr. and Apple LISA. This is not to say that the adherence to the methods and models put forth in this book or any other will guarantee success or even the absence of failure. Rather, the methods and models proposed here do attend to the entire life cycle of the system and provide a process that makes sense, can be tailored to various levels of detail as dictated by the complexity of the system being addressed, and attend to all the details that many engineers during years of practice in systems engineering have determined to be useful.

The concepts of design and integration are critical to the methods addressed in this chapter and the book. The word *design* is used by many professions (artists, architects, all disciplines of engineering) and is claimed by each.

The American Heritage Dictionary [Berube, 1991] defines design as:

> de-sign (di-zin') v. - signed, - sign·ing, - signs. —tr. 1. To conceive in the mind; invent: *designed his dream vacation*. 2. To form a plan for: *designed a marketing strategy for the new product*. 3. To have a goal or purpose; intend. 4. To plan by making a preliminary sketch, outline, or drawing. 5. To create or execute in an artistic or highly skilled manner. –intr. 1. To make or execute plans. 2. To create designs. –n. 1. A drawing or sketch. 2. The invention and disposition of the forms, parts, or details of something according to a plan. 3. A decorative or artistic work. 4. A visual composition; pattern. 5. The art of creating designs. 6. A plan; project. 7. A reasoned purpose; intention. 8. Often designs. A sinister or hostile scheme: *He has designs on my job* …

All but the third and eighth definitions for noun usage will apply at various times in this book. *Design* during the engineering of a system, as discussed in this book, is the preliminary activity that has the purpose of satisfying the needs of the stakeholders; it begins in the mind of the lead engineer but has to be transformed into models employing visual formats in a highly skilled manner for success to be achieved. While this book addresses the engineering methods and models used during the design process, there is always an element of artistry that is required for the design process and the system to be successful.

Integration brings all the detailed elements of the overall design together through a process of testing (or qualification) to achieve a valid system for meeting the needs of the stakeholders. Engineers of appropriate disciplines perform integration according to the specifications defined by the design of the system's engineers. The integration process involves testing or qualification of both the elements of the system and the system itself to ensure that the system meets the ultimate needs of the stakeholders.

This chapter first provides an overview of the issues and processes associated with the engineering of a system. This overview addresses the phases of the system's life cycle, describes the importance of performing the engineering of a system well, provides a definition for the engineering of a system, introduces the key process model for the engineering of a system called the Vee model, describes the richness of decisions that are inherent in the engineering process, and discusses the diversity of expertise required for this engineering process. Section 1.3 describes process models that have been adopted by the software engineering community. Architectures play a key role in the engineering of systems and are introduced next. Requirements, Section 1.5, play a major role in the engineering of a system because they serve the role of defining the engineering design problem and capturing the key information needed to describe design decisions. The life cycle of the system is next examined in more detail. Finally, the Vee model for engineering a system is described in more detail.

The key method addressed in this chapter is the process used to perform the engineering of systems. Supplementing this discussion of the engineering method are discussions of the key concepts needed to understand the method at an introductory level. This method is presented as a process model; models and modeling are discussed in detail in Chapter 3, so the reader is asked to accept the notion of the process discussion as a discussion of a model until more detail on models can be provided in Chapter 3.

1.2 OVERVIEW OF THE ENGINEERING OF SYSTEMS

The development process in systems engineering is commonly viewed [Forsberg and Mooz, 1992; Lake, 1992] as a decomposition (or design) process followed by a recomposition (or integration) process (see Sidebar 1.1). During the decomposition process, the stakeholders' requirements are analyzed and defined in engineering terms and then partitioned into a set of specifications (or specs) for several segments, elements, or components. It is critical that this design process be *broad in perspective* so that nothing is left out and every contingency is considered. Systems engineers must be "big picture" people. Depth is only achieved by many iterations through the design process, as many as are needed, until the system's specifications are sufficiently detailed for individual configuration items (CIs) to be built or purchased. This design process defines what the system must do, how well the system must do it, and how the system should be tested to verify and validate the system's performance. To do this, the systems engineers must maintain a very clear focus on the objectives that the system's stakeholders (users, owners, manufacturers, maintainers, trainers, etc.) have defined for the system.

One of many possible representations of the life cycle of a system is shown in Figure 1.1, beginning with the identification of the need for the system and progressing through the retirement of the system. Some of the phases of the life cycle are accomplished in parallel, as the diagram tries to depict; exactly which phases occur in parallel depends upon the type of system, the organization, and the context. For additional details, see Driscoll [2007].

As shown in Figure 1.1, the design includes the preliminary system design as well as parts of the identification of need and concept definition. Parts of the identification of need and concept definition include the development of a basic idea and the first embodiment of the idea; these two initial activities are often called invention and are usually not part of the engineering of a system. The invention has a heavy technological and scientific focus. The last portions of the identification of need and concept design phases, plus preliminary system design, address the initial or follow-on commercialization of the idea based upon a specific statement of stakeholders' needs.

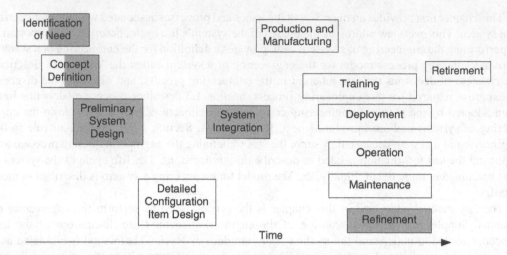

FIGURE 1.1   Phases of the system life cycle.

formed a company (R-W) in 1954 to perform systems engineering for the Air Force's ICBM program. In 1956, R-W and the Air Research and Development Command (under the direction of Bernard Schriever) reached a legal definition of systems engineering: (1) The solution of interface problems among all weapon system subsystems to insure technical and schedule compatibility of the systems as a whole. (2) The surveillance over detailed subsystems and overall weapon design to meet Air Force required objectives. (3) The establishment and revision of program milestones and schedules and monitoring of contractor progress in maintaining schedules, consistent with sound technical judgment and rapid advancement of the state of the art [Johnson, 2002]. R-W later became TRW.

Hall [1962] asserts that the first attempt to teach systems engineering as we know it today came in 1950 at MIT by George W. Gilman, Director of Systems Engineering at Bell [1953]. The first book on Systems Engineering was written by Goode and Machol in 1957, titled *System Engineering – An Introduction to the Design of Large-Scale Systems*.

Hall [1962] defined systems engineering as a function with five phases: (1) system studies or program planning; (2) exploratory planning, which includes problem definition, selecting objectives, systems synthesis, systems analysis, selecting the best system, and communicating the results; (3) development planning, which repeats phase 2 in more detail; (4) studies during development, which includes the development of parts of the system and the integration and testing of these parts; and (5) current engineering, which is what takes place while the system is operational and being refined.

The RAND Corporation was founded in 1946 by the United States Air Force and created *systems analysis*, which is certainly an important part of systems engineering.

The Department of Defense entered the world of systems engineering in the late 1940s with the initial development of missiles and missile defense systems [Goode and Machol, 1957].

Paul Fitts addressed the allocation of the system's functions to the physical elements of the system in the late 1940s and early 1950s [Fitts, 1951].

There is a special bibliography on the Wiley website for this book devoted to historical references.

The products of the design process serve as inputs to the hardware and software design of detailed CI design. The CIs then reenter the systems engineering process during system integration for integration testing, verification, and validation. Further adjustments to the design occur during the refinement phase. The life cycle phases associated with the engineering of the system are shaded in Figure 1.1. The term *concurrent engineering* simply means that the systems engineering process should be done with all of the phases (and their associated requirements) of the system life cycle in mind [Prasad, 1996]. This notion of concurrent engineering is a key concept addressed in this book.

The importance of systems engineering is highlighted by examining a generally accepted relationship between the phases of the system life cycle and the commitment versus the incursion of costs. The time associated with the system's life cycle is plotted on the x-axis; note that the time increments are notional and should not be interpreted as equal to the relative length of the four stages being addressed. See Prang [1992] for an illustration based on computer boards. (Prang is also referenced in Schelber [1995].) Figure 1.2 shows the major phases of the system life cycle on the horizontal axis. The curves represent the cost committed, based upon engineering design decisions, and the cost incurred, based upon actual expenditures. As can be seen, about 80% of the cost of the system is committed by the end of design and integration, while only about 20% of the actual cost for the system has been spent. Obviously, mistakes made in the front end of the system life cycle can have substantially negative impacts on the total cost of the system and its success with users and bill payers.

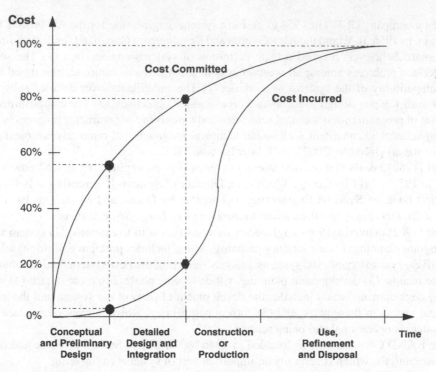

FIGURE 1.2 Cost commitment and incursion in the system life cycle.

There have been many definitions of systems engineering put forward since the 1950s when systems engineering became a profession. Table 1.1 provides several of these definitions. There are two important trends to note over the 20-year span of these definitions. First, the role of management in the systems engineering process is made explicit in the definitions from the 1990s. Second, the three pillars of engineering success (cost, schedule, and technical performance) from the 1970s evolved into concerns over the life cycle, namely concurrent engineering.

The American Heritage Dictionary [Berube, 1991] defines engineering as:

> *The application of scientific and mathematical principles to practical ends such as the design, construction, and operation of efficient and economical structures, equipment, and systems.*

The following definitions of engineering and the engineering of systems are adopted here:

Engineering: discipline for transforming scientific concepts into cost-effective products through the use of analysis and judgment.

Engineering of a System: engineering discipline that develops, matches, and trades off requirements, functions, and alternate system resources to achieve a cost-effective, life-cycle-balanced product based upon the needs of the stakeholders.

INCOSE has formally defined model-based systems engineering:

Model-Based Systems Engineering (MBSE): the formalized application of modeling to support system requirements, design, analysis, verification, and validation activities beginning in the

TABLE 1.1   Definitions of Systems Engineering

Source	Definitions of Systems Engineering
Mil-Std 499A [1974]	The application of scientific and engineering efforts to: (1) transform an operational need into a description of system performance parameters and a system configuration through the use of an iterative process of definition, synthesis, analysis, design, test, and evaluation; (2) integrate related technical parameters and ensure compatibility of all related, functional, and program interfaces in a manner that optimizes the total system definition and design; (3) integrate reliability, maintainability, safety, survivability, human, and other such factors into the total technical engineering effort to meet cost, schedule, and technical performance objectives.
Sailor [1990]	Both a technical and management process; the technical process is the analytical effort necessary to transform an operational need into a system design of the proper size and configuration and to document requirements in specifications; the management process involves assessing the risk and cost, integrating the engineering specialties and design groups, maintaining configuration control, and continuously auditing the effort to ensure that cost, schedule, and technical performance objectives are satisfied to meet the original operational need.
Sage [1992]	The design, production, and maintenance of trustworthy systems within cost and time constraints.
Forsberg and Mooz [1992]	The application of the *system analysis and design process* and the *integration and verification process* to the logical sequence of the *technical aspect of the project life cycle.*
Wymore [1993]	The intellectual, academic, and professional discipline, the primary concern of which is the responsibility to ensure that all requirements for a bioware/hardware/software system are satisfied throughout the life cycle of the system.
Mil-Std 499B draft [1993a]	An interdisciplinary approach encompasses the entire technical effort to evolve and verify an integrated and life-cycle-balanced set of system people, product, and process solutions that satisfy customer needs. Systems engineering encompasses: (1) the technical efforts related to the development, manufacturing, verification, deployment, operations, support, disposal of, and user training for system products and processes; (2) the definition and management of the system configuration; (3) the translation of the system definition into work breakdown structures; and (4) development of information for management decision-making.
INCOSE [2019]	Systems Engineering *is a transdisciplinary and integrative approach to enable the successful realization, use, and retirement of engineered systems, using systems principles and concepts, and scientific, technological, and management methods.*
	The terms "engineering" and "engineered" are used in their widest sense: "the action of working artfully to bring something about." "Engineered systems" may be composed of any or all of people, products, services, information, processes, and natural elements.

conceptual design phase and continuing throughout development and later life cycle phases [INCOSE, 2007].

Definitions for digital engineering, digital threads, and digital twins are emerging:

Digital Engineering (DE): an integrated digital approach that uses authoritative sources of systems' data and models as a continuum across disciplines to support life cycle activities from concept through disposal [ODASD (SE), 2017].

Digital Thread: (1) the use of digital tools and representations for design, evaluation, and life cycle management. (2) A data-driven architecture that links together information generated from across the product life cycle and is envisioned to be the primary or authoritative data and communication platform for a company's products at any instance of time. (3) More narrowly, the lowest level design and specification for a digital representation of a physical item.

Digital Twin: (1) a related yet distinct concept to digital engineering. The digital twin is a high-fidelity model of the system that can be used to emulate the actual system [SEBoK Editorial Board, 2023]. (2) An integrated multiphysics, multiscale, probabilistic simulation of an as-built system, enabled by a Digital Thread, that uses the best available models, sensor information, and input data to mirror and predict activities/performance over the life of its corresponding physical twin [DoD, 2018].

A DE ecosystem is an interconnected infrastructure, environment, and methodology that enables the exchange of digital artifacts from an authoritative source of truth across engineering disciplines. MBSE is a subset of DE. Figure 1.3 shows the design and integration process as a "Vee" with the emphasis of this model of the engineering process for a system being on the activities that the engineers perform. The left or decomposition side of the Vee coincides with the three phases at

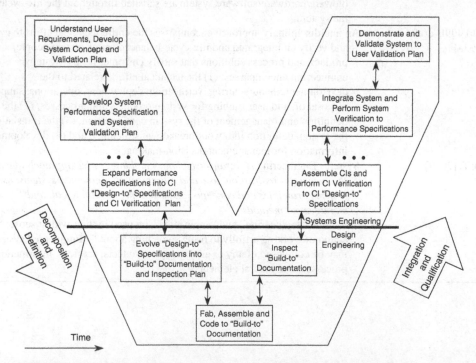

FIGURE 1.3 Systems engineering "Vee." (Adapted from Forsberg and Mooz [1992].)

the beginning of the life cycle from Figure 1.1. Time proceeds from left to right in Figure 1.3, just as it did in Figure 1.1. The process is initiated at the top left of the Vee with the definition of the operational need of the stakeholders. The focus of the decomposition and definition process (or design) is the movement from an operational need to system-level requirements to specifications for each component to the specifications (or specs) for each CI. Since time is moving from left to right in Figure 1.3, parallel work on high- and low-level design activities is not only permitted but encouraged. The iterative nature of this design process, from high-level issues such as stakeholders' requirements to low-level issues such as component and CI design, is accomplished by moving vertically in the Vee over short increments of time. This vertical movement during the design process is critical to success and has been observed in studies of expert designers [Guindon, 1990]. Note, this Vee model does not emphasize the interaction with the stakeholders even though that interaction is assumed to occur in order to enable the engineering processes depicted in the Vee model.

The horizontal line, drawn just under the middle intersection of the Vee in Figure 1.3, depicts the hand off of the final products of the design process, the CI specs, to the discipline (or design) engineers, those engineers whose orientation is electrical, mechanical, chemical, civil, aerospace, computer science, and the like and whose job it is to produce a physical entity. This dividing line can be drawn higher or lower to signify decreasing or increasing overlap between design and integration activities. As the dividing line is drawn in Figure 1.3, the sloping lines of the middle portion of the Vee can be extended until they meet the dividing line, with the resulting very modest overlap between design and integration. If the dividing line is raised above the intersection of the sloping lines of the Vee, there would be no intersection of design and integration. This complete separation of design and integration is often sought in practice to enhance contractual relationships between procurer and supplier of the system; however, this separation negatively impacts the schedule and cost associated with the development of the system. There is significant integration and qualification activity that should take place during design, as is discussed in Chapter 11. In many systems engineering activities, the horizontal dividing line between systems engineering and the discipline of engineers is drawn significantly lower than shown in Figure 1.3.

The right-hand side of the Vee depicts the integration and qualification activities of the engineering of a system. Integration involves the assembly of the CIs into components, the assembly of lower-level components into higher-level components, and the assembly of high-level components into the system. All this assembly involves testing (or qualification) of the newly assembled system elements to determine whether the assembled element meets the set of requirements (or spec) that the design phase had established for that element; this qualification is called verification. Finally, after the system is verified against the system requirements, the system must be validated. After validation, the stakeholders determine whether the system is acceptable. Naturally, there are problems throughout this process that require modifications to be made either to the design of the elements of the system or to the requirements that were developed during the design. Recall that time is running from left to right in Figure 1.3; the Vee process allows for the low level of verification of CIs to happen in parallel with some high-level validation and even acceptance activities.

A sample of the movement from operational need to CI specs is given for a race car in Table 1.2. The first column states the operational need or mission requirement: Win the Indianapolis 500. Associated with this need are stakeholders' requirements concerning the pretrial average speed and the average speed during the race with the expected number of yellow flags and pit stops (note the numbers in Table 1.2 are notional and are not accurate reflections of race conditions). System-level requirements can then be derived that are more meaningful during engineering. As an example, the key system-level requirement involves the g–g space of a vehicle [Milliken and Milliken, 1995]. Race cars, when driven by experienced drivers, are always changing velocity in speed or direction. (Recall that speed is the velocity you are traveling in your direction of travel. But when traveling around a curve, you also have a component of velocity perpendicular to your direction of travel.)

TABLE 1.2   Race Car Example of Requirements and Tests

Operational Need or Mission Requirements – Partially Validated by Operational Test (Proven by Real-World Experience)	System-Level Requirements – Verified by System-Level Tests	Component-Level Requirements – Verified by Component-Level Tests
• Win the Indianapolis 500 • Pretrial average speed of 215 mph • Average speed in the "500" of 190 mph	• Top speed of X mph • Acceleration in all directions, "g–g" space • Average standard pit time of Y seconds	• Engine horsepower of x Btu • Body's drag coefficient of y • Range per tank of gas of z mi

Therefore, the acceleration ability of the car in both longitudinal and lateral directions (see Fig. 1.4) is critical in the design process. Figure 1.4 portrays the g–g curve for a single car driven by three racers (charts a–c); the bottom right space (chart d) is the inferred g–g space of the vehicle. Finally, each of these system-level requirements is "flowed-down" to component-level requirements, such as the engine's horsepower and the drag coefficient of the body of the race car. (Note the true values of these parameters are closely guarded secrets of racing teams.) This process continues until the requirements for CIs are defined, establishing a hierarchy of requirements, from mission or need down to the CIs.

The system integration process starts during the decomposition and definition (or design) process. As part of design, the integration and qualification plans are developed. The purpose of qualification is the verification and validation of the system's design. *Verification* addresses the following question: Does the component, element, segment, or system meet its requirements, or have we built the component, ... , system right? On the other hand, *validation*, which is often combined with acceptance testing, demonstrates that the system satisfies the users' needs, or have we built the right system? Note that as verification moves farther from the CIs and closer to the system, it is not possible to conduct enough testing to prove anything statistically. Demonstration is often the best that can be done. It is expected, though not desired, that there will be issues and problems that arise as part of this qualification process. Decisions must be made concerning the relaxation of requirements versus design changes to specific CIs and components. During the design phase, integration activities should be planned to maximize the effectiveness of qualification within the resources and time available. These planned activities are then carried out during integration, with adaptations as needed. There should have been some thought given during design about what the most likely adaptations would be so that the integration phase has sufficient, built-in flexibility.

To be successful, the engineering design of systems must embrace the notion that many decisions are made during the development process. This is not a controversial position to take. However, adopting the notion that these decisions should be made via a rational, explicit process is not consistent with much of the current practice in the engineering of systems. Table 1.3 lists a sample of the many categories of development decisions. Chapter 14 provides a method for addressing these decisions. An important philosophical point in decision-making is that decisions should be made with the best information available at the time, realizing that the outcomes associated with the decision remain uncertain when the decision is made. Therefore, distinguishing between a good decision and a good outcome is important. The material in this book will also distinguish between the level of detail needed to make decisions in the engineering of a system and the level of detail needed to ensure proper implementation of the system's components and CIs.

In order to accomplish this difficult job of engineering a system, people with many different specialties must be involved in the systems engineering team. The stakeholders are central to the

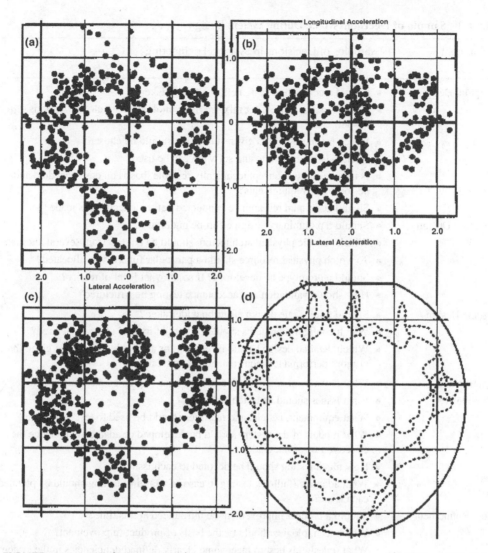

FIGURE 1.4 "*g–g*" design region for a racecar. (Adapted from Milliken and Milliken [1995].)

success of this effort and need to be represented on the systems engineering team. Discipline engineers with knowledge of the technologies associated with the system's concept are needed to provide the expertise needed for design and integration decisions throughout development. Discipline engineers not only come from traditional engineering fields such as electrical, mechanical, and civil but also from the social sciences to address psychological, informational, physical, and cultural issues. In addition, systems engineers who model and estimate system-level parameters such as cost and reliability fall in the category of discipline engineers. Analysts skilled in modeling and simulation, more and more of which is done on the computer rather than with scaled-down mock-ups of the system, are also important members of this team. Engineers skilled in the processes (or methods) of systems engineering form the nucleus of this collection of skills. *These processes and associated models are the nucleus of this book.* Finally, managers that are in charge of meeting cost and schedule milestones need to be present. These five disciplines are depicted in the Venn diagram in Figure 1.5.

TABLE 1.3 Sample of Decisions Made During System Design

Development Phase	Examples of Decisions in Systems Engineering
Conceptual Design	• Should a conceptual design effort be undertaken? • Which system concept (or mixture of technologies) should be the basis of the design? • Which technology for a given subsystem should be chosen? • What existing hardware and software can be used? • Is the envisioned concept technically feasible, based on cost, schedule and performance requirements? • Should additional research be conducted before a decision is made?
Preliminary Design	• Should a preliminary design effort be undertaken? • Which specific physical architecture should be chosen from several alternatives? • To which physical resource should a particular function be allocated? • Should a prototype be developed? If so, to what level of reality? • How should validation and acceptance testing be structured?
Full-scale Design	• Should a full-scale design effort be undertaken? • Which configuration items should be bought instead of manufactured? • Which detailed design should be chosen for a specific component given that one or more performance requirements are critical?
Integration and Qualification	• What is the most cost-effective schedule for implementation activities? • What issues should be tested? • What equipment, people, and facilities should be used to test each issue? • What models of the system should be developed or adapted to enhance the effectiveness of integration? • How much testing should be devoted to each issue? • What adaptive (fallback testing in case of a failure) testing should be planned for each issue?
Product Refinement	• Should a product improvement be introduced at this time? • Which technologies should be the basis of product improvement? • What redesign is best to meet some clearly defined deficiencies in the system? • How should the refinement of existing systems be implemented given schedule, performance, and cost criteria?

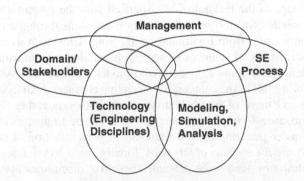

FIGURE 1.5 Expertise required on the engineering team for a system.

Sidebar 1.2 describes Joe Shea, who was hired by the National Aeronautics and Space Administration (NASA) in 1961 to take charge of systems engineering for the Office of Manned Space Flight.

SIDEBAR 1.2

"It was 1943 when he graduated (from high school), wartime, and Shea heard about a special Navy program that would send him to college ... Then the Navy sent him to M.I.T., and after that to the University of Michigan. ...

For the next several years Shea moved back and forth between Michigan, where he eventually obtained his engineering doctorate, and Bell Labs. It was an educational odyssey that took him from engineering mechanics to electrical engineering to theoretical mathematics to physics to inertial guidance. "The nouns change but the verbs remain the same" became one of Shea's sayings as he went from one specialty to another.

Then in 1956 Shea found out how it all fit together. At the age of 29, Shea was named systems engineer for the radio guidance project connected with the Titan I. "I didn't know what 'systems engineer' meant," Shea said, but he learned quickly, traveling around to the subcontractors on the Titan I, becoming a member of the small fraternity of engineers who were coming of age in this new field. At night after work they gather at a bar near the plant where they had been working that day. They didn't even drink that much, Shea recalled, they were so busy talking – about testing, grounding, vibrational spectrums, weights, stability, electrical interfaces, guidance equations, all the myriad elements of the system that some lucky guy, like a systems engineer, got to orchestrate.

By 1959, Shea had acquired enough of a reputation within the ballistic missile fraternity for General Motors to hire him to run the advanced development operation for its A.C. Sparkplug Division, which was trying to wedge its way into the missile business. Shea was in charge of preparing a proposal for the inertial guidance contract for the Titan II. After the proposal won, Shea went back to administering the advanced development office. But a year later, in September 1960, the contract he had won was six months behind and Shea was called away to rescue it.

Shea began to discover that he had a knack for leading. His was not a gentle style, but if he was tough on people who fell short, he was generous and loyal to those who didn't ... It didn't make any difference what your specialty was. Shea's maxim was that if you understood it, you could make him understand it – and once he did, you never had to explain it again. The only problem was keeping up.

It was about this time that Shea discovered the uses of what he would come to call his "controlled eccentricity." When he was still at Bell, his wife had bought him a pair of red socks as a joke. One day in a meeting he absent-mindedly put his feet up on the table, getting some laughs and loosening up the meeting. So Shea started wearing red socks, not all the time, but to important meetings. Eventually the socks were accepted as a good-luck charm to wear to presentations. Even senior management at General Motors, where putting one's feet on a desk was discouraged and wearing red socks was unthinkable, got used to the idea ...

Armed with his red socks and his puns and an emerging sense of how good he was getting to be at this sort of engineering, Shea set out to rescue the lagging Titan contract. He moved into the plant, and for five days a week, all three shifts, he was there, catching catnaps on a cot set up in his office. It was a pattern he would repeat later, during Apollo. The reasons were partly motivational – people work harder when they see the boss working all three shifts.

"But it also lets you find out everything that's going on," Shea said. "Things I'd find out at night, I'd get corrected during the daytime." Shea began handing out red socks as an award for good performance. His enthusiasm and energy were infectious.

Shea pulled it off, making up the six months." [Murray and Cox, 1989, pp. 121–123]

1.3 APPROACHES FOR IMPLEMENTING SYSTEMS ENGINEERING

We have just provided a description of what happens inside the process associated with the design of an engineered system and are about to describe several approaches for organizing that process. But let us step back a minute to look at the bigger picture, as summarized in Figure 1.6. The system that we have been tasked to design exists in a broader system, called the metasystem. This metasystem contains other systems and is purposefully pursuing some objectives. There is likely a sustainment system that is part of this metasystem that is providing supplies and support to one or more of the systems that comprise the metasystem.

The approach we take to modeling systems, systems of systems, and their metasystems is based on the input-process-output (I-P-O) paradigm [Wymore, 1993]. I-P-O is foundational for rigorous modeling across the life cycle of structural and executable integrity and fidelity to the level acceptable to stakeholders. Systems controllability, observability, and identifiability critical to assuring stability can be assessed using the I-P-O paradigm.

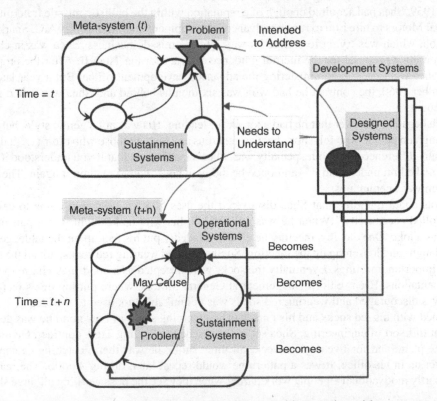

FIGURE 1.6 Characterizing the broader systems' design problem. (Adapted from Martin [2004].)

Some group has identified a problem with the achievement of the objectives being attained by the metasystem and has tasked a development system (organization) to design a system that will replace or upgrade one or more of the systems in the metasystem. In order to understand how to design this new or upgraded system, the people in the development system must understand the metasystem, or they will have little chance of success. Understanding the metasystem includes the interaction between the system to be replaced and other systems in the metasystem as well as the context or environment in which that metasystem operates. We will refer many times in this book to the creation of metasystem (or mission) requirements and an operational concept as approaches to achieving this understanding of the metasystem.

At some later time, after the metasystem has gone through many changes for which the development system must be tracked and adjusted, the designed system will be deployed and become an operational system within the changed metasystem first studied. Not only will the context of the metasystem have changed, but many of the systems inside the metasystem will have changed. In fact, there may well be other development systems working on some of these other systems in the metasystem, including the sustainment system. The introduction of the operational system may in fact introduce new problems into the metasystem. Such potential problems should be imagined as part of the development process and avoided or minimized via the design.

A final caution to the reader is that the development system (an organization of systems engineers and other engineers and experts) must design itself to have any chance of success. This design of the development system must emphasize adaptability to the inevitable change going on in the metasystem, as described in Figure 1.6 as well as another metasystem in which the development system exists.

I The Traditional, Top-Down Systems Engineering (TTDSE) process has evolved from the 1950s. Software engineers have evolved several approaches, starting with a waterfall process, moving to spiral development, and currently focused on object-oriented design. Object-oriented (OO) software design gained popularity in the early 1990s, shortly after object-oriented programming languages became available. Software engineering has led the agile engineering approach from the software perspective, and systems engineering is adapting agile engineering from the systems perspective; this adaptation is not a simple substitution replacing the word *software* with the word *systems* as everything must tailored within the context of the different disciplines and domains.

1.3.1 TTDSE

TTDSE (described in the overview in Section 1.2 and shown in Figure 1.7) is a process for systems engineering that begins a thorough analysis of what the problem is that needs to be solved; this is usually done with an analysis of the current metasystem (the system of interest and its peers [external systems]) performing one or more missions for the primary stakeholders. The result of this analysis is a statement of the *problem to be solved*. Based on this statement of the problem to be solved, several potential, competing concepts for implementing the system of interest are defined; this set of concepts should initially include a very broad range of ideas, some of which are relatively inexpensive while others are very expensive. Next, there will be an analysis of the competing concepts, resulting in the selection of the most favorable concept for implementation. Note, this analysis could really be many analyses. (Note, this book does not address the problem definition and evaluation of concepts. This material is covered by most texts on problem-solving for defining the problem to be solved. See Checkland [1993], Klir [1985], and Warfield [1990]. Decision analysis [Chapter 14] addresses the evaluation of concepts.)

On the basis of this selection, an operational concept and system-level requirements are defined for that solution concept. These two products (operational concept and system-level requirements) are a statement of the *problem being solved*. Next, a layered (or onion-peeling) iterative process

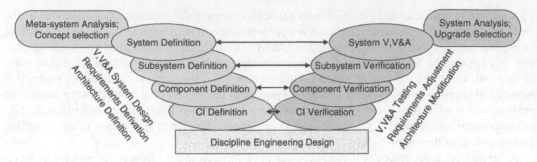

FIGURE 1.7 Traditional, top-down systems engineering.

begins for creating an architecture, deriving requirements, and refining the needed test system and associated data collection requirements. This layered process can have as many layers as are needed; the bottom layer addresses the CIs that the discipline engineers will design. Each layer repeats the same process (defined in detail in Chapters 6–10 of this book). Systems engineers commonly perform a great deal of analysis and modeling during each layer of this process; trade studies are often conducted to examine alternate ways to proceed or solutions that optimize some objective (e.g., cost, reliability, weight) while minimizing the impact on all other objectives.

Once the CIs have been designed and delivered for integration, the verification, validation, and acceptance testing process begins. Each layer of the decomposition process is verified against the associated derived requirements. During the process, requirements may be adjusted or the architecture and design of the system may be modified as needed. At the system level, validation against the concept of operations and acceptance testing (as defined by the stakeholders) is conducted. Chapter 11 defines this process. If a positive result is obtained, the system is deployed, and systems engineering continues by analyzing the usage of the system for needed modification and selecting upgrades that will be implemented in the future. The actual upgrading of the system should follow the same process as defined by the Vee-like structure in Figure 1.7.

TTDSE is primarily a process for designing the many pieces of a system in such a way that many different organizations can be tasked to design one or several pieces, and all the pieces can be integrated easily and effectively to achieve the desired system. Other references for TTDSE are Blanchard and Fabrycky [1998], Hatley and Pirbhai [1988], Sage [1992], and Wymore [1993].

1.3.2 The Waterfall Model of Software Engineering

One of the earliest concepts of the software engineering process was called the "waterfall" model by Boehm [1976] but introduced by Royce [1970]. The waterfall model (Fig. 1.8) is characterized by the sequential evolution of typical life cycle phases, allowing iteration only between adjacent phases. The waterfall model is known and discussed throughout the software and systems engineering communities and was the basis for Military Standard 2167A for software development. The major problem with the waterfall process is that iteration between phases that are widely separated is all too common.

1.3.3 The Spiral Model of Software Engineering

The spiral model (Fig. 1.9), developed in the 1980s [Boehm and Papaccio, 1988] and then modified several times [Boehm, 1986, 1988], addressed the need to shorten the time period between the users' statement of requirements and the production of a useful product with which the users could interact.

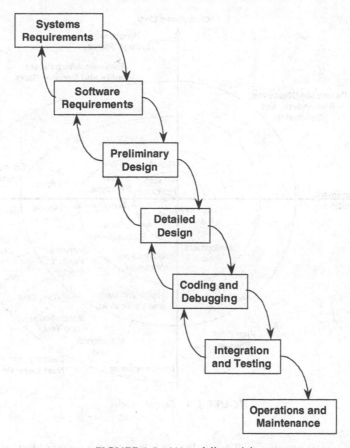

FIGURE 1.8 Waterfall model.

Too many systems and software implementations were being produced and rejected because the development life cycle took too long; valid requirements at the beginning of the cycle were no longer valid at the time of delivery. In addition, new systems were degraded because the vestiges of learning about the system domain tainted the early designs.

The spiral model has four major processes, starting in the top left of Figure 1.9 and moving clockwise: design, evaluation and risk analysis, development and testing, and planning with stakeholder interaction and approval. These four processes are repeated as often as needed. The radial distance to any point on the spiral is directly proportional to the development cost at that point. The spiral model views requirements as objects that need to be discovered, thus putting requirements development in the last of the four phases as part of planning. The early emphasis is on the identification of objectives, constraints, and alternate designs. These objectives and constraints become the basis for the requirements in the fourth step. There is also a major emphasis on evaluation and risk analysis as part of the management activities. This management activity is to identify which requirements are most important to discover early in order to minimize problems associated with cost, schedule, and performance. The development effort is composed of prototyping activities, which provide mock-ups of the software or system that will enable the stakeholders to define their requirements. This third step ends with evaluation and testing. The fourth step involves documenting requirements gleaned from the intense prototyping interaction with the users during the current trip around the spiral and planning the next trip around the spiral. The number of iterations around the spiral is variable and defined

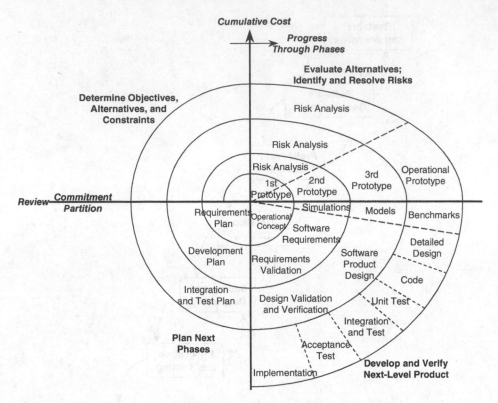

FIGURE 1.9   Spiral model.

by the software or systems engineers. The final cycle integrates the stakeholders' needs into a tested and operational product.

Shortly after the spiral model was introduced, various authors [e.g., Boar, 1984] spoke of rapid prototyping as a development process. The rapid prototyping process is meant to produce early, partially operational prototypes. The use of these operational prototypes by stakeholders generates new and improved requirements, as well as provides the stakeholders with increased functionality via early releases of the system. Thus, one could view rapid prototyping through the spiral process model in which the prototypes were partially operational.

1.3.4   Object-Oriented Design

OO design followed from OO programming in the 1970s. OO design is a bottom-up process that begins by defining a set of objects that need to be part of the system to achieve the system-level functionality desired. Objects are thought to be basic building blocks that can perform functions (methods) and contain information. Key properties of OO design are inheritance and information hiding. Inheritance means that a general object can be specialized by adding special characteristics; the specialized object will "inherit" all the properties (methods and data) not overridden by the specialization. Information hiding means that an object does not need to know how another project is producing the information being sent to it, just what that information is. Systems engineers have referred to this idea as modularity for years. Besides being the basic building blocks of a system, objects are seen to promote reusability, testability, and maintainability. For more information, see Ambler [1997].

1.3.5   Composability Engineering

Composability is a systems design approach made to increase agility and accelerate application development by reusing existing assets and reassembling them in unique ways to satisfy specific user requirements. Composability is the opposite of TTDSE. The principles of composability are modularity, autonomy, and discoverability. It is common for SoS to be engineered from the composition of its identified constituent systems [SEBoK, 2022].

1.3.6   Agile Systems Engineering

Agile systems engineering [SEBoK, 2023] is an iterative, incremental approach to continually modeling, analyzing, developing, and trading options in the engineering of systems. Program leadership, systems engineering, and all team members work using an *agile mindset* that is a combination of beliefs and actions from agile values that include focusing on delivering working capabilities often, trusting the knowledge workers to find the best solution, improving the product and process through regular demonstrations and retrospections, planning often to implement lessons learned.

1.4   MODELING APPROACHES FOR SYSTEMS ENGINEERING

Modeling techniques for designing systems were created as early as the early 1950s. These techniques address the connection of system components, the decomposition of system functions, and the dynamic behavior of the system. The modeling techniques and tools have continued to evolve, including the Unified Modeling Language (UML) for software, the extension of UML 2.0 to the Systems Modeling Language (SysML) (Object Management Group [OMG]), the Object Process Methodology (OPM) [Dori, 2016; ISO, 2015], and the Life cycle Modeling Language (LML) [Dam, 2014; LMO, 2022]. The principles, models, and methods in this text are tool agnostic, with relevant examples implemented in the different tools available on the companion website. The UML was created by several of the OO software gurus under the aegis of the OMG, who had developed their own approaches to modeling and decided an integrated approach was needed. The US Department of Defense Architecture Framework (DoDAF) was developed within the Command, Control, Communications, Computers, Intelligence, Surveillance, and Reconnaissance (C4ISR) community and then extended to all of DoD. DoDAF has been adapted and extended internationally to the Unified Architecture Framework (UAF). These architecture frameworks have been instantiated in the systems engineering tools.

1.4.1   Modeling Approaches for TTDSE

The first modeling approach of TTDSE was the block diagram. Each block represented a system component. Lines between the blocks represented the exchange of information, energy, or physical entities. Next, N-squared (N^2) diagrams were created to capture a high-level view of the flow of information, energy, and physical entities among the components at a given level of abstraction for the system (see Lano [1990b]). Next, the N^2 diagram was transformed to a functional perspective, the components being exchanged for major system functions. Function flow block diagrams were then developed to capture the dynamics of the system's behavior. Meanwhile, software designers were creating data flow diagrams to model software systems. Manufacturing designers created the Structured Analysis and Design Technique (SADT), which was later transformed into the Integrated Computer-Aided Manufacturing (*I*CAM) *Defi*nition or IDEF0. Data flow diagrams, N^2 charts, and IDEF0 diagrams all capture the same basic time-lapsed flow of information, energy, and physical items among functions. State transition diagrams (or state machines) were developed and enhanced

TABLE 1.4   Diagram Types for UML 2.0

Structure Diagrams	Behavior Diagrams	Interaction Diagrams
Class	Activity	Collaboration – communication
Component	State machine	Interaction overview
Composite structure	Use case	Sequence diagram
Deployment		Timing
Object		
Package		

by several engineering disciplines to capture dynamic behavior; these techniques have been applied to some TTDSE efforts. Finally, Petri nets have been developed to model the dynamics of systems. Many TTDSE practitioners use some subset of these modeling techniques. All of these techniques are covered in later chapters of this book.

1.4.2   UML

The UML is a specification language for modeling objects that is approved by the Object Manage Group. UML 2 was adopted in 2004 and is often described as a graphical modeling language. Critical ideas underlying OO modeling are multiple views at varying levels of abstraction, object, class, inheritance, and extensibility. All useful approaches to systems and software engineering use modeling approaches that enable modeling a system at multiple levels of abstraction. An object is a basic building block of OO programming that can receive messages, process data, and then send messages to other objects. An object can be viewed as a component or actor that has the resources to receive, process, and send data. A class in OO terminology is a grouping of related variables or functions; this is a key to addressing a system at multiple levels of abstraction. Inheritance (now often called generalization) is the process of creating instances of a class based upon specializations of class parameters; this is often the key to software reuse. Extensibility is a way of extending the UML modeling language. For example, stereotypes permit extending elements of UML to a specific problem domain.

UML 2.0 contains 13 different diagram categories that can be aggregated into three diagram types; see Table 1.4. Structure diagrams address those issues or elements that are part of the system being modeled. Concepts for structure diagrams include actor, attribute, class, component, interface, object, and package. Behavior diagrams examine the activities that must happen in the system being modeled. Behavior diagram concepts include activity, event, message, method, operation, state, and use case. Interaction diagrams (considered by some to be a subset of behavior diagrams) address the flow of data and control among the elements in the system being modeled. Concepts for interaction diagrams include aggregation, association, composition, depends, and generalization (or inheritance).

Some important ideas in UML are the use case diagram, which is a high-level view of the use cases, the class diagram which describes the relationship between structural elements of the system and the external domain, and a set of object diagrams that are more definitive than the class diagram about the structural elements of the system and their relationships over time. Sequence diagrams are a representation of scenarios or use cases, something that traces back decades. The key design elements are the software objects.

The use case diagram and sequence diagrams define the requirements in a qualitative way; there are seldom any quantitative performance requirements, and there are no non-functional requirements. Similarly, there is no top-level functional analysis; each object contains operations that can be performed and data that can be used for those operations. UML is primarily a graphical modeling language for creating abstractions or generalizations so that the resulting software system will be more flexible and adaptable.

This UML process is more of a bottom-up design process in which the components of the software are derived from more specific software objects that are designed to be adapted from existing code or coded from scratch. Useful references on UML are Ambler [2004] and Eriksson and Penker [2000]. Software engineers believe the appropriate model of their design is the code itself so very little modeling and analysis is performed during this process.

1.4.3 DoDAF

The DoDAF provides three integrated views needed for a system architecture; each of the three views is composed of subviews using graphical, tabular, and textual descriptions. A data model is defined that defines entities and relationships among the data elements that are part of these integrated views. This effort began in 1995, produced versions 1 and 2 of the C4ISR Architectural Framework in 1996 and 1997 (respectively), and yielded versions 1 and 2 of the DoDAF in 2003 and 2009, respectively. The Ministry of Defence (MOD) of the United Kingdom and the North Atlantic Treaty Organization (NATO) have adopted similar architecture frameworks: MODAF and NAF, respectively.

In DoDAF 1, there were three top-level views: operational, systems, and technical. The operational view addresses the organizational and human context in which the system will be utilized. The system's view switches to the physical and functional world, starting outside the system and moving inside the system. The technical view addresses standards and conventions. In DoDAF 2, seven viewpoints were adopted: capability, data and information, operation, project, services, standards, and systems. These viewpoints are oriented toward supporting DoD decision-makers associated with systems engineering activities. The *capability viewpoint* serves the needs of capability portfolio managers. Business activities are supported by the *data and information viewpoint*. The *operational viewpoint* is used to describe the tasks and activities, operational elements, and operational resource flows. The *project viewpoint* enables the description of contributions by programs, projects, portfolios, or initiatives. The *services viewpoint* is used to describe the satisfaction of DoD functions via services, as opposed to systems. The set of rules governing the management, interaction, and interdependence of parts or elements is defined by the *standards viewpoint*. The *systems viewpoint* describes the systems and other interconnections providing DoD functions.

Each viewpoint has several products, which capture a subset of the concepts, associations, and attributes relevant to the view. It is important to conceive of the DoDAF as containing a central database of all the entities and relationships. Each product of each view is then a representation of a subset of that central database. The developers of the DoDAF continue to strive to make this structure useful to decision-makers and systems engineers. References include Lcvis and Wagenhals [2000] and Dam [2006].

1.4.4 SysML

There has been a push among some systems engineers for an approach to systems engineering that is less text-based and, therefore, more model-based. The arguments against text-based processing are its inefficiencies in finding errors and stress points, testing both performance and timing behavior in one or more competing designs, and providing actionable information for trade studies and design reviews. Ultimately, there is a need to examine performance issues and conduct tests before the first prototype is completed. Software engineers, for the most part, seem to have no problem with waiting until the code is written to find out that there are major timing and latency problems. Hardware has

TABLE 1.5   Diagram Types for SysML v1

Structure Diagrams	Behavior Diagrams	Interaction Diagrams	Requirement (new)
Class – renamed to be	Activity (modified)	~~Collaboration - Communication~~	Requirement (new)
Block definition	State machine	~~Interaction overview~~	
Internal block	Use case	Sequence diagram	
~~Component~~		~~Timing~~	
~~Composite structure~~			
~~Deployment~~			
~~Object~~			
Package			
Parametric design (new)			

traditionally taken much longer to redesign, so systems engineers prefer to get the bad news early. This emphasis has led to MBSE efforts, the most visible of which is SysML.

SysML version 1 (v1) is a visual modeling language adapted from UML 2.0 and enhances the traditional top-down systems engineering process. SysML extends the modeling language of TTDSE; this extension should make the traditional approach to systems engineering less prone to errors and more efficiently implemented. Table 1.5 shows which UML 2.0 diagrams have been dropped (strikethrough), adopted (new), or modified (modified) for SysML. The first thing to notice in Table 1.5 is that there is a new column for requirements with a single diagram type. The column with the most changes is the first column for structure diagrams. Here, the class diagram has been renamed to capture two different concepts associated with the physical architecture: block definition and internal block connectivity of parts. A new diagram was created for modeling performance, called the parametric diagram. The package diagram was kept as is from UML 2.0. Within the category of behavior diagrams, the activity diagram has been modified while the state machine and use case diagrams have been kept as is. Finally, most of the interaction diagrams have been dropped; the only remaining interaction diagram is the sequence diagram. The implication of these changes is that SysML v1 places much greater emphasis on behavior compared to interaction than UML does.

These diagram concepts will be introduced in later chapters of this book.

The real challenge for SysML (and every other model-based approach) is to include easily understood descriptions of the system design and the associated requirements for non-engineering stakeholders. References for SysML v1 are Bock [2006], Friedenthal and Moore [2014], and Delligatti [2013]. Stakeholder critiques of SysML v1 have led INCOSE and the OMG to evolve SysML v1. SysML v2 introduces a metamodel based on formal semantics and not constrained by UML while preserving most UML modeling capabilities. SysML v2 has flexible graphical, tabular, and textual view and viewpoint specifications and execution, as well as a standardized application programming interface (API) for the system model to be interoperable with other tools.

1.5   INTRODUCING THE CONCEPT OF ARCHITECTURES

Levis [1993] has defined an analytical systems engineering process (for the left side of the Vee process) that begins with the system's operational concept and includes the development of three separate architectures (functional, physical, and allocated) as part of this decomposition. The functional

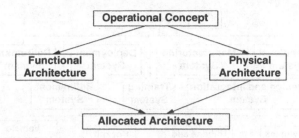

FIGURE 1.10   Architecture development in the engineering of a system. (Adapted from Levis [1993].)

(or logical) architecture defines what the system must do, that is, the system's functions and the data that flows between them. The physical architecture represents the partitioning of physical resources available to perform the system's functions. The allocated architecture (see Fig. 1.10) is the mapping of functions to resources in a manner that is suitable for discrete-event simulation of the system's functions and is analogous to Alford's [1985] approach with behavior diagrams. Figure 1.10 suggests that the functional and physical architectures are developed independently of each other and then combined to form the allocated architecture. This suggestion is inaccurate; rather, the two architectures are developed in parallel but with close interaction to ensure that the allocated architecture is meaningful when the functional and physical architectures are combined. Chapters 7–9 address these three architectures and their development in detail and discuss the interactive development of them.

Critical to this multiple-architecture approach is the balancing of information among them. To be complete, three separate models must be developed: data, process, and behavior models. The functional architecture includes the first two (data and process) models and the initial behavioral model, as discussed in Chapter 7. The behavioral model should be finished and exercised as part of the allocated architecture; see Chapter 9. Each of these three models must be integrated to define the three architectures properly.

Figure 1.11 shows an organization chart representation of the physical architecture of the F-22 fighter. Note that this physical architecture includes more than the F-22; the training and support systems are included as well. For a life-cycle-balanced (concurrent engineering) definition of the F-22, the physical architecture should have been decomposed, as shown in Figure 1.12.

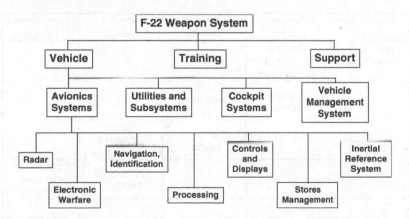

FIGURE 1.11   Sample physical architecture (F-22 Type A Spec). (Adapted from Reed [1993].)

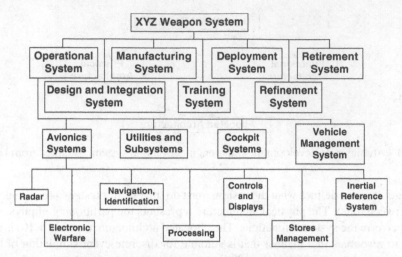

FIGURE 1.12 Life cycle physical architecture.

Graphical techniques, such as Figures 1.11 and 1.12, are invaluable because they serve as an excellent communication medium; communication is one of the most important functions of systems engineers. A physical architecture subdivides the problem into manageable parts, permitting and encouraging an iterative process and providing excellent documentation.

Figure 1.13 depicts the systems engineering design process in terms of requirements and architectures in a similar manner as the waterfall process, a sequential decomposition of requirements and the allocated architecture (functions mapped to physical resources) by moving from left to right and top to bottom. A question often asked by new students is: What is the difference between a requirement and a specification? A *requirement* is one of many statements that constrain or guide the design of the system in such a way that the system will be useful to one or more of its stakeholders. A *specification* is a collection of requirements that completely define the constraints and performance requirements for a specific physical entity that is part of the system. The systems engineering design process involves defining all the system's requirements and then bundling them by segments and refining them into a specification for each of the system's segments, elements, components, and CIs.

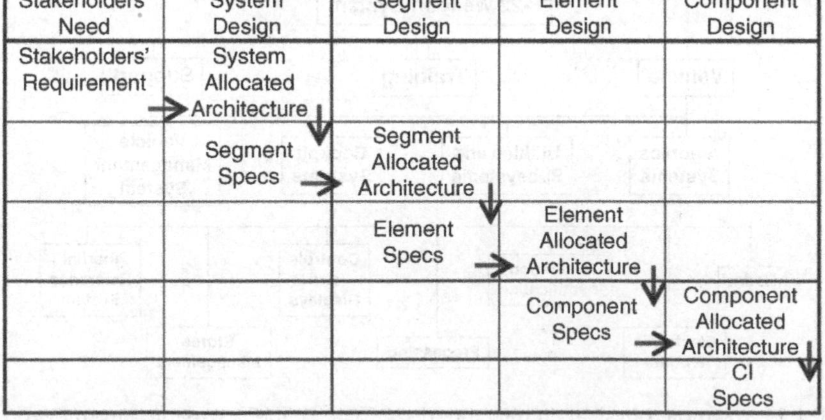

Stakeholders' Need	System Design	Segment Design	Element Design	Component Design
Stakeholders' Requirement	System Allocated Architecture			
	Segment Specs	Segment Allocated Architecture		
		Element Specs	Element Allocated Architecture	
			Component Specs	Component Allocated Architecture
				CI Specs

FIGURE 1.13 Design decomposition of architectures and specs.

1.6 REQUIREMENTS

Requirements for a system address the needs and objectives of the stakeholders. Just as there is a hierarchy associated with the physical components of the system, there is a hierarchy of requirements. At the top of the hierarchy are mission requirements, which relate to needs associated with missions or activities that are important to one or more groups of stakeholders. These *mission requirements* typically involve the interaction of several systems, one or more of which include individuals or groups of people, and are therefore stated in the context of the operation of the system in question with these other systems, called the metasystem or supersystem or SoS. Mission requirements represent stakeholder preferences for the increased ability to perform their activities with the introduction of the system in question at a lower cost arid in a faster time than the existing capability.

Stakeholders' requirements are statements by the stakeholders about the system's capabilities that define the constraints and performance parameters within which the system is to be designed. Systems engineers take these high-level stakeholders' requirements and derive a consistent set of more detailed engineering statements of requirements as the design progresses. For the purposes of this introduction, requirements are divided into constraints and performance indices. Some constraints are simple; for example, the system must be painted a specific shade of green. Other constraints are the minimally acceptable level associated with a performance requirement. A performance requirement defines a desired direction of performance associated with an objective of the stakeholders for the system. For an elevator system (which is used throughout this book) a performance requirement might be to minimize passengers' waiting time. For any performance requirement, there must also be a minimum acceptable performance constraint or threshold and a design goal associated with the index; this threshold dictates that no matter how wonderful a design's performance is on other objectives, performance below this threshold on this requirement makes the design unacceptable. This is a very strong statement of needs, and so minimal acceptable thresholds must be established very carefully.

Every major organization, governmental or commercial, has established its own guidelines for system or product development. The names and organizations of the several requirements documents vary somewhat but cover similar material. Table 1.6 summarizes the common major requirements documents that are produced during the beginning of the design phase. The Problem Statement (or Mission Element Need Statement in the military) gets the process rolling and identifies a problem for which a solution In the form of a system (new or improved) is needed. This document supports and documents a decision-making process to start a system development effort. The Systems Engineering Management Plan (SEMP) then defines the systems engineering development system.

Stakeholders' requirements are found in the Stakeholders' Requirements Document (StkhldrsRD). This document is produced with or by the stakeholders and is written in their language(s). Systems engineers need to be involved in a substantial way in this activity, although not all systems engineers share this view. Experience has shown that if this document is left to the stakeholders, the document will be very incomplete. The systems engineers can play a major facilitation role among the various groups of stakeholders as well as bring an assortment of tools to bear on a difficult problem, the creation of this document. These tools (a major focus of this book) ensure greater completeness and consistency. The methods and tools presented here are equally applicable to the rest of the systems engineering process.

The systems engineer then begins restating and "derivin" requirements in engineering terms, called *system* requirements, so that the systems engineering design problem can be solved. This derivation of the StkhldrsRD becomes the Systems Requirements Document (SysRD).

It is critical that the requirements in all of these documents address "what" and "how well" the system must perform certain tasks. Requirements do not provide solutions but rather define the problem to be solved.

TABLE 1.6 Typical Requirements Documents

Document Titles	Document Contents
Problem Situation or Mission Element Need Statement and Systems Engineering Management Plan (SEMP)	• Definition of stakeholders and their relationships • Stakeholders' description of the problem and its context • Description of the current system • Definition of mission requirements • Definition of the systems engineering management structure and support tools for developing the system
Stakeholders' Need or Stakeholders' Requirements Document (StkhldrsRD)	• Definition of the problem needing solution by the system (including the context and external systems with which the system must interact) • Definition of the operational concept on which the system will be based • Creation of the structure for defining requirements • Description of the requirements in the stakeholders' language in great breadth but little depth • Trace of every requirement to a recorded statement or opinion of the stakeholders • Description of trade-offs between performance requirements, including cost and operational effectiveness
System Requirements Document (SysRD)	• Restatement of the operational concept on which the system will be based • Definition of the external systems in engineering terms • Restatement of the operational requirements in engineering language • Trace of every requirement to the previous document • Justification of engineering version of the requirements in terms of analyses, expert opinions, stakeholder meetings • Description of test plan for each requirement
System Requirements Validation Document	• Documents analyses to show that the requirements in the SysRD are consistent, complete, and correct, to the degree possible • Demonstrates that there is at least one feasible solution to the design problem as defined in the SysRD

The Systems Requirements Validation Document defines requirements associated with the verification, validation, and acceptance of the system during integration. These requirements are high-level requirements that state the needs of the stakeholders for qualifying the design of the system. These requirements form the basis of the problem definition for creating the qualification system that will be used during integration. In addition to defining the high-level qualification requirements, this document should demonstrate that if the systems engineering process continues, an acceptable solution is possible. Unfortunately, this "existence proof" of a feasible solution is seldom produced in practice, leading to a major downfall of many systems engineering efforts. Namely, the realization many months (or years) later that not all of the requirements can be satisfied, and the stakeholders must relax the requirements that the engineers promised could be met.

TABLE 1.7 Comparison of the Relative Cost to Fix Software in Various Life Cycle Phases

Source	Phase Requirements Issue Found			
	Requirements	Design	Code	Test
Boehm [1981]	1	5	10	50
Hoffman [2001]	1	3	5	37
Cigital [2003]	1	3	7	51
Rothman [2002]		5	33	75
Rothman-Case B [2002]			10	40
Rothman-Case C			10	40
Rothman [2002]	1	20	45	250
Pavlina [2003]	1	10	100	1000
McGibbon [2003]		5		50
Mean	1	7.3	25.6	177
Median	1	5	10	50.5

Systems engineers have always desired to demonstrate the importance of requirements and getting the requirements right, for example, complete, consistent, and correct. In the mid-1970s, three organizations (GTE [Daly, 1977], IBM [Fagan, 1974], and TRW [Boehm, 1976]) conducted independent studies of software projects. These studies addressed the relative cost to fix a problem based upon where in the system cycle the problem was found. Boehm [1981] and Davis [1993, p. 25] compared the results of the three studies (see the first row of Table 1.7). The costs have been normalized so that the relative cost to repair an average problem found in the coding phase is 10 units. These results stood for 20 years. The next eight rows of Table 1.7 show results from recent studies, summarized in Haskins et al. [2004]. As can be seen, the results have held up well. Getting the requirements right is a very difficult task and, therefore, a task that is fraught with errors. An error that is caught during requirements development can be fixed for about 10% of the cost associated with an error caught during coding. Errors caught during maintenance in the operation of the system cost about 20 times that of an error caught during coding and 200 times the cost of an error caught during requirements development. Unfortunately, many of these errors are not caught until late in the life cycle, causing the expenditure of significant money.

1.7 SYSTEM'S LIFE CYCLE

There are many ways to define a system's life cycle. However, the common phases associated with a system are development, manufacturing, deployment, training, operations and maintenance, refinement, and retirement. Systems engineers have activities in all these phases, but the primary phases of concern to the systems engineers are development and refinement. Stakeholders use and maintain the system in the operation and maintenance phase. A common mistake is to envision these phases as distinct and separate in time. In fact, it is common (though not required) to have four distinct periods: development only, pre-initial operational capability development and testing, operational use and refinement, and retirement. All but the first period have multiple phases occurring in parallel, as shown in Figures 1.14–1.17.

In the development period, the systems engineering team receives resources from the bill payer and begins the development of the system. This period involves heavy interaction with the stakeholders as the requirements process is begun, and the architectures and models for simulation and

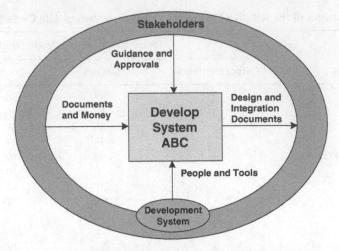

FIGURE 1.14   Development period.

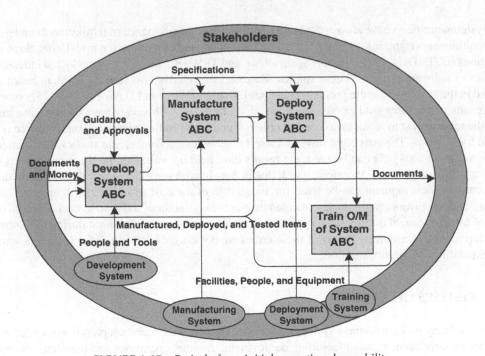

FIGURE 1.15   Period of pre-initial operational capability.

analysis are initiated. However, this period ends when the manufacturing, deployment, and training teams begin preparation for the system.

Development, manufacturing, deployment, and training activities are pursued concurrently during the second period, after concurrent design occurred in the first period. Specifications flow from the development process to the other three. Manufactured, deployed, and training equipment flow to development for testing. Interaction continues with the stakeholders as final testing occurs, leading to the acceptance of the system by the stakeholders. This period ends just as the first operational systems are being delivered to the users.

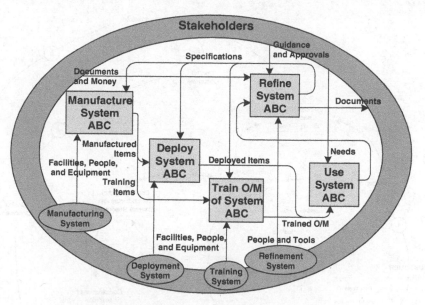

FIGURE 1.16 Period of operational use and refinement.

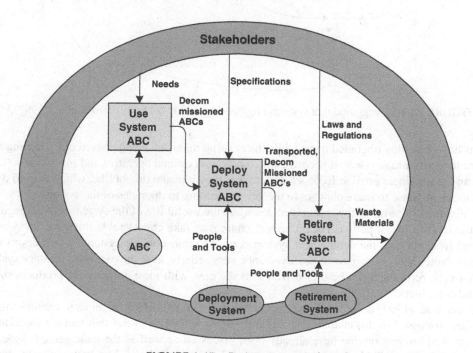

FIGURE 1.17 Retirement period.

The third period begins as users receive the first operational items. This period also contains continued production of the system, as well as deployment of and training on the system. Refinement of the design begins here. Manufactured items are sent to the deployment system, which delivers them to users. One of the most difficult problems to solve adequately from the perspective of the

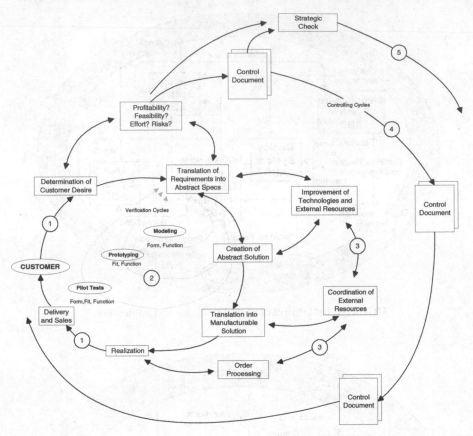

FIGURE 1.18   Cycle model of systems engineering. (Adapted from Wenzel et al. [1997].)

users is how to deploy upgraded items while the existing items are being phased out. Training items are sent to the training system (if needed), which produces trained operators and maintainers (O/M). Users and maintainers provide feedback about what they like and do not like, which is used during the refinement phase to make changes to the design, leading to upgrades of the system.

Finally, the bill payer of the system decides when the useful life of the system is over, beginning the initiation of the last period. The retirement phase may take considerable time. As the system is removed from service, the deployment system is used to transport the system from users to waste facilities. Note that this retirement process can be very orderly, as is the case with military systems. Alternatively, retirement can be user-driven as is the case with most commercial products such as cars and computers.

Wenzel et al. [1997] describe the cycle model (see Fig. 1.18), which attempts to capture many of the issues discussed in this chapter. The cycle model stresses five cycles that include the elements of design and integration that have already been discussed as well as the management aspects of systems engineering. Table 1.8 describes these cycles in some detail. The first cycle satisfies the key elements of stakeholder satisfaction, beginning with the determination of the need and ending with the delivery of the system to satisfy those needs. The development functions in this first cycle include requirements development and the creation of the system design. The second cycle (verification) addresses the modeling, prototyping, and testing that must be part of the development process; these cycles within the verification cycle enable the requirements and the solution to be refined and verified. The third cycle enables management to insert technologies and external resources into both

TABLE 1.8 The Cycles of the Cycle Model

Design and Integration Cycles	Management Cycles
1. *Core Cycle*: Realization of stakeholder needs, followed by requirements development, design, manufacturing, and product delivery.	3. *Technologies and External Resources Cycle*: Insertion of the appropriate technologies and resources into the systems engineering process.
2. *Verification Cycle*: Analysis, simulation, prototyping, integration, and testing.	4. *Controlling Cycle*: Configuration management of the design process and multiple product releases and updates.
	5. *Strategic Check Cycle*: Management assessment and approval of product development.

the development and the manufacturing processes to improve the chances of stakeholder satisfaction, subject to the constraints faced by management. The controlling cycle provides configuration management throughout development and enables product releases and updates throughout the system's life cycle. Finally, top-level management and stakeholder review and approval are included in the final cycle.

1.8 DESIGN AND INTEGRATION PROCESS

Recall the design and integration of Vee as identified by Forsberg and Mooz [1992]. The Vee model defines five major functions for the design or decomposition phase, as shown in Figure 1.19. Note that these functions must be repeated for each stage of the decomposition process. A modification of the more detailed design functions, as put forth by Forsberg and Mooz [1992], is shown in Figure 1.20. This figure also shows how the Forsberg and Mooz [1992] functions are grouped to be comparable to the five analytical systems engineering functions.

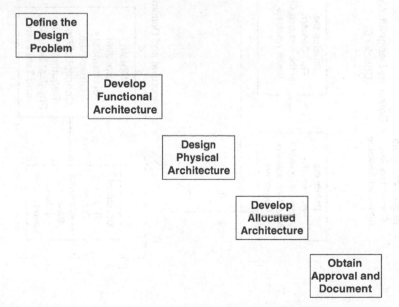

FIGURE 1.19 Five major functions of the engineering design of a system.

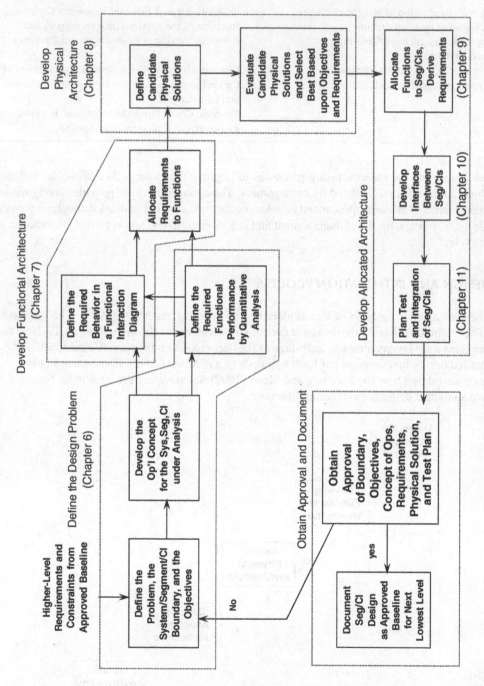

FIGURE 1.20 Detailed functions of systems engineering design.

Develop Physical Architecture (Chapter 8)

Develop Functional Architecture (Chapter 7)

Develop Allocated Architecture

Define the Design Problem (Chapter 6)

Obtain Approval and Document

Define Candidate Physical Solutions

Evaluate Candidate Physical Solutions and Select Best Based upon Objectives and Requirements

Allocate Functions to Seg/CIs, Derive Requirements (Chapter 9)

Develop Interfaces Between Seg/CIs (Chapter 10)

Plan Test and Integration of Seg/CIs (Chapter 11)

Allocate Requirements to Functions

Define the Required Behavior in a Functional Interaction Diagram

Define the Required Functional Performance by Quantitative Analysis

Develop the Op'l Concept for the Sys, Seg, CI under Analysis

Obtain Approval of Boundary, Objectives, Concept of Ops, Requirements, Physical Solution, and Test Plan

Higher-Level Requirements and Constraints from Approved Baseline

Define the Problem, the System/Segment/CI Boundary, and the Objectives

Document Seg/CI Design as Approved Baseline for Next Lowest Level

yes

No

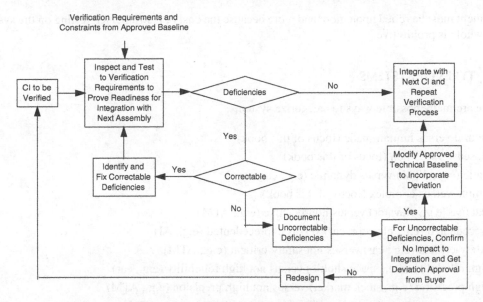

FIGURE 1.21   Functions of the systems engineering integration process.

The five detailed functions that comprise the design phase must address up to five different dimensions of data; see van den Hamer and Lepoeter [1996]: (1) system variants when the system is a member of a product family (e.g., personal computers, automobiles), (2) system versions when the system is a product that evolves over time (e.g., operating systems), (3) views of the system (e.g., data, process), (4) hierarchical detail or onion peels (e.g., system, subsystem), and (5) status of the data (e.g., stable and approved versus tentative or draft).

For many systems, five modeling views [Karangelen and Hoang, 1994] are critical for capturing the totality of a system: environment, data or information, process, behavior, and implementation. The *environmental* view captures the system boundary, the operational concept, and the objectives of the system's performance. The *data* or *information* view addresses the relationships among the data elements that cross the system's boundary and those that are internal to the system; this view can be critical for information and software systems but incidental to mechanical systems. The *process* view examines the functionality of the system and is used to create the functional architecture. The *behavior* view addresses the control structures in which the system's functions are embedded. The *implementation* view examines the marriage of the physical architecture with the process and behavior views; the allocated architecture represents the implementation view. In later chapters, these views and the tools that are used to execute them will be addressed.

Figure 1.21 shows a modification of the Forsberg and Mooz [1992] integration functions. Most of this activity is dedicated to the verification that the integrated components, elements, and segments meet the derived requirements (specifications) of the systems engineering process. The final iteration of the integration functions is devoted to the validation of the system; is this system the system the stakeholders wanted? Will they accept the system? The answer to this question is substantially determined by the extent to which the systems engineers have kept the stakeholders involved throughout the process. The greater the involvement, the more the stakeholders understand what trade-offs were made and why.

There are four primary methods for testing the system to complete the verification and validation process: instrumented test using calibrated equipment, analysis and simulation using equations and computers, demonstration or functional test using human judgment, and examination of documentation using human judgment. As integration moves from CIs and approaches the system level, human judgment must be relied upon more and more because the cost of instrumented testing on the system as a whole is prohibitive.

1.9 TYPES OF SYSTEMS

There are many possible ways to categorize systems:

natural versus human-made (focus of this book)
closed versus open (focus of this book)
static (e.g., ATM) versus dynamic (e.g., car)
simple versus complex (focus of this book)
reactive (e.g., elevator) versus non-reactive (e.g., ATM)
precedented (e.g., elevator, car) versus unprecedented (e.g., AI)
safety-critical (e.g., car) versus not safety-critical (e.g., ATM)
high reliability (e.g., space shuttle) versus not high reliability (e.g., car)
high precision (e.g., stock market) versus not high precision (e.g., ATM)
human-centric (e.g., bank operations) versus non-human (e.g., ATM)
high durability (e.g., telephone system) versus not high durability (e.g., ATM)
software-intensive (e.g., current auto) versus not software-intensive (e.g., elevator)
human-machine team (e.g., military aircraft) versus non-human-machine team (e.g., elevator)
    (Human is defined to be part of military aircraft systems and not part of an elevator system)

Yet the process described in this book should work for all "human-made" systems, with some tailoring. Clearly, a great deal more engineering and systems engineering is required for an unprecedented system (the Shuttle) than for a precedented one (a new automobile).

Magee and de Weck [2004] propose a two-dimensional classification structure of systems that was derived from the work of several other authors. The two dimensions include the character (energy, matter, etc.) of the major output of the system as well as the type of operation or process being employed to produce this major output. The major outputs of Magee and de Weck were broadened to include:

- *Matter (M)*: physical objects, including organisms that exist unconditionally
- *Energy (E)*: stored work that can be used to power a process in the future
- *Information (I)*: anything that can be considered an informational object
- *Value (Monetary) (V)*: monetary and intrinsic value objects used for exchange

Magee and de Weck [2004] also broadened the list of operands or process manipulators to include:

- *Transformation Systems*: transform objects into new objects
- *Distribution Systems*: provide transportation, i.e., change the location of objects
- *Storage Systems*: act as buffers in the network and hold/house objects over time
- *Market Systems*: allow for the exchange of objects mainly via the Value layer
- *Control Systems*: seek to drive objects from some actual state to a desired state

Table 1.9 provides an example for each of the 20 combinations in the Magee and de Weck structure.

TABLE 1.9 System Classification by Magee and de Weck [2004].

Major Process and Operand	Major Output			
	Matter	Energy	Information	Value
Transform or process	Manufacturing plant	Power plant	Computer chip	Mint
Transport or distribute	Package delivery company	Power grid system	Telecommunication network	Banking network
Store or house	Dam	Dam	Public library	Bank
Exchange or trade	Internet auction company	Energy market	News agency	Stock trading market
Control or regulate	Health care company	Energy agency	International Standards Organization	Monetary regulator

1.10 SUMMARY

Engineering involves the practice of applying scientific theories to the development, production, deployment, training, operation and maintenance, refinement, and retirement of a system or product and its parts. The engineering discipline that addresses the creation of a system that meets the needs of defined stakeholders is systems engineering. The engineering of a system involves both the design of the system's components and CIs and the integration of those CIs and components into a qualified system acceptable to the stakeholders across the life cycle of the system.

The Vee model of the engineering of a system defines the design and integration processes of TTDSE and forms the basis for this book. These processes are iterative. As illustrated in Figure 1.22, design starts as a top-down process and is analogous to peeling an onion to uncover the specifications associated with increasingly detailed components of the system. However, the trade-offs and decisions associated with the design process are so complex and intertwined that there is significant movement between low-level and high-level design issues. The key to successful design is the isolation of design decisions using sound engineering principles so that this movement between low- and high-level design issues is consistent with the needs of the development process. There are logical arguments for decreasing development costs by spending the money to conduct a reasonable, systematic engineering effort of the total system.

Multiple types of architectures are introduced to differentiate between what the system does (its functions) and what the system is (its resources) and how the functions are allocated to the resources to enhance the cost-effectiveness of the system in the eyes of the stakeholders. The functional and physical architectures are developed in parallel to enhance their integration into the allocated architecture.

Requirements are used to define the design problem being solved at various levels of detail. Mission requirements define the problem in terms most meaningful to the stakeholders, terms that relate to enabling the stakeholders to accomplish tasks better, faster, and cheaper. Stakeholders' requirements are the next level of detail that constrain specific characteristics of the system so as to achieve the mission requirements. Derived requirements relating to the system and specific components are even more detailed constraints upon the system. In addition to the requirements related to the system, qualification system requirements must be developed to address the verification, validation, and acceptance of the system during integration.

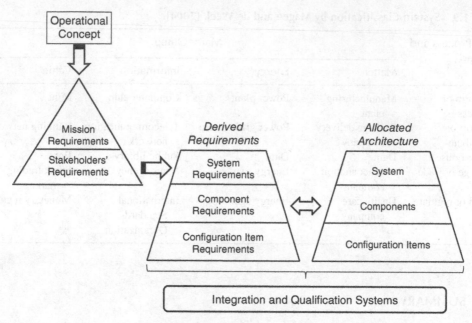

FIGURE 1.22   Summary of TTDSE.

The integration process receives less attention than the design process and is often viewed as the yin (weaker and passive side) of development, design being the yang (stronger and active side). However, integration cannot be passive after an active design process. Rather, design and integration must proceed in harmony; integration, if done well, actually improves as well as checks the design process.

There are at least five ways that good systems engineering adds value. First is defining the problem clearly and well and then finding a good solution that balances the needs of varying segments of stakeholders and the multiple engineering disciplines. Second, systems engineers serve as a communication interface between stakeholders and engineers. Finding showstoppers that are present in the design and getting them fixed is the third value-adding element. Finding design errors early when these errors are still relatively cheap to fix is the fourth. Fifth, systems engineers help identify high-risk elements of the design and develop risk mitigation strategies.

CASE STUDY: HUBBLE TELESCOPE TESTING DECISIONS

Lyman Spitzer of Princeton University (1946) suggested that a telescope in space would eliminate the atmospheric effects that blurred images seen on Earth. The National Academy of Sciences proposed launching a telescope into space in 1972. NASA began the Hubble Space Telescope in 1977. After many project mishaps, the Hubble was ready to be launched in 1986. However, the explosion of the shuttle *Challenger* delayed the launch by four years. In April 1990, the Hubble was launched; on May 20, the moment of truth arrived. At first, the scientists were thrilled with the data that was arriving from space; after further work, though, the scientists noticed a spherical aberration. The Hubble provided a resolution of three times that available with telescopes on the ground, but the originating requirement for Hubble had been ten times Earth-based telescopes. In June 1993, the shuttle *Endeavor* carried a repair team to the hobbled Hubble. The astronauts spent three days of painstaking efforts to install a corrective "contact lens," replace the original Wide-Field and Planetary Camera, and replace the original solar panels to eliminate jitter twice each orbit as the satellite crossed from daylight to darkness. These repairs cost over $50 million.

When the first images from Hubble were examined, the scientists knew that Hubble needed some adjustment. Several focusing tests were proposed. The telescope was taken completely out of focus and then brought slowly back into focus; this is a common approach to check for errors in any optical device. Meanwhile, another scientist wrote a software program to simulate the images from a telescope with a spherical aberration in its mirrors. The test images were amazingly similar to the simulated images, leading to a devastating conclusion.

The Hubble telescope is a two-mirror reflecting telescope, a special type of Cassegrain telescope called a Ritchey-Chretien telescope. The primary mirror (96 inches) and secondary mirror were to be hyperbolic in shape; the manufacturing process is to grind the mirror as close as possible to this shape and then polish the mirror to remove all possible aberrations within the specified tolerances. During the grinding and polishing process, tests were conducted with a computer-controlled optical device, a reflective null corrector consisting of two small mirrors and a tiny lens. Unfortunately, the spacing between the lens and the mirrors was off by 1.3 millimeters. The aberration, 0.001 arcseconds from the design specification, resulted in an error 100,000 times the size of the desired 1/50 the wavelength of light.

Why was a mistake this large not detected? Photos taken during the manufacturing process in 1981 showed the flaw, but the flaw was not noticed in the photos or other testing. A knife-edge test was conducted on the main mirror. This sophisticated and complex test produced results showing that the null corrector results were incorrect. Either Perkin-Elmer (the prime contractor) thought these results invalid and did not report them to NASA, or NASA managers ignored them on the grounds that the knife-edge test results were not correct. Two other tests could have been conducted but were not. Eastman-Kodak was a competing contractor and had built an identical primary mirror. The primary mirrors could have been swapped, and the null corrector tests rerun. The second test was an end-to-end test conducted on the assembled mirrors and other components. This test was deemed too expensive; NASA claimed the test would have cost more than $100 million but soon had to back down when independent estimates were 10 times lower, and the Air Force could possibly have conducted tests using existing equipment.

This testing situation was aggravated and explained by management conflicts and mistakes within NASA and by cost overruns. NASA devised a management structure that included two centers, Goddard and Marshall. Marshall was given primary responsibility even though Goddard had more experience in systems of this type. Lockheed Aerospace was awarded the prime contract. Eastman-Kodak and Perkin-Elmer competed for the job of the primary mirror. Eastman-Kodak had more experience, but Perkin-Elmer provided a lower bid. Eastman-Kodak was given a contract to produce a backup primary minor, a risk mitigation strategy that could have been proven very insightful if the flaw in the Perkin-Elmer mirror had been detected [Fienberg, 1990; Petersen and Brandt, 1995; Sinnott, 1990].

PROBLEMS

1.1 Compare and contrast the waterfall, spiral, cycle, and Vee models of the systems engineering process. In particular, what (e.g., functions performed, time sequence of functions, outputs produced, interaction with stakeholders) is the same in each of these processes and what is different? Are there some categories of systems for which one process would be better than the others? Use outside references to gain more information on the waterfall and spiral models.

1.2 Describe your personal experience with a system whose capability disappointed you. In your opinion, was this disappointment a design mistake made by the system's designers or the result of a trade-off decision that had to be made during the system's design? For example, a keyboard that is too small to be as usable as you would like on a laptop computer is the result of a trade-off decision. However, a keyboard with a poor touch for typing is a design mistake. Consider the following examples:

Example	Design Mistake	Trade-off Decision
Alarm Clock/Radio – the requirements are to show the time, to provide radio reception and listening capabilities, and to serve as an alarm clock. On this particular unit, there are two buttons on the top to adjust the time. The buttons can be depressed easily, both on purpose and accidentally. Accidental depressions will cause the alarm to activate at the wrong time.	This is a design flaw. Requirements development should have established this as a design issue. Testing should have identified the problem.	
Alarm Clock/Radio – the sleep timer, timed play and record, and clock display are only available via the remote control. If the remote is lost, these features cannot be changed.	This may have been a design flaw if not consciously addressed.	This may have been a trade-off decision if placing controls on the unit was too costly.
Digital Audio System – the user wants a repeat button that causes the repetition of a track from a CD; the current repeat button replays the entire CD. The user also wants a means to fast-forward or rewind a few seconds of a track on a CD.	A repeat button for a track on a CD is quite common, so this was probably a design mistake.	Fast-forwarding or rewinding a few seconds could have been a trade-off decision.
Stereo System – The components of the system can be turned on separately, but there is only one Power Off button that controls the entire system.	This is a design flaw; if the components can be turned on separately, they should be able to be turned off separately.	

1.3 More often than desired, engineers are required to estimate quantities related to some aspect of a system because the necessary data is not available. Systems engineers often have to estimate quantities related to the metasystem. There has been quite a bit of attention to estimation in K-12; a common example is to estimate the number of gas stations in the 48 continental states of the United States.

a. How would you go about this? What are several ways to estimate this quantity? Besides information about how many people there are in the United States or how many cars there are in the United States, what other information do you know that might be related to the number of gasoline stations?

b. Search the web and make a list of ways that other people have tackled this problem. Does this list give you any new ideas? What are they?

Chapter 2

Overview of the Systems Engineering Design Process

2.1 INTRODUCTION

This chapter provides a quick tour of many of the major concepts found in Chapters 6–11. This tour is quite valuable as part of an academic course on the engineering of systems because this chapter provides the context for the detailed discussion to follow. However, the advanced reader may wish to skip this chapter. Section 2.2 addresses the processes for design and for integration and qualification. Included here are definitions for key terms such as system, function, and external system. Section 2.3 describes many of the key concepts of the design and integration processes. Included in the design process are the concepts of operational concept, external system model with corresponding external systems diagram, objectives hierarchy, requirements, functions, items, components, and interfaces. Verification, validation, and acceptance are discussed as part of the integration and qualification process. Section 2.4 introduces the Systems Engineering Modeling Language (SysML®) standard from the Object Management Group (OMG). Section 2.5 introduces the GENESYS software, which is a systems engineering tool used in selected portions of this book to enable the student to practice and learn many engineering concepts discussed here.

2.2 DESIGN PROCESS

This section begins by defining some key terms that set the stage for discussing the engineering of a system. Then a more detailed discussion of the two legs of the Vee process, design, and integration (and qualification), is presented in more detail than in Chapter 1.

2.2.1 Key Terms

As part of this overview of the design process, we must establish some important definitions:

System: set of components (subsystems, segments) acting together to achieve a set of common objectives via the accomplishment of a set of tasks. The system may be described as a service.

The Engineering Design of Systems: Models and Methods, Fourth Edition. Dennis M. Buede and William D. Miller
© 2024 John Wiley & Sons, Inc. Published 2024 by John Wiley & Sons, Inc.
Companion website: www.wiley.com/go/engineeringdesignofsystems4e

The system may also be a system of constituent systems, that is, a system of systems. The constituent systems are systems in their own right.

System Task or Function: set of functions that must be performed to achieve a specific objective.

Human-Designed System

- specially defined set of segments (hardware, software, physical entities, humans, and facilities) acting as planned,
- via a set of interfaces, which are designed to connect the components,
- to achieve a common mission or fundamental objective (i.e., a set of specially defined objectives),
- subject to a set of constraints,
- through the accomplishment of a predetermined set of functions.

System's External Systems: set of entities that interact with the system via the system's external interfaces. Note in Figure 2.1, the external systems can impact the system and the system does impact the external systems. The system's inputs may flow from these external systems or from the context, but all of the system's outputs flow to these external systems. The external systems, many or all of which may be legacy (existing) systems, play a major role in establishing the stakeholders' requirements.

System's Enabling Systems: set of supporting systems that facilitate the life cycle activities of the system. The enabling systems provide services needed by the system during life cycle stages, for example, the modeling, development, supply chain, production, training, and logistics systems. The system and its enabling systems are integrated in digital engineering as the digital thread over the life cycle.

System's Context: set of entities that can impact the system but cannot be impacted by the system. The entities in the system's context are responsible for some of the system's requirements. A caveat is that the context may change over the life cycle of the system and therefore should be modeled and tracked; the change may even be stimulated by the system. Changes in context over the system life cycle must be addressed so that the system remains fit for purpose.

Ecosystem: integration of the system (or system of systems), enabling systems, external systems, and context to establish a closed universe to rigorously model and engineer the system of interest. The ecosystem is both spatial and temporal and is modeled over the life cycle of the system to assure the rigor and fidelity acceptable to the stakeholders subject to the availability of resources.

Fit for Purpose: Used informally by the U.S. National Institute of Standards and Technology (NIST) to describe a process, configuration item, service, etc., that is capable of meeting its objectives or service levels. Being fit for purpose requires suitable design, implementation, control, and maintenance.

System Model: "Models represent various aspects of a system, but the only complete, full, or perfect representation of the system is the system itself. This is particularly true for cyber systems and other non-deterministic systems where non-deterministic system response modeling approaches are not well defined. In modeling a system, the system model is composed of multiple models of aspects of that system, and each approximates reality" [INCOSE, 2022a,b]. See Figure 2.1.

2.2.2 Design

As discussed in Chapter 1, design includes identification and modeling of the ecosystem, decomposition, and definition of both the requirements, or statement of the design problem, and

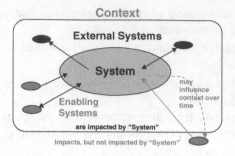

FIGURE 2.1 Depiction of the system, external systems, enabling systems, and context. The system may influence the context over time.

the system architectures: functional and physical, and allocated representations of the system. The allocated architecture addresses which physical resources of the system are going to perform which functions, but also includes a mapping of all the requirements to the physical resources of the system. In addition to addressing the system that must be operational for users, the design process should also address all relevant enabling systems needed during the life cycle of this system: the development system, of which the systems engineers are part; the manufacturing system, if needed; the deployment system, if needed; any training systems that are needed; a refinement system for system upgrades; and the retirement system, if needed. Finally, the qualification systems for each of these systems need to be addressed.

There are nine functions below that capture the complete systems engineering process throughout the system life cycle. The first two (0a and 0b) are performed at the very beginning and are really part of a general problem-solving process. The last seven functions (1–7) are the focus of this book and are considered the core repetitive functions of the systems engineering design process:

0a *Define the problem to be solved.*
0b *Define and evaluate alternate concepts for solving the problem.*
1. Define the system-level design problem being solved.
2. Identify the ecosystem to the rigor and fidelity acceptable to stakeholders.
3. Develop the system functional architecture.
4. Develop the system physical architecture.
5. Develop the system allocated architecture.
6. Develop the interface architecture.
7. Define the qualification system for the system.

All nine of these functions are shown in Table 2.1 with their major inputs and outputs. Chapters 6–11 address the last seven functions, respectively. As can be seen by the respective inputs and outputs of these functions, these last seven functions cannot be conducted in series but must be concurrent. The resource that performs these functions is the systems engineering team.

The control flows of these design functions are adaptable to top-down, middle-out, composable, reverse engineering, and agile system design approaches.

The first of the repetitive design functions must create a definition of the problem being solved for which the next six develop a set of designs (across the system's life cycle). The seven functions that comprise this first design function, defining the design problem (stakeholders' requirements), are as follows:

1. Develop an operational concept.
2. Define the system boundary with the ecosystem.

TABLE 2.1 Functions of the Design Process

Design Function	Major Inputs	Major Outputs
Define problem to be solved	Concerns and complaints by stakeholders Available data from stakeholders	Definitions of measures of effectiveness and desired ranges Constraints
Identify the ecosystem	Contextual, enabling systems, external systems	Enabling systems, external systems, and possibly contextual
Develop and evaluate alternate concepts for solving problem	Ideas for concepts from all interested parties	Recommended concept(s) Objective hierarchy and value parameters for meta-system
Define system-level design problem being solved	Stakeholders' inputs	Stakeholders' requirements Operational concept
Develop system functional architecture	Stakeholders' requirements Operational concept	Functional architecture
Develop system physical architecture	Stakeholders' requirements	Physical architecture
Develop system allocated architecture	Stakeholders' requirements Functional architecture Physical architecture Interface architecture	Allocated architecture
Develop interface architecture	Draft allocated architecture	Interface architecture
Develop qualification system for the system	Stakeholders' requirements Systems requirements	Qualification system design documentation

3. Develop the system objectives hierarchy.
4. Develop, analyze, and refine the requirements (both stakeholders' and system).
5. Ensure requirements feasibility.
6. Define the test system requirements.
7. Obtain approval of system documentation.

These seven functions are shown with their major inputs and outputs in Table 2.2. There is an important distinction between the stakeholders' requirements and the system requirements. Stakeholders' requirements are those requirements that the system's stakeholders agree define their needs. As such, the stakeholders' requirements are written in the common language of the stakeholders (e.g., English and Chinese). The system requirements are a translation of the stakeholders' requirements into the appropriate engineering terminology (e.g., foot-pounds, bits, and decibels). Systems requirements may be derived using formal methods, such as Boolean logic statement amenable to formal verification methods. Chapter 6 presents more details about this process.

The detailed processes for developing the functional, physical, allocated, and interface architectures are not presented here because many of the concepts for these architectures are not appropriate for this overview. The decomposition of the last function of design, developing the qualification system for the system, is a replication of the design process for the system but with the focus on the elements of the qualification system.

The design process, as presented here, is not a formal process in the sense that success can be proved; designs can be proved to be correct, and so forth. Some researchers have developed formal processes, primarily in software engineering. These formal processes have succeeded in embedded

TABLE 2.2   Functions of the System-Level Design Problem

Design Function	Major Inputs	Major Outputs
Develop operational concept	Stakeholders' inputs	Operational concept
	Objectives hierarchy and value parameters for meta-system	System concept
		Input–output traces
	Recommended concept	Meta-system MOEs
Define system boundary with the ecosystem	Operational concept	System boundary
		System's inputs and outputs
Develop system objectives hierarchy	Operational concept	System-level objectives hierarchy
	Stakeholders' inputs	
Develop, analyze, and refine requirements (stakeholders' and system)	Operational concept	Stakeholders' and systems' requirements
	System boundary, input and output objectives hierarchy	
	Stakeholders' inputs	
Ensure requirements feasibility	Stakeholders' and systems' requirements SE team's inputs	Design feasibility
Define the test system requirements	Stakeholders' and systems' requirements Stakeholders' inputs	Test system requirements
Obtain approval of system documentation	Stakeholders' and systems' requirements	Stakeholder' and systems' requirements documents

systems and microprocessors, as well as testing, but have not succeeded very frequently in software development and are relatively rare in the engineering of systems. An example of such a formal process for engineering design is Suh's [1990] axiomatic design process. Suh defines two major concepts – functional requirements and design parameters. He posits two axioms: (1) independence axiom: maintain the independence of the functional requirements and (2) information axiom: minimize the information content of the design.

While Suh introduces hierarchical decomposition in his axiomatic process, there is not sufficient richness of concepts in his process to handle the complexity of the engineering issues associated with the development of a system. First, as will be discussed in Chapter 6, the set of functional requirements is a derived entity that has no inherent meaning to the stakeholders; input and output requirements are statements that relate to stakeholders needs. Second, Suh's process does not provide a sufficient process to develop and enable validation of the requirements. Finally, the interaction of functions, components, and interfaces, as described in this book but is missing in some richness in Suh's approach, is needed to deal with the generation and analysis of design options, as well as guide the qualification of the system in terms of the stakeholders' needs.

2.2.3   Integration and Qualification

The second half of the Vee model, integration and qualification, is primarily a bottom-up process that comprises integrating the most basic building blocks of the system and verifying that these lower-level components meet the specifications, or sets of requirements, that were developed for them during design. However, before this integration and verification process begins, a validation of the requirements development process should take place to attempt to demonstrate that the low-level design solutions are still consistent with the stakeholders' needs. At the end of qualification come the important steps of validating that the system that has been designed and verified does in fact agree with the operational concept and is acceptable to the stakeholders. The stakeholders include the bill

TABLE 2.3   Functions of the Integration and Qualification Process

Design Function	Major Inputs	Major Outputs
Conduct early validation	Stakeholders' inputs Operational concept Stakeholders' requirements Derived requirements	Validated requirements Validated operational concept
Conduct integration and verification testing	Configuration items (CIs) Components Derived requirements	Verification testing document Verified components and system
Conduct system validation testing	Verified system Stakeholders' requirements Stakeholders' inputs	Validation testing document Validated system
Conduct system acceptance testing	Validated system Acceptance test plan Stakeholders' inputs	Acceptance testing document Accepted system

payer, the users, the maintainers and supporters, the manufacturers, the trainers, the deployers, the refiners, and the retirers. The integration and qualification process can be divided into four segments:

1. Conduct early validation
2. Conduct integration and verification testing
3. Conduct system validation testing
4. Conduct system acceptance testing

Table 2.3 presents these four functions that comprise integration and qualification with their major inputs and outputs. Each of these functions is described in more detail in Chapter 11.

2.3   KEY SYSTEMS ENGINEERING CONCEPTS

This section provides a detailed discussion of key constructs for design and integration. An operational concept, ecosystem model, objectives hierarchy, and requirements are the essential elements of the definition of the design problem. Functions and items comprise the functional (or logical) architecture. Components are the building block for the physical architecture. Interfaces combined with the functional and physical architectures are key to understanding the allocated architecture. Qualification is verification, validation, and acceptance during integration. Figure 2.2 provides an entity relation diagram showing many of these concepts (or entities) and their relationships. Different systems engineering modeling tools may use different verbs to specify these relationships, and the relationship verbs in a modeling tool may be changed in different versions of the tool.

2.3.1   Operational Concept

An *operational concept* is a vision for what the system is (in general terms), a statement of mission requirements, and a description of how the system will be used. A set of scenarios describes how the system will be used by defining the system's interaction with other systems. For example, the operational concept for an elevator system may begin with a description of cars that carry people and equipment moving vertically in shafts. The mission requirements would discuss the desired times that passengers would wait from the time they requested service until they arrived at their

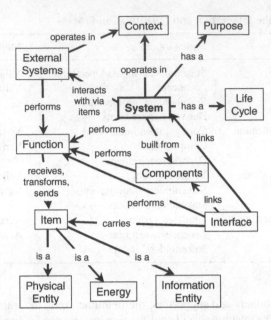

FIGURE 2.2 Many of the concepts and their relationships.

TABLE 2.4 Sample Operational Concept Scenarios for an Elevator

(1) Passengers (including mobility, visually, and hearing challenged) request up service, receive feedback that their request was accepted, receive input that the elevator car is approaching and then that an entry opportunity is available, enter the elevator car, request a floor, receive feedback that their request was accepted, receive feedback that the door is closing, receive feedback about the floors at which the elevator is stopping, receive feedback that an exit opportunity is available at the desired floor, and exit the elevator with no physical impediments.

(2) Passenger enters the elevator car, as described in (1), but finds an emergency situation before an exit opportunity is presented and notifies the police or health authorities using communication equipment that is part of the elevator. Elevator maintenance personnel create an exit opportunity.

(3) A maintenance person needs to repair an individual elevator car; the maintenance person places the elevator system in "partial maintenance" mode so that the other cars can continue to pick up passengers while the car(s) in question is (are) being diagnosed, repaired, and tested. After completion, the maintenance person places the elevator system in "full operation" mode.

destination. Next, the operational concept would use scenarios such as those in Table 2.4 to define how the elevator would be used during the elevator's operational phase.

2.3.2 Ecosystem Model

Defining the boundaries of a system is critical but often neglected. An *ecosystem model* is used to establish the bounds of the system and communicate the results of this bounding process. This model can be created from the scenarios in the operational concept for the system and should be completely consistent with those scenarios.

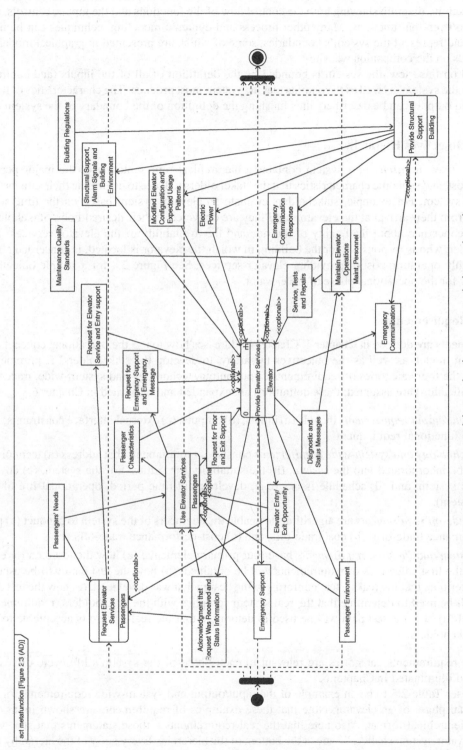

FIGURE 2.3 Ecosystem model view for the operational phase of an elevator.

Figure 2.3 shows an ecosystem model view for the operational phase of the elevator using a SysML activity diagram showing both the control flow of the functions and the inputs, controls, and outputs to/from the functions. Many other process and dynamic modeling techniques can be used to draw and represent the system's boundaries, some of which are presented in graphical modeling techniques on the companion website.

Based on this view, the system is bounded by the definition of all of the inputs (and controls) that enter the system, as well as the outputs that the system must produce. The characteristics of these inputs and outputs can be described, thus finishing the definition of the boundary of the system.

2.3.3 Objectives Hierarchy

The *objectives hierarchy* of a system contains a hierarchical representation of the major performance, cost, and schedule characteristics that the stakeholders will use to determine their satisfaction with the system. For example, stakeholders evaluate an elevator system based on the time spent waiting from their arrival at the elevator until they are delivered at their desired floor. Stakeholders are also concerned about the quality of the ride and the availability of the elevator services. The stakeholder, who is responsible for the building in which the elevator is located, is concerned about the monthly operating cost of providing elevator services. See Figure 2.4 for a sample objectives hierarchy for the operational phase of the elevator.

2.3.4 Requirements

Requirements are defined in Chapter 1. Chapter 6 addresses how to use the operational concept and ecosystem model, as well as the objectives hierarchy, to develop the stakeholders' requirements. For now, the four categories of requirements (input/output, technology and system-wide, trade-off, and qualification) are assumed; these definitions are expanded and motivated in Chapter 6.

1. *Input/output requirements* include (a) inputs, (b) outputs, (c) external interface constraints, and (d) functional requirements.
2. *Technology and system-wide requirements* consist of requirements that address (a) technology to be incorporated into the system, (b) the suitability (or "ilities") of the system, (c) cost of the system, and (d) schedule issues (e.g., development time period, operational life of the system).
3. *Trade-off requirements* are algorithms to enable the engineers of the system to conduct (a) performance trade-offs, (b) cost trade-offs, and (c) cost-performance trade-offs.
4. *System qualification requirements* have four primary elements: (a) how the test data for each of the first categories of requirements will he obtained, (b) how the test data will be used to determine that the real system conforms to the design that was developed, (c) how the test data will be used to determine that the real system complies with the stakeholders' requirements, and (d) how the test data will be used to determine that the real system is acceptable to the stakeholders.

These requirements categories are relevant to each phase of the system's life cycle discussed above and illuminated in Chapter 6.

Consider Table 2.5 to be an example of the input/output and system-wide requirements for the operational phase of an elevator. Note that these examples of requirements are shown in an outline or hierarchical format. Also note that the real requirements – those statements that start with "The Elevator system shall ... " – are at the bottom of the hierarchy. Every entry of the hierarchy that has another level below it is not really a requirement, but a group of requirements.

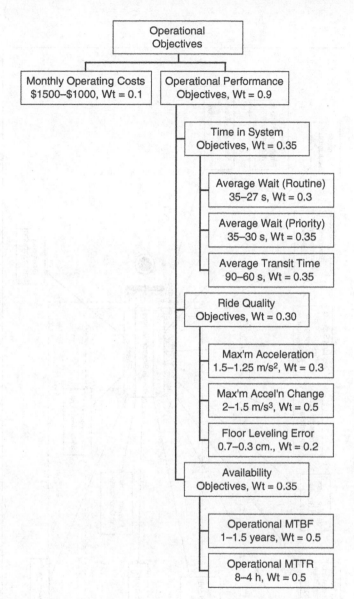

FIGURE 2.4 Fundamental objectives hierarchy for operational phase of elevator.

2.3.5 Functions

A *function* is a transformation process that changes inputs into outputs. A system is modeled as having a single, top-level function that can be decomposed into a hierarchy of subfunctions. The system's top-level function transforms all the inputs to the system into all of the outputs of the system. See Figure 2.5 for the elevator example using a SysML activity diagram. This system function can be taken directly from the activity diagram of the ecosystem model.

The top-level function is decomposed into subfunctions as part of the development of the functional architecture. Each subfunction transforms a subset of the inputs from the outside (plus some other internally generated inputs) into a subset of the outputs (plus some other internally generated

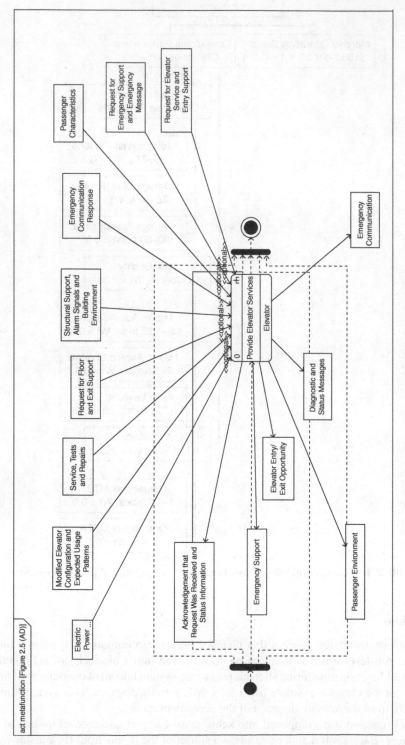

FIGURE 2.5 Top-level function of the elevator.

TABLE 2.5 Sample Elevator Requirements for the Operational Phase

---

4.3 Stakeholders' Requirements
    4.3.5 Operational-Phase Requirements
        4.3.5.1 Input/Output Requirements
            4.3.5.1.1 Input Requirements
                4.3.5.1.1.1 Emergency Support Inputs
                    4.3.5.1.1.1.1 The system shall support manual overrides
                    4.3.5.1.1.1.2 The elevator system shall allow passengers with a designated pass key to assume
                          complete control of an elevator car
            4.3.5.1.2 Output Requirements
                4.3.5.1.2.1 Passenger Environment Outputs
                    4.3.5.1.2.1.1 The elevator car shall have adequate illumination
            4.3.5.1.4 Functional Requirements
                4.3.5.1.4.1 The elevator shall accept passenger requests and provide feedback
                4.3.5.1.4.2 The elevator shall move passengers between floors safely and comfortably
                4.3.5.1.4.3 The system shall control elevator cars efficiently
                4.3.5.1.4.4 The system shall enable effective maintenance and servicing
        4.3.5.2 System-wide and Technology Requirements
            4.3.5.2.1 The system MTBF shall be greater than 1 year. The design goal is 1.5 years. Failure is
                defined to be a complete inability to carry passengers
            4.3.5.2.2 The system MTTR shall be less than 8 hours. The design goal is 4 hours. Repair means the
                system is returned to full operating capacity

---

outputs). This decomposition cannot be found in a book or dictated by the stakeholders; the decomposition is a product of the engineers of the system and is part of the architectural design process that is attempting to solve the design problem established by the requirements. The decomposition can be carried out as deeply as needed to define the transformations that the system must be able to perform. Figure 2.6 shows the first-level decomposition of the elevator system function represented with an activity diagram.

2.3.6 Items

Items are the inputs or triggers that are received by the system, the outputs that are sent by the system to other systems, and the inputs or triggers that are generated internally to the system and sent to other parts of the system to assist in the transformation process for which the system is responsible. Items can be physical entities that have mass and energy, or they can be information that is somehow transformed into items with mass or energy to be transmitted from one physical element to another. Items may be input to Functions as either inputs or triggers. In a trigger-type connection, the receiving Function does not execute until the "trigger" item has "arrived." For input type connections, the Function is executed independent of the Item's "arrival." The external inputs/triggers and outputs of the elevator are shown in Figure 2.5. Both the external and internal items of the elevator, at the first level of functional decomposition, are shown in Figure 2.6.

2.3.7 Components

A *component* of a system is a subset of the physical realization (and the physical architecture) of the system to which a subset of the system's functions has been (will be) allocated. A component could be the integration of hardware and software, a specific piece of hardware, a specific segment

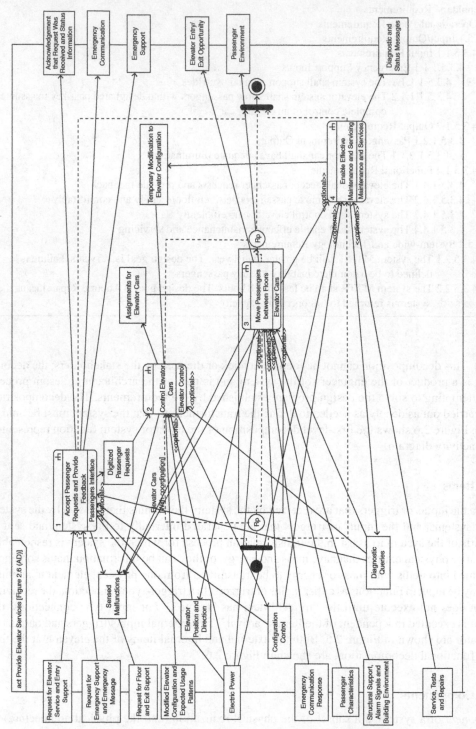

FIGURE 2.6 First-level decomposition of the elevator system function.

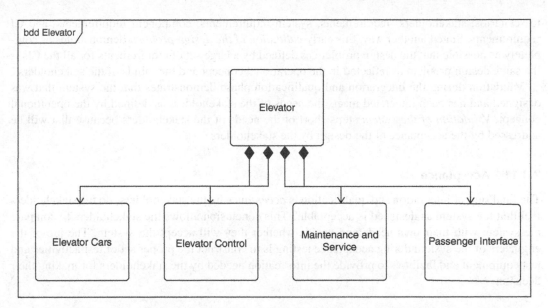

FIGURE 2.7 Physical architecture of the elevator system.

of the system's software, a group of people, facilities, or a combination of all of these. As with the requirements and functions, there is often a hierarchical structure to the components that comprise the system. The first-level and second-level decompositions of the elevator into components are shown in Figure 2.7.

2.3.8 Interfaces

An *interface* is a connection resource for hooking to another system's interface (an external interface) or for hooking one system's component to another (an internal interface). Interfaces have inputs, produce outputs, and perform functions. An interface can be as simple as a wire or conveyor belt or as sophisticated as a global communication system (which is a system in its own right). Interfaces are represented as Links that may be decomposed. The Links transfer Items between Functions.

2.3.9 Verification

As discussed in Chapter 1, verification addresses whether the system was built right. In practical terms, *verification* is the determination that each configuration item (CI), component, and the system meets the requirements for it during the design phase. These requirements will be input/output or technology/system-wide requirements. Inspection, testing, analysis and simulation, or demonstration can establish this verification. Inspection and testing are most common at the CI level. Demonstration and analysis/simulation are most common at the system level.

2.3.10 Validation

Validation addresses whether the right system has been developed. Validation gets back to the illusive needs of the stakeholders. One aspect of validation that should be performed during the design phase of development attempts to demonstrate that the design problem is evolving properly from a high-level statement in the operational concept that correctly reflects the needs of the stakeholders to scenarios, stakeholders' requirements, system requirements, component requirements, and CI requirements. Stated another way, this early *validation of the design problem* demonstrates as completely as possible that the design problem as defined by a large set of requirements for all the CIs is the same design problem as reflected in the operational concept and the minds of the stakeholders.

Validation during the integration and qualification phase demonstrates that the system that was designed and has been integrated meets the needs of the stakeholders as defined by the operational concept. *Validation of the system* stops short of the needs of the stakeholders because that will be addressed by the acceptance of the design by the stakeholders.

2.3.11 Acceptance

The final step of integration and qualification is *acceptance* by the stakeholders; do the stakeholders feel that the system as designed is acceptable? This conclusion allows the stakeholders to compare the system with their own needs and decide whether they will accept the system. The job of the engineers of the system during acceptance testing is to construct the proper set of test activities and test equipment and facilities to provide the information needed by the stakeholders for making their decision.

2.4 INTRODUCTION TO SysML

As described in Chapter 1, SysML is a visual modeling language for conducting systems engineering. Sections 2.2 and 2.3 have introduced you to some of the key terms and basic process for performing traditional, top-down systems engineering (TTDSE). Figure 2.8 shows the processes of TTDSE associated with the left-hand side of the Vee model on the left-hand side of the figure. Included with these process elements of TTDSE are modeling activities that should utilize a graphical modeling language. The right-hand side of Figure 2.8 shows the diagram types of SysML. Double-headed arrows in Figure 2.8 show which modeling elements from TTDSE are addressed by which diagram types of SysML. Naturally, there are many-to-many and many-to-one relations shown. There are also three modeling activities in TTDSE that are not supported by SysML (requirements taxonomy, creativity and the morphological box for physical architecture alternatives, and risk and trade studies). Each of these topics is addressed in Chapters 6, 8, and 14, respectively. Also, there are two diagram types (use case and package) in SysML that are not related to the elements of TTDSE from the first edition of this book; these two diagram types are covered in Chapter 3.

This figure should convey to you that SysML is a nearly complete interconnected set visual modeling diagram that enables a substantial improvement in model-based systems engineering. All elements of the design process that we cover in this book are covered. Since the test system is another system, SysML can be equally applied to the design and operation of the test system, just as it can be applied to the design of the systems engineering system of engineers, domain experts, technologists, and managers.

2.5 USE OF GENESYS SYSTEMS ENGINEERING TOOL

Part of the educational material provided with this book is an academic version of a system engineering tool, called GENESYS, a one-off of the original CORE software. (You may download the academic version of GENESYS from http://www.vitechcorp.com.) The rest of this chapter provides an overview of concepts embedded in GENESYS. Instructions for using GENESYS can be obtained from the user's manual and guided tour available with GENESYS.

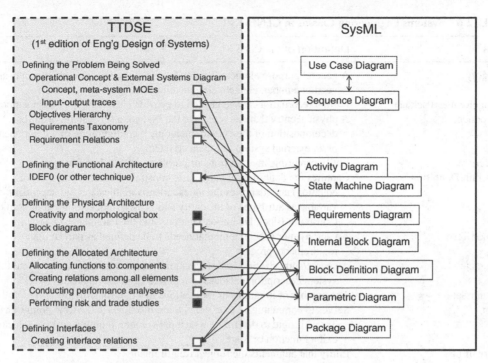

FIGURE 2.8   Comparison of TTDSE and SysML.

At its simplest, GENESYS is comprised of classes (e.g., requirements, functions, and items), examples or elements of those classes (e.g., specific requirement), and relations. The most basic user activities of GENESYS are entering and editing elements of the classes and establishing relations between elements of classes. Other important activities include viewing products of the design data, saving your work, and obtaining reports that document the design contained in the database. The automated tutorial demonstrates these functions for a geospatial library system as a sample problem. GENESYS is compliant with the SysML standard and renders all SysML diagram types in addition to rendering legacy diagrams.

2.5.1   Classes

Table 2.6 lists the systems engineering relevant classes in GENESYS. The GENESYS documentation describes the complete set of pre-established classes and additional classes can be customized. These classes contain both the major elements of the systems engineering design process that are discussed in this chapter and a number of supporting classes. For a given system, the job of the engineers of the system is to define specific elements of the system for each of these major classes (e.g., stakeholders' requirements, functions, components, and items).

2.5.2   Relations

As part of the engineering design process, requirements must be related to functions and components using the specify relation, functions allocated to components, and inputs and outputs assigned to interfaces. Table 2.7 defines the relations available in GENESYS to define relationships within the design and integration classes. These relations are fully compatible with the mathematical definition

TABLE 2.6 Systems Engineering Classes in GENESYS

Class	Definition of the Class
Category	A general purpose element that can be used to represent such concepts as version number, element classification, etc.
ChangeRequestPackage	Formal record of a change needed to provide change management support
Component	A physical entry that can represent the system, a subsystem, or further decomposition of the system, including a configuration of a component; or an external system or the metasystem
Concern	Entity capturing any questions or problems with a Requirement
ConstraintDefinition	Captures the definition of parametric constraints as an expression (or equation) and identifies the independent variables and the dependent variable as attributes of the entity and is used in development of both the Constraint Block Definition Diagram and the Parametric Diagram
Defined Term	An acronym or special term that needs to be defined as part of the requirements process
Document	A source/authorization for information from stakeholders entered into the system description database or reported from the database
Domain Set	The number of iterations or replications in a control structure
Event	Serves to communicate to external State machines at the time point of a
Exit	TransitiLogic to determine which path among multiple paths is selected to exit a Function or State
ExternalFile	Entity that augments the subject Requirement
Function	A process that accepts one or more inputs (items) and transforms them into outputs (items). A function should have a completion criterion for each exit
Item	Physical entities or data that flows within and between functions. An item is an input to or output from a function and may be data, material, or energy
Link	The physical implementation of an interface
MitigationActivity	Action performed to reduce probability of occurrence or consequence/impact of an uncertainty element
Mode	Collection of States
Note	Informal comments, additional information, or queries regarding the characteristics of a particular entity in a design model
Package	Arbitrary clustering of model entities to communicate groupings and in interrelationships of interest
Port	Identification of the place where entities can connect to and interact with a specified component block
PortDefinition	Port providing a PortDefinition shown on SysML Flow Internal Block as the ball notation connected to the Port
Requirement	A requirement extracted from the source documentation for a system, or a refinement of a higher-level requirement. Requirements should be refined until only a single, testable statement of a system's feature remains
RequirementGroup	Grouping of Requirements
Resource	A characteristic (e.g., power, channels, instructions processed per second) of one or more components that are used, captured, or generated and can be depleted during the operation of the system
Risk	The uncertainty of attaining/achieving a product performance level or program milestone

(continued overleaf)

TABLE 2.6 (*continued*)

Class	Definition of the Class
ServiceSpecification	Service attributes for an internal service developed throughout the operational and system process or service attributes for an external service (one which is an external in the system context) provided by the service provider
State	Sometimes used as the highest-level functional breakout to define a set of functions that system performs at a point in time, for example, start-up, normal operation, recovery operations, and shutdown
Text	Entity augmenting a State for the purpose of further enhancing the meaning or representation of the State
Transition	The movement of one State to another State occurs through a Transition, which triggers an Event
UseCase	Scenario in which a system receives an external request as an input and responds to it. The requirement to be met by the qualification system, the
VerificationRequirement	level at which it must be met, the method of qualification, and current qualification status
VerificationRequirement Group	A clustering of Verification Requirements

of relations in Chapter 4 and can be depicted graphically by directed graphs as discussed in Chapter 5. For each relation, there is an opposing relation that reverses the direction of the relation. Table 2.7 shows common systems engineering relations (and their opposing relations in parentheses) and defines each relation by identifying which class is on the left side (tail of the arrow) and which class is on the right side (head of the arrow). The GENESYS documentation describes the complete set of pre-established relations and additional relations can be customized.

One subtlety that has been ignored so far is the relating of requirements to functions and items, or the system. Input/output requirements are defined in such a way that each such requirement is directly relatable to both specific functions and items. Technology and system-wide requirements are those requirements that cannot be related to specific functions or items but must be satisfied by the system. As a result, each input/output requirement is traced to (or specifies) the lowest-level function that receives the relevant input or produces the relevant output, all the functions that are above that function in the functional decomposition, and the item directly relevant to that requirement. (Note that the third category of input/output requirements is function requirements; these requirements specify the top-level system function because they define the decomposition of that function.) Similarly, each technology and system-wide requirement specifies the system. Relating requirements to functions and the system through the specify relation is important because this activity initiates the process of creating a set of requirements for the system to satisfy and provides the material for subsets of the requirements to be associated with specific components. These subsets of requirements become the specifications that each CI design team must meet. (Note that the requirements related to functions ultimately are assigned to the system and its components when each function is allocated to one or more components for execution.)

2.5.3 Documents

GENESYS enables you to design your document. However, the outline of the document that is used throughout this book is the System Description Document (SDD), which can be found under the reports available from GENESYS in the Project Explorer. This SDD outline (see Table 2.8) has

TABLE 2.7   Common Systems Engineering Relations

Relation [Classes] (Opposing Relation)	Definition	Application
refines [requirements] (refined by) decomposes [functions, items, links, state] (decomposed by)	The left side is a subset of the right side. For example, the requirement on the left side incorporates the one on the right, and the function on the left side is decomposed by the one on the right	Hierarchies of requirements, functions, items, components
input to/triggered by (inputs/triggers)	The item on the left side is an input or trigger to the function on the right side	Development of the functional architecture
output from (outputs)	The item on the left side is an output from the function on the right side	Development of the functional architecture
"incorporated by" ("incorporates")	The state/mode on the left side incorporates the function on the left side	Development of the functional architecture
"built from" ("built in")	The left side (system or component) is comprised of the system or component on the right side	Development of the physical architecture
"exhibited by" ("exhibits")	The state/mode on the left side is exhibited by the component on the right side	Development of the physical architecture
"allocated to" ("performs")	The function on the left side is being assigned to the component on the right side for the purpose of execution	Development of the operational architecture
"basis of/specifies" ("specified by")	The requirement on the left side specifies the function, state/mode, item, component, interface, or link on the right side	Development of the functional and operational architectures
"transfers" ("transferred by")	The link on the left side transfers the item on the right side	Development of the interface architecture
"connects to" ("connected to")	The link on the left side connects to the component or system on the right side	Development of the interface architecture

options for Component, Package, or Project and contains an initial section in which a general description of the system would be provided; the operational concept would be found here if GENESYS captured this material. The requirements are found in Section 2 is on risks.

Section 3 enables the systems engineering team to capture the design concerns and decisions, which are usually addressed as part of the allocated architecture. Risk management is addressed in Section 4; this is the place that key uncertainties are defined, and the potential impact of bad outcomes is also defined. Use cases are documented in Section 5. States and Modes are described in Section 6. The functional architecture is defined in Section 7 as both a process model and a behavioral model; GENESYS uses an N^2 model view for the system's process and function flow block diagram or activity diagram model for the system's behavior(both models are covered in the material on Graphical Modeling Techniques on the Companion website). The item dictionary (or data model) is found in Section 8. The resources consumed to operate the system are defined in Section 9, and the physical architecture is defined in Section 10. The logical and physical interfaces developed by the system engineering team are described in Section 11. Verification is defined in Section 12. Section 13 provides a requirements traceability matrix (RTM). Section 14 documents acronyms, and Section 15 is the glossary.

Note that SysML does not have an N^2 diagram type; SysML tool vendors provide an N^2 table view in their tools.

TABLE 2.8   Outline of System Description Document

0.1 Cover Page
  1 Overview
  2 Requirements
  3 Concerns and Decisions
  4 Risks
  5 Use Cases
  6 States and Modes
  7 Functions
  8 Item Dictionary
  9 Resources
  10 Components
  11 Interfaces and Links
  12 Verification
  13 Requirements Traceability Matrix (RTM)
  14 Acronyms
  15 Glossary

2.6   SUMMARY

This chapter has given definitions and provided discussions on the most important concepts in the engineering of systems. The operational concept of the system provides the theme for the system as viewed by the stakeholders and defines scenarios depicting how its users will employ the system and how the system will interact with other systems. The ecosystem model defines the interaction in terms of inputs and outputs with other systems and is consistent with the operational concept. The objectives hierarchy of the system lays out the performance, cost, and schedule objectives that the stakeholders have for the system; this objectives hierarchy provides a satisfaction index for the stakeholders for alternate system designs. The requirements of the system provide constraints and performance ranges for the system in terms of inputs and outputs and its system-wide and technology-related characteristics. The requirements also state the trade-offs that the stakeholders are willing to make in the development of the system and the constraints and performance ranges associated with testing the system. These first four concepts deal with defining the design problem.

Three additional concepts (functions, components, and interfaces) are part of the design process. Functions are those activities performed by the system (and all other systems) to transform inputs into outputs. Components are the physical entities of the system that perform the system's functions. Resources are the consumables used to operate the system. Interfaces and links connect components, external interfaces and links connect components of the system to components of other systems, and internal interfaces and links connect components of the system to each other.

Throughout this entire process, from the operational concept through requirements, it is important to remember that the engineers have to concern themselves not only with the operational system that the users of the system want but also with the systems relevant to every stage of the life cycle of the system (e.g., the development system, the manufacturing system, and the retirement system).

IDEF0 is a legacy process modeling technique that is described in detail in Chapter 3 and used throughout this book to model the processes to engineer systems. The elevator case study model is represented with SysML diagram views. The software product GENESYS used extensively in this book was described in terms of its classes and the relationship between those classes for systems engineering. GENESYS' data structure is a one-off from the earlier CORE software tool that originated from a data modeling technique called entity–relationship (ER) diagrams, which is discussed in more detail by the graphical modeling techniques on the companion website.

PROBLEM

2.1 Use the requirements in Table 2.5 to define the elevator's requirements. Use the activity diagram views of the elevator in Figures 2.4 and 2.5 to define the functional decomposition of the elevator system and to identify the external inputs and outputs, as well as those that are internally generated and consumed. Use the first-level decomposition in Figure 2.6 to define the physical decomposition of the elevator components.

Enter all the above information on the elevator as the system into GENESYS (or tool of choice): enter the requirements shown in Table 2.5, enter the functions shown in Figures 2.5 and 2.6, enter the items shown in Figure 2.5, and enter the components shown in Figure 2.6 as elements of the corresponding classes. Then establish the relevant relations associated with Table 2.7. These relations include hierarchies for the requirements and functions as well as the system with its components. Use the "specify" relation to connect the requirements to the appropriate function, item, or system, designate items as inputs or outputs of the relevant functions, and allocate each function to the system or appropriate component.

Chapter 3

Modeling and SysML Modeling

3.1 INTRODUCTION

This chapter serves two major purposes: it describes models and their role in the engineering of systems and introduces several modeling techniques associated with both legacy modeling and SysML. The legacy modeling techniques introduced in this chapter are use case diagrams, sequence diagrams, IDEF0 (Integrated Definition for Function Modeling), N^2 charts, Function Flow Block Diagrams (FFBDs), enhanced Function Flow Block Diagrams (EFFBDs), state machine diagrams, and block diagrams. SysML also uses use case diagrams, sequence diagrams, and state diagrams as well as requirements diagrams, block definition diagrams, interface block diagrams, activity diagrams, and parametric diagrams. IDEF0 is a process modeling technique that is not part of SysML but is utilized throughout this book. SysML activity diagrams are equivalent to legacy FFBDs and EFFBDs.

Models, abstractions of reality, are critical in the engineering of systems. These models start as very high-level representations that address what needs the system should meet and then progressively define how the system will meet these needs. These models contain increasingly more mathematical and physical details of the system as the design portion of the development phase ends. The various engineering disciplines create even more detailed mathematical and physical representations of the configuration items (CIs) before the final prototype of each CI is produced for testing and integration. During the qualification of the system design, these CI prototypes are tested with a test system that itself is comprised of many models of the system's components, other systems and the context with which the system interacts, models of scenarios that depict how the system will be used, and analysis and simulation models for creating and analyzing the test results. In fact, models are so pervasive in the engineering of systems that engineers must always remind themselves not to confuse *reality* with the *models of reality* that are being created, tested, and used.

Every modeling technique is a language used to represent some part of reality so that some questions can be answered with greater validity than could be obtained without the model. All languages have a set of symbols or signs, known as *semantics*, that are used like we use letters and numbers to form expressions. Similarly, every language has a *syntax* that defines proper ways of combining the

The Engineering Design of Systems: Models and Methods, Fourth Edition. Dennis M. Buede and William D. Miller
© 2024 John Wiley & Sons, Inc. Published 2024 by John Wiley & Sons, Inc.
Companion website: www.wiley.com/go/engineeringdesignofsystems4e

symbols to form thoughts and concepts. Section 3.2 summarizes the descriptive versus normative purposes of models and then categorizes models as physical, quantitative, qualitative, and mental.

SysML version 1 is a modeling language that is an extended subset of the Unified Modeling Language (UML) for software engineering. SysML matches somewhat closely with Traditional Top-Down Systems Engineering (TTDSE). SysML version 2 breaks the reliance on UML to add a data object relationship meta-model demanded by users. In Section 3.3, we introduce SysML, including use case and sequence diagrams for high-level metasystem interactions, activity diagrams for dynamic behavior modeling, block diagrams for structural modeling, and parametric models for modeling equations. Note, SysML tool vendors also address requirements using tables with textual representations in addition to the requirements diagram. Users find the requirements table views more fit for purpose because of the relatively large space taken by requirements diagrams. Chapter 6 of this text addresses textual representations of requirements.

Use case diagrams capture the various systems that comprise the metasystem, one of which is the system of interest. The use case diagram also identifies numerous scenarios in which the systems in the use case diagram interact during the relevant life cycle phase of the system of interest. Each of these scenarios is then defined in more detail in a sequence diagram. Section 3.4 defines these diagrams and gives examples.

Process models address how outputs are transformed from inputs via some function, activity, or task. There are numerous process modeling techniques in use today, one of which is IDEF0. Other process modeling techniques (data flow diagrams and N^2 charts) are described in graphical modeling techniques on the companion website. Process models are graphical representations that provide qualitative descriptions to explain how inputs are transformed into outputs. These process models can be used at both shallow and detailed levels of abstraction. IDEF0, presented in Section 3.5, is a popular modeling technique because it has rich and standardized semantics and syntax.

FFBDs, EFFBDs, and activity diagrams capture dynamic behavior in a representation that can be simulated. They are discussed in Section 3.6.

Block diagrams are used within SysML to capture the interconnections between pairs of components within the physical architecture so that interfaces between these pairs of components can be defined. Section 3.7 presents this material.

Requirements diagrams and requirements tables are introduced in Section 3.8.

Parametric diagrams, used to capture variable relationships in systems of equations for simulating system performance, in combination with block definition diagrams, are discussed in Section 3.9.

Exercise Problem 3.1 introduces a process model of the TTDSE process using the IDEF0 model. Selected pages of this IDEF0 model are used in Chapters 6–11 to describe the methods that comprise this engineering process, while the elevator system model is rendered as SysML activity diagrams and block definition diagrams.

3.2   MODELS AND MODELING

A *model* is any incomplete representation of reality, an abstraction. Models can be physical representations of reality. A subscale aircraft is used in a wind tunnel to test the aerodynamics of the real aircraft; this subscale aircraft does not contain the instrument panel used by the pilot or the seats in which passengers sit because they are not relevant (we think) for testing the aerodynamics of the aircraft. Similarly, models can be mathematical. A random number generator can be used to model the propensity of a coin to turn up heads or tails in a flip. Similarly, we can develop either an analytic or a simulation model of an aircraft's aerodynamics or an information system's response to user inputs. The wind tunnel data taken from the physical model of the aircraft can be used to refine the simulation data. The simulation data can be used to guide additional wind tunnel tests.

Qualitative models are also quite useful. The set of requirements for a system is an example of a qualitative model that serves as a model of the system's performance and capabilities. Finally, each of us has several mental models that we use in everyday life. However, in every case, the *essence of a model is the question or set of questions that the model can reliably answer* for us.

Before describing the types of models, discussing the types of questions that can be answered is important. The questions can be divided into three categories: descriptive (or predictive), normative, and definitive. A *definitive* model addresses the question of how an entity should be defined; this is the major category of questions that will be addressed in this book. The focus is building a definition of how the system is being designed in terms of its inputs and outputs, functions, and resources. A *descriptive* model attempts to predict answers to questions for which the truth may or may not be obtained in the future. Descriptive models are the most used in science and engineering. Executable models, which will be discussed in Chapter 12, are descriptive models because they predict the behavior of the system's design in specific situations, given the modeled design definition of the system. *Normative* models address how individuals or organizational entities ought to think about a problem and guide decision-making. A normative model for decision-making, deciding about the engineering of a system, is developed in Chapter 14.

Every modeling technique requires a language to establish a representation of reality. Models should be used to provide an answer to one or more questions; these answers should provide greater validity or insight than is possible without the model. Any language has semantics, a set of symbols or signs, which form the basis of representations in the language. In addition, every language has a syntax that defines proper ways of combining the symbols to form thoughts and concepts.

Definitive models require a rich language, both in terms of semantics and syntax, since these models are used to establish an interpretation of some aspect of reality and communicate that interpretation to a broad range of people and possibly computers. This language must be understandable to its audience. Unfortunately, richness and understandability often conflict with each other. That is, making a modeling language richer usually makes it less understandable. A third aspect, formality, is useful for proving that certain characteristics exist or do not exist; formality tends to conflict with both richness and understandability.

Descriptive models are measured by their power or richness for addressing a wide range of problems, understandability to both wide and narrow audiences, and accuracy or precision with which they can be used to define the relevant entity. Descriptive models can sometimes be tested as to their predictive accuracy in various situations. This predictive accuracy must be understood by those using the descriptive model because the ability to predict accurately when the model that is being used cannot be known exactly. Nonetheless, talking about descriptive models as being right or wrong is fruitless – all models are wrong. Rather, the model's usefulness in terms of predictive accuracy in general and the cost of building and using the model are very relevant.

Normative models, on the other hand, cannot be tested but are judged on their understandability and appeal across all disciplines in which they can be used. A normative model for making decisions cannot be tested because the world can never be examined in the same conditions with and without the use of the normative model. Rather, the normative model is tested by decision-makers based upon the model's ability to reflect the intuitions of the decision-makers or provide logical arguments that refute this intuition.

One possible taxonomy of models is shown in Table 3.1. This taxonomy begins by breaking models into physical, quantitative, qualitative, and mental models. A *physical model* represents an entity in three-dimensional space and can be divided into full-scale mockup, subscale mockup, breadboard, and electronic mockup. Full-scale mockups are usually used to match the interfaces between systems and components as well as to enable the visualization of the physical placement of elements of the system. The design of the Boeing 777 replaced the physical mockups with a very detailed three-dimensional electronic mockup. Subscale models are commonly used to examine a specific

TABLE 3.1 Taxonomy of Models

Model Categories	Model Subcategories	Typical Systems Engineering Questions
Physical	Full-scale mockup	How much?
	Subscale mockup	How often?
	Breadboard	How good?
		Do they match?
Quantitative	Analytic	How much?
	Simulation	How often?
	Judgmental	How good?
Qualitative	Symbolic	What needs to be
	Textual	done?
	Graphic	How well?
		By what?
Mental	Explanation	All of the above!
	Prediction	
	Estimation	

issue such as fluid flow around the system. A breadboard is a board on which electronic or mechanical prototypes are built and tested; this phrase was legitimized in dictionaries in the mid-1950s but is not used as much now.

Quantitative models provide answers that are numerical; these models can be either analytic, simulation, or judgmental models. Simulation models can be either deterministic or stochastic, as can analytic and judgmental models. Similarly, these models can be dynamic (time-varying) or static snapshots (e.g., steady state). An analytic model is based upon an underlying system of equations that can be solved to produce a set of solutions; these solutions can be developed in closed form. Simulation methods are used to find a numeric solution when analytic methods are not realistic, such as when friction in some form is introduced as an element of the model. When the equations involve the movement through time of numerous variables, we say the simulation is dynamic, involving differential or difference equations. However, a simulation need not involve time; the model may address spatial issues. Simulations that include uncertainty are often called "Monte Carlo" simulations; Monte Carlo simulations involve the repetitive solution of the same set of equations based upon different samples of the underlying probability distributions for the uncertainty specified in the equations. Judgmental models provide representations of real-world outcomes based solely on expert opinions. Explicit judgmental models are not used as often as the other types discussed here, but many analysts have found them to be an extremely useful precursor to other quantitative modeling activities.

Qualitative models provide symbolic, textual, or graphic answers. Symbolic models are based on logic or set theory, samples of which are provided in Chapter 4. Textual models are based on verbal descriptions; many models of the social sciences use textual models in which a model is described in one or more paragraphs. Many requirements documents in systems engineering are examples of textual models of the system's ultimate performance. Graphical models use either elements of mathematical graph theory or simply artistic graphics to represent a hierarchical structure, the flow of items or data through a system's functions, or the dynamic interaction of the system's components. This use of artistic graphics as a modeling approach is often given the pejorative name of "view graph" engineering. Most engineers view graphical models as one step above textual models. If graphical models can be based on mathematical graph theory, then these qualitative models can be powerful additions to the systems engineers' toolkit.

Finally, we need to address the *mental models* that we all carry around inside of us as abstractions of thought. The concept of a mental model arose in at least three separate communities relatively independently. Craik [1943] introduced mental models to cognitive psychology as our foundation for reason and prediction. Little was done with Craik's concept of a mental model until the early 1980s when Johnson-Laird [1983] and Gentner and Stevens [1983] published two books on the subject. The research in cognitive psychology has moved from the question of whether people do have mental models to the question of how best to capture and utilize these mental models for educational and other pursuits. The second field in which mental models became popular and useful was that of manual control, comprised of both psychologists and engineers. Early authors in the manual control field [Veldhuyzen and Stassen, 1977; Jagacinski and Miller, 1978; Rasmussen, 1979] addressed the use of mental models by system operators for controlling and predicting system performance. The third field to adopt mental models [Alexander, 1964; Pennington, 1985] is our field of engineering and architectural design. Alexander [1964] discussed mental pictures as representations of the problem definition and alternate solutions. Do you have a mental model of the street network in your neighborhood of your residence?

Engineers need to develop a mental model of the system on which they are working to be successful. Modelers who are developing qualitative, quantitative, or physical models clearly must develop a mental model of the model they are developing. The advantage of these non-mental models is that there is a much clearer communication mechanism; mental models fall in benefit in terms of enabling communication among people. People engaged in the same conversation may have a very different mental model, but due to the imprecise nature of natural language, they often feel that they can agree with each other at the end of a conversation even though their models of reality are quite different.

This book emphasizes the *qualitative aspects of systems engineering*. As a result, this chapter introduces the qualitative modeling approaches in SysML and IDEF0. The next two chapters introduce the mathematics of set theory and graph theory, which provide some mathematical underpinnings and limitations of these modeling approaches. Chapter 14 introduces decision analysis as the quantitative method for framing the design decisions discussed throughout this book.

SIDEBAR 3.1: QUANTITATIVE MODELS AND MENTAL MODELS

It is tempting to think that a quantitative model is more objective than a mental model and, by extension, that a more complex quantitative model is more objective than a less complex quantitative model. Certainly, more complex models are more explicit than less complex models. Also, the data inputs to these complex models are more specific and objective appearing. However, we must always remember that any quantitative model is developed via a mental process of one or more people and is the product of their mental models. Therefore, it is a mistake to ascribe objectivity to models. Complex mathematical models often have subjective assumptions throughout their equations and data.

The *purpose* of developing a model is to answer a question or set of questions better than one can without the model. Often, models are used to check each other; non-mental models should always be used to check mental models. This checking process is a two-way street; each model can be assumed to have certain strengths (answers known to be valid within some degree of accuracy or precision). These strengths can be used to help verify the abilities of the other model. Ultimately, a model is developed to provide answers in an area for which we feel we cannot get reliable answers any other way. However, we are commonly looking for more than just an answer; we want to understand "why" the answer is what it is, that is, obtain insight into how the real world works. Qualitative models are typically created to achieve agreement among individuals (shared visions) and to communicate that agreement to other people. Quantitative and physical models are better mechanisms to provide insight.

The more specific a question or set of questions that a model must answer, the easier it is to develop a model that can be useful. Models that are expected to answer a wide range of questions or generic questions well are the most difficult to develop and the least likely to provide insight into the logic for the answer. The easiest questions to answer are those for which we are looking for a relative comparison of alternate options: Which aircraft design weighs the most? How much more does one design weigh than another? The hardest questions involve providing an absolute answer: How much does this aircraft weigh?

The most *effective process* for developing and using a model is to begin by defining the questions the model should be able to answer. (This is analogous to defining the requirements for a system.) Then, the model should be developed, tested, and refined. The model should be validated and shown to be answering the right questions. Finally, there should be some verification process to show that the model is providing the right answers for known test cases. Now, we are ready to use the model for unknown test cases.

Often, there may be existing models that we believe are appropriate for use. In this case, we should begin by defining the questions to be answered. Then we can decide which model to use, perhaps with some enhancements. There should again be a period of verification for the chosen model in relevant cases before usage begins.

The incorrect approach to modeling is to begin by building or revising a favorite model before we know what questions need to be answered. People enthralled with the modeling process rather than the question-answering process employ this approach far too often. Modeling enthusiasts are more interested in the intrinsic properties of the model than in the model's ability to answer important questions. Note the more complex the model, the harder it is to obtain the insight we are seeking as to why the answer is what it is. This is why many experienced model builders opt for the most parsimonious (simplest) model that will provide a reasonably accurate answer.

Before using a model, it is important to establish the validity of the model. Model validity is difficult to establish and must first be defined. Recall from Chapter 1 that a system's validity addresses whether we have built the right system. By extension, model validity concerns whether we have built the right model. The validity of a model has several dimensions: conceptual, operational, and data. *Conceptual* validity addresses the model representation, that is, the theories employed, and the assumptions made. Conceptual validity addresses whether the model's structure is appropriate to answer the questions being asked. For a qualitative model, conceptual validity is the most important. For a quantitative model, the *operational* validity is key; that is, does the model's output behavior represent that of the real world for the questions being asked? Finally, *data* validity addresses whether the appropriate inputs were employed in building, testing, and using the model. Data validity for a qualitative model addresses whether the right individuals were involved in creating the model and whether they obtained access to the best set of information about the real world during the creation process. For quantitative models, the selection of a modeling technique may ride on what type of information will be available for running the model. When input data is scarce, judgmental models are often selected. In summary, establishing a model's validity must be tied to the model's ability to answer the questions that the model was designed to address.

Models have many potential uses in systems engineering: *creation* of a shared vision, *specification* of the shared vision, *communication* of the shared vision, *testing of* the shared vision, *estimation* or *prediction* of some quantitative measure associated with the system, and *selection* of one design option over other design options. The shared vision could be the inputs and outputs of the system, the system's requirements, the system's architecture, or the test plan for validating the system's design. As can be seen, all but the last two uses involve a qualitative activity. *This is the basis for emphasizing the use of qualitative models as adjuncts to our mental models in this book.* Quantitative models remain important, but qualitative models are not given their due value in engineering.

3.3 SysML MODELING

In Table 1.5 four topic areas are defined for SysML version 1 modeling [Friedenthal and Moore, 2014]: structure, behavior, interaction, and requirements. This is the decomposition provided by the Object Management Group, Inc. (OMG), which produces the specifications for SysML.

Another way of viewing these categories that is more consistent with the organization of this book would be:

1. metasystem modeling with use case and associated sequence diagrams as well as requirements relations with requirements diagram,
2. behavior modeling of the system's activities or processes (including both static and dynamic modeling) using activity and state machine diagrams,
3. structural modeling of the system's components including block definition and internal block diagrams,
4. parametric modeling of performance characteristics of the system, and
5. the process and structure that the systems engineering team is taking using package diagrams.

Section 3.4 introduces use case diagrams and sequence diagrams. This material is presented in terms of the very important modeling of the system's interaction with other systems (or the metasystem) that is often not done. Chapter 6 revisits this material and provide more context about how to use these diagrams.

Systems and software engineers have devised many ways to model processes or activities, at all levels of granularity – metasystem, system, and down to and through components. The activity or process modeling category within SysML includes state machines, a modeling technique that combines some properties of Petri nets (see the graphical modeling techniques on the companion website), and activity diagrams. Not directly mentioned are the legacy static, time-lapsed representations of dynamic processes such as IDEF0, data flow diagrams and N^2 charts. This text will continue to stress the value of static modeling techniques such as IDEF0 as a stepping-stone for getting to the more complex behavior models as well as for capturing the inputs and outputs that are passed from function to function in the behavioral model. The process modeling approach primarily employed in this book is IDEF0, which is described in Section 3.5. State machine models are discussed in the graphical modeling techniques on the companion website. FFBDs, EFFBDs, and activity diagrams are described in detail here in Section 3.6. Table 3.2 provides a categorization of process models into static and dynamic, as well as into SysML versus non-SysML approaches. The graphical modeling techniques on the companion website cover most of the techniques not addressed here in Chapter 3. Chapter 7 returns to this material and provides a process for building these types of models.

Structural modeling diagrams (block definition and internal block) are described and illustrated in Section 3.7. These diagrams have a long history in systems engineering and have finally been formally defined and standardized by SysML. Chapter 8 provides more detail on how to build these kinds of models.

Systems engineers have historically built many types of performance models of their system design to estimate the final performance capabilities of the design prior to fabrication of the initial prototypes. Performance modeling using a combination of block definition and parametric diagrams

TABLE 3.2 **Representation of Process Modeling Techniques**

	Static View	Dynamic View
SysML	–	State machines
		Activity diagrams
Non-SysML	Data flow diagrams	FFBDs and EFFBDs
	Control flow diagrams	Behavior diagrams
	N^2 diagrams	Petri nets
	IDEF0 diagrams	Statecharts
		ROOMcharts

is introduced in Section 3.9. Chapters 9 and 15 introduce the types of performance modeling commonly used in the engineering of systems.

3.4 METASYSTEM MODELING

To describe a modeling language, we have to describe the way in which the language is used to communicate to its readers/listeners. To describe a language, we need to identify the semantics (signs and symbols) and the syntax (composition of signs and symbols) of that language. The following definitions for semantics and syntax are taken from *The American Heritage Dictionary* [Berube, 1991].

Semantics: study of relationships between signs and symbols and what they represent.

Syntax: way in which words are put together to form phrases and sentences.

This and each of the following sections will describe the semantics of the language tool. The syntax will then be described formally or via examples.

It is critical that there be a team of engineers and domain experts that is performing the systems engineering process. This team can create a huge problem for itself by diving right into the design of the system without first learning about the other systems with which the focus of the design activity is to interact. This is probably the most common and most major problem encountered in the engineering of systems.

Chapter 6 introduces the operational concept, which includes scenarios or use cases that are supposed to describe how the system's interest will interact with humans and other systems throughout its life cycle. It is in the operational concept that the use case diagram and many sequence diagrams would be used to describe these scenarios. Developing these sequence diagrams is a major part of getting ready to develop the system's requirements. This section provides the fundamental semantics and syntax for using use case diagrams, sequence diagrams, and requirements diagrams within SysML.

The purpose of the use case diagram is to provide a higher level of how all of the individual use cases or usage scenarios combine within the operational concept to describe how the stakeholders think the system will be operated. The use case diagram originated in software engineering and is now commonly employed within the engineering of systems.

The semantics of a use case diagram contains:

1. labeled stick figures or drawing/picture for each class of humans or external systems
2. labeled ovals to define each use case

3. solid lines connecting stick figures and ovals
4. labeled dashed lines connecting ovals.

The syntax is that there is a diagram for each relevant phase of the system's life cycle, e.g., operations and training. For a given phase of the life cycle, the appropriate classes of humans (e.g., operators and maintainers) and other external systems are each given a stick figure or are represented by a drawing or picture. Then, all the possible interaction sequences among the system and these classes of humans and external systems are categorized and labeled as ovals. These interaction sequences are later defined one at a time in a sequence diagram. There is a significant amount of iteration between the first draft of the use case diagram, the defining of individual sequence diagrams, the improvement of the use case diagram, and so on.

Figure 3.1 provides an example of a use-case diagram for an elevator. The stick figures are (1) passenger class of humans, (2) maintenance workers, (3) building personnel, (4) a centralized service center including humans and other technology assets, and (5) the building. The high-level operational scenario is of course to use and maintain the elevator. There are four extensions of this basic scenario: responding to a fire, keeping the doors open, rescuing people from a stopped elevator, and ensuring that the load on the elevator is within a safe range. These extensions provide more detail about the basic scenario in specific situations that may or may not occur. There are two other ovals on this use case diagram for updates to the basic scenario that must always be present: providing electric power and maintaining a comfortable environment. Finally, there is one use case (fix the elevator) that does not involve passengers. In fact, this would be a basic scenario with extensions and inclusions if we were designing a real elevator.

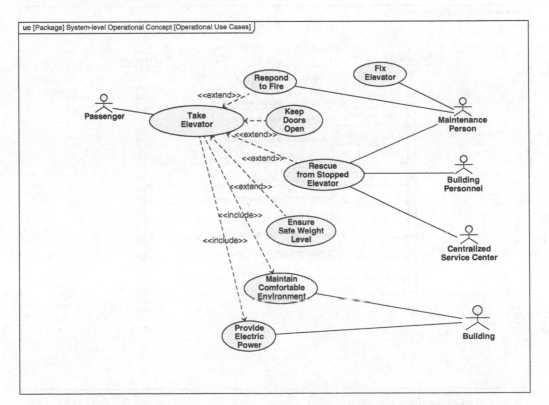

FIGURE 3.1 Exemplary use case diagram. (Developed with MagicDraw.)

For each labeled oval in the use case diagram, there should be a sequence diagram that defines the interactions among it and the other systems (including people, facilities, etc.) that the use diagram depicts as relevant. The semantics of a sequence diagram are:

- a labeled vertical line
- a labeled horizontal arrow that connects two or more vertical lines.

One labeled vertical line represents the system of interest. Each vertical line represents an external system with which the system interacts (exchanges inputs and outputs) during the use case. There must be at least two vertical lines. Time is assumed to go from the top to the bottom of the vertical lines. The labeled horizontal arrows represent the flow of items (information, energy, or physical entities) between the systems that the horizontal arrows connect. These items move in the direction of the arrow.

The syntax of sequence diagrams dictates that earlier flows in the use case appear above later flows, but time is not represented in appropriately scaled time units. Figure 3.2 provides a simple example of an elevator system in which a potential passenger calls an elevator to go up or down.

One contentious issue in sequence diagrams is what the labels of the horizontal arrows should represent. Many authors and practitioners label the arrows with the function being performed by the system of interest. In this book, we adopt the convention of labeling the horizontal arrows with a name that represents the item being transferred from one system to another. The reason for this

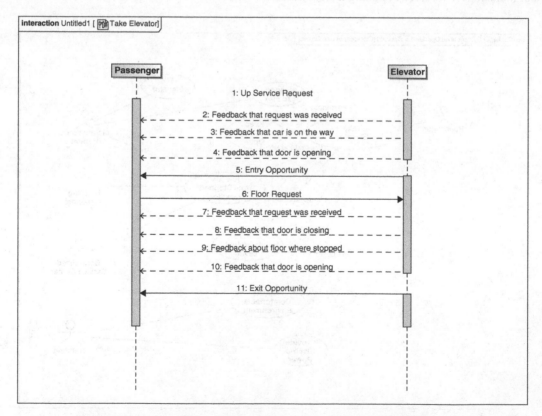

FIGURE 3.2   Exemplary sequence diagram. (Developed with MagicDraw.)

convention is described in detail in Chapter 6 and is associated with the contention that functional requirements should be written about inputs and outputs rather than functions.

SIDEBAR 3.2: USE CASE DIAGRAMS AND TEXT-BASED USE CASES

An alternative to use case diagrams is structured text-based use cases [Cockburn, 2001]. The structure includes the following attributes: (1) Primary Actor, (2) Scope, (3) Level, (4) Stakeholders and Interests, (5) Preconditions, (6) Minimal Guarantee, (7) Success Guarantee, (8) Main Success Scenario (steps 1 through *n*), and (9 Extensions (keyed to individual main success scenario steps). These structured text-based use cases provide richer information than use case diagrams. The combination of use case diagrams and sequence diagrams covers most of the attributes of text-based use cases.

Finally, SysML provides a basic representation for defining requirements and a broad set of representations for relating requirements to other requirements and system concepts. Rather than detail this part of SysML, we will use the capabilities in GENESYS described in Chapter 2.

3.5 STATIC BEHAVIORAL PROCESS MODELING WITH IDEF0

While IDEF0 was not included in SysML as a modeling technique, it is used throughout this book. First, it provides a very useful graphical representation of the interaction of the functional and physical elements of a system. IDEF0 is definitely not a sufficient modeling representation for the engineering of systems since it is *not* precise enough to define a unique dynamic representation of the system's design. In fact, it is not even a necessary modeling language since other languages have been successfully used for decades in its place. However, IDEF0 has gained wide acceptance and standardization and has been used successfully for decades as an approach to start the modeling process.

The IDEF acronym comes from the U.S. Air Force's Integrated Computer-Aided Manufacturing (ICAM) program that began in the 1970s. IDEF is a complex acronym that stands for ICAM Definition. The number, 0, is appended because this modeling technique was the first of many techniques developed as part of this program. In 1993, the U.S. Department of Commerce (National Institute of Standards and Technology [NIST]) issued Federal Information Processing Standard (FIPS) Publication 183 [1993a] that defines the IDEF0 language and renames the acronym, Integrated Definition for Function Modeling. In 2008, FIPS PUB 183 was withdrawn but is still online at https://nvlpubs.nist.gov/nistpubs/Legacy/FIPS/fipspub183.pdf. In 2012 ISO/IEC/IEEE 31320-1:2012 identified the basic components of Integration Definition 0 (IDEF0) syntax (the drawn, visual elements of the language and how they may be used together) and IDEF0 semantics (what it means when the visual elements are used together in specific, allowable ways), specifies the rules that govern the use of these modeling components, and describes the types of diagrams used in an IDEF0 model. ISO/IEC/IEEE 31320-1:2012 identifies and discusses the model pages with which each diagram in an IDEF0 model is associated and discusses specific features found in an IDEF0 diagram. It describes IDEF0 reference expressions and IDEF0 diagram feature references. It also presents an abstract formalization of the IDEF0 language.

The roots of IDEF0 can be traced to the structured analysis and design technique (SADT), which was developed and tested by Doug Ross at SofTech, Inc. from 1969 to 1973.

A sample of the modeling languages developed as part of the IDEF family is:

- *IDEF0*: a major subset of SADT; focus is a functional or process model of a system.
- *IDEF1*: focus is an informational model of the information needed to support the functions of a system.
- *IDEF1X*: focus is a semantic data model using relational theory and an entity–relationship modeling technique.
- *IDEF2*: focus is a dynamic model of the system.
- *IDEF3*: focus is both a process and object state-transition model of the system.

3.5.1 IDEF0 Semantics or Elements

An IDEF0 model is comprised of two or more IDEF0 pages. The two semantical elements of an IDEF0 page are functions and flows of material, energy or information. The semantics of IDEF0 are general enough to be applicable for framing models in SysML, OPM, and the Lifecycle Modeling Language (LML) that are introduced in Chapter 1.

A *function* or activity is represented by a box and described by a verb-noun phrase and numbered to provide context within the model (see Fig. 3.3). A function in this context is a transformation that turns inputs into outputs.

Inputs to be transformed into outputs enter the function box from the left, controls that guide the transformation process enter from the top, mechanisms (physical resources that perform the function) enter from the bottom, and outputs leave from the right.

A *flow* of *material, energy,* or *data* is represented by an arrow or arc that is labeled by a noun phrase (see Fig. 3.4). The label is a noun phrase and represents a set or collection of elements defined by the noun phrase. The label is connected to the arrow by an attached line, unless the arc leaves the page, in which case the label is placed on the appropriate edge of the page.

3.5.2 IDEF0 Diagram Syntax

An IDEF0 model has a purpose and viewpoint and is comprised of two or more pages, each page being a syntactical element of the model. The IDEF0 model:

- Answers definitive questions about the transformation of inputs into outputs by the system.
- Establishes the boundary of the system on the context page. This boundary is explicated, if needed, as a meta description.

FIGURE 3.3 Syntax for an IDEF0 function.

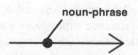

FIGURE 3.4 Syntax for an IDEF0 flow of material or data.

- Has one viewpoint; the viewpoint is the vantage or perspective from which the system is observed.
- Is a coordinated set of diagrams, using both a graphical language and natural language.

The A-0 page is the context diagram, which defines the inputs, controls, outputs, and mechanisms (ICOMs) for the single, top-level function, labeled A0. The context page establishes the boundaries of the system or organization being modeled by defining the inputs and controls entering from external systems and the outputs being produced for external systems.

Other pages in the IDEF0 model represent a decomposition of a function on a higher page, except for the external system diagram page, which is described later. The number of subfunctions for any IDEF0 function is limited to six, or possibly seven, for purposes of a readable display on a page. The decomposition of a parent function preserves the ICOMs of the parent. There can be no more, no less, and no differences. Every function must have a control, and input is optional. Functional boxes are usually placed diagonally on the page, with the more control-oriented functions being on the top left and the functions responsible for producing the major outputs being on the bottom right. Arcs are decomposable, just as functions are. Feedback is modeled by having an output from a higher-numbered function on a page flow upstream as a control, input, or mechanism to a lower-numbered function.

Arc decomposition and joining is necessary to minimize the number of arcs on the upper pages of a model, enhancing the readability or communicability. Arc decomposition and joining are handled by branching and joining, respectively. The labeling conventions for joins and branches are shown in Figure 3.5. If an arc is labeled before a branch and not labeled after the arc branches into two or more segments (as shown in the first of four examples in Fig. 3.5), then the arc before the branch carries on after the branch. Similarly, if an arc is labeled after two or more arcs join (see the second example in Fig. 3.5), then the label after the join also applies to the arcs before the join. If the label before a branch (after a join) does not apply to one or more of the arcs after the branch (before the join), then the arcs that deviate must have their own label. These labels of the exception branches must be subsets of the labels before the branch (after the join), as shown in the bottom two examples of Fig. 3.5.

Three different types of feedback are possible within an IDEF0 page: control, input, and mechanism. (The general topic of feedback will be discussed in more detail in Chapter 7.) Feedback in an IDEF0 diagram enables data or physical resources to be sent against the flow, down and to the right, so that closed-loop control can be used to improve key performance issues. The semantical protocols for showing these three types of feedback are shown in Figure 3.6. A control arc indicating feedback

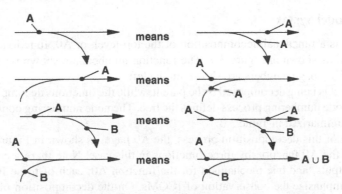

FIGURE 3.5   Labeling conventions for branches and joins.

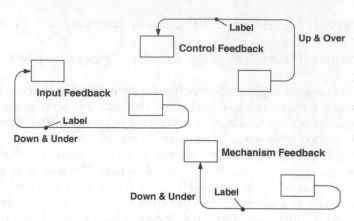

FIGURE 3.6 Feedback semantics within an IDEF0 page.

must go up and over the functions involved, coming down on the function for which it is a control. Input feedback is indicated by an arc that goes down and under the functions involved, coming up and into the function for which it is an input from the left. Finally, mechanism feedback must also be achieved by an arc that goes down and under the function for which it is a mechanism.

A major difficulty with IDEF0 models is determining whether an item should be an input or control. The primary distinction is that inputs are items that are transformed or consumed in the functional process associated with the production of its outputs. Controls, on the other hand, are not transformed or consumed, but rather are information or instructions that guide the functional process. Typical examples of controls are a blueprint and recipe instructions (e.g., bake at 375°F for one hour, use a 9.5- by 12-inch baking pan). Nonetheless, there are many times when it is very difficult to determine whether an item is an input or control. In these cases, the decision is the author's, with the provision that every function must have at least one control while inputs are optional.

Readers of an IDEF0 model are often surprised to see a function with a control and output, but no input. This seems to suggest a counterexample to the conservation of mass and energy in physics. Remember, though, that outputs of a function in an IDEF0 model do not have to have mass or energy but can be information. A common example of a function that can produce an output without an input is a function that produces a time mark for other parts of the system. This function receives control whenever the time mark is needed and uses its timekeeping resources to produce the time mark as an output.

3.5.3 IDEF0 Model Syntax

An IDEF0 model is a functional decomposition of the top-level, or A0, function. The decomposition is a hierarchy, as shown in Figure 3.7. The function numbers are shown on the right, and the corresponding IDEF0 page numbers are shown on the left.

The function that is being decomposed is the parent, while the functions decomposing it are called its children. The node numbering process defines the tree. The node numbering convention, as shown in Figure 3.7, is summarized in Table 3.3.

As an example of this decomposition process, the A0 page, as shown in Figure 3.8, defines the decomposition of the A0 function by three functions in this case. Note there are two inputs, three controls, three outputs, and one mechanism for the function A0; each of these ICOMs is given a generic label to emphasize the conservation of ICOMs. On the decomposition of A0 into A1, A2, and A3, there are again two external inputs, three external controls, three external outputs, and an

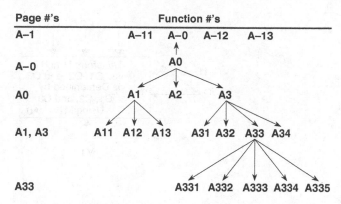

FIGURE 3.7 IDEF0 functional decomposition.

TABLE 3.3 IDEF0 Page Hierarchy

Page Number(s)	Page Content
A-1	Ancestor or External System Diagram
A-0	Context or System Function Diagram (contains A0)
A0	Level 0 Diagram with first-tier functions specified
A1, A2, ...	Level 1 Diagrams with second-tier functions specified
A11, A12, ..., A21, ...	Level 2 Diagrams with third-tier functions specified
...	...

external mechanism. Note that I1, C2, C3, and M1 branch on this A0 page. In addition, the joining of outputs from Al and A2 produces 02. Several internal items are produced, some of which branch and join.

IDEF0 models can also address the interaction of the system with other systems. This interaction is modeled on the A-1 page, which takes the A0 function and places it in context with other systems or organizations. This representation is often critical to understanding the relationship of the system being addressed to the system's outside world and establishing the origination of inputs and controls and the destination of outputs.

An IDEF0 model also has a data dictionary. An IDEF0 model should have a glossary page that defines the special words and acronyms in the labels and functions of the model. The data dictionary defines the arc decompositions. These decompositions reflect the arc branches and joins in the model. The dictionary also describes which functions use/produce which data elements.

3.5.4   IDEF0 Advanced Concepts

Advanced concepts to be discussed in this section are loops, tunneling, functional activation rules, exit rules, and call arrows.

IDEF0 allows the use of loops to show memory storage and feedback (see Fig. 3.9). A loop is showing that there is feedback involved in the decomposition of the function shown with the loop. Usually, the loop is not needed because the feedback will be seen on the decomposition. If the function is not going to be decomposed, it may be wise to show the loop. There are very few instances in which a loop is appropriately shown.

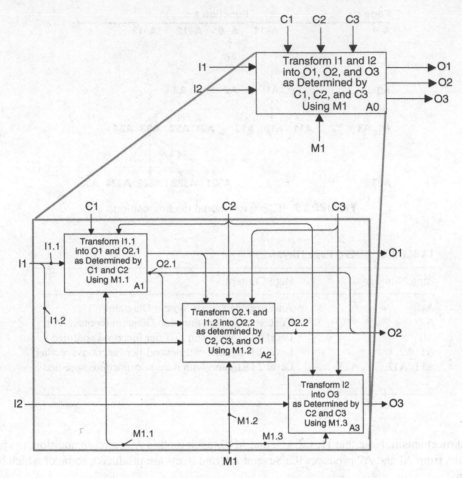

FIGURE 3.8 Functional decomposition in an IDEF0 model.

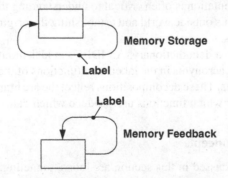

FIGURE 3.9 Memory syntax in IDEF0.

Tunneling is a technique within IDEF0 to hide an input, control, output, or mechanism in part of the model. The use of parentheses around either the head or tail of an arrow depicts a tunnel in IDEF0. Parentheses around the head of an arrow that is entering a functional box indicate that the input, control, output, or mechanism associated with that arrow will not be seen on the decomposition of that function; that is, the ICOM is going underground and may or may not reappear. If the ICOM does reappear, it will have parentheses around its tail to depict that it is exiting the ground. The rationale for tunneling is that certain ICOMs are not particularly relevant for understanding the functional model at specific levels of detail and, therefore, should not clutter up these pages of the model.

Each function is activated when sufficient inputs and controls are present to produce the relevant outputs given those inputs and controls. This functional activation is typically defined as a set of rules. A rule is a set of "if ... , then ... " statements, or preconditions and postconditions. Boolean algebra is used to specify these rules. These activation rules are embedded in each function; a "for exposition only," or EEO page, is often used to articulate the activation rules of a particular function or set of functions.

For each function, there are one or more exit criteria that determine when the function has completed its execution. Typically, the exit criterion is associated with the production of one or more outputs. If more than one output may be produced by a given function, then it is critical to state the exit criteria.

The final advanced concept is that of a call arrow. A call arrow is an arrow that breaks all the rules of ICOMs that have been presented so far and is seldom used *in* the author's experience. The call arrow exits the bottom of an activity's box and points toward the bottom of the page; see FIPS Publication 183 [1993a] for an example. The label attached at the end of the call arrow signifies another box that may be part of the IDEF0 model, or part of another IDEF0 model. The call arrow indicates that there is no decomposition of the activity from which the call arrow is exiting but that there is a decomposition of the activity at the box associated with the label of the call arrow. The advantage of the call arrow is that fewer pages need to be part of the IDEF0 model if several of the boxes have the same decomposition.

3.5.5 Systems Engineering Use of IDEF0 Models

A major emphasis in this book is the development of a functional architecture for the system that defines what functions the system must perform to transform the system's inputs into its outputs. An IDEF0 model, minus the mechanisms, can be used to define the *functional architecture*.

As part of the development of the allocated architecture the system's functions are allocated to the system's components and CIs. This allocation of functions is captured by adding the mechanisms to the functional architecture, producing a description of the *allocated architecture*.

3.6 DYNAMIC BEHAVIORAL PROCESS MODELING WITH EFFBDs

FFBDs were traditionally used in conjunction with N^2 diagrams as the original approach to functional decomposition in systems engineering. (In this book, we are substituting IDEF0 for N^2 diagrams; N^2 diagrams are covered in Chapter 12 for the interested reader.) Later, FFBDs were extended and enhanced to become EFFBDs. The extended FFBDs added more types of dynamic control logic. The enhanced FFBDs included some items in the models for better explication and understanding. This section first presents the full set of control logic of EFFBDs. Then, it shows how the items will be added.

An EFFBD model contains all the information in an IDEF0 model plus sufficient information to create a unique discrete event simulation of the dynamic behavior of the system. This is quite an added benefit over the IDEF0 model, but it also requires additional sophistication to create. The view adopted here is that the IDEF0 model is a stepping-stone to the completed EFFBD model for beginning systems engineers. Many experienced systems engineers can skip the IDEF0 model and create the EFFBD directly. However, there are many other experienced systems engineers who view the IDEF0 modeling process as an important learning and communication process for the stakeholders.

An EFFBD model has pages just as an IDEF0 model does. In fact, one could take an IDEF0 model, add control logic to each page, and end up with an EFFBD model. So, the EFFBD model provides a hierarchical decomposition of the system's functions with a control structure that dictates the order in which the functions can be executed at each level of the decomposition. The control structure and arrival sequence of "triggers" (special control inputs) determine this order. This makes the syntax and semantics of an EFFBD model identical to that of an IDEF0 model.

The only semantical difference between an IDEF0 and an EFFBD page is that the EFFBD has control symbols and lines that are not present in IDEF0. These control symbols and lines will be the main emphasis of this section.

In the *original, or basic*, FFBD syntax, there were four types of control structures that were allowed: series, concurrent, selection, and multiple-exit functions. A set of functions defined in a series control structure (see Fig. 3.10) must all be executed in that order. In fact, the second function cannot begin until the first function is finished, and so on. (Note that in the diagrams shown in this chapter, the two nodes at each end with missing center panels on the top and bottom of the functional rectangles are functions that are outside of the decomposition of the system function.) Control passes from left to right in FFBDs along the arc shown from outside (depicted by a function in a box with broken top and bottom lines) and activates the first function. When the first function has been completed (i.e., the function's exit criterion has been satisfied), control passes out of the right face of the function and into the second function, and so on. (Note that the little solid squares in the upper left corner of functions 1 and 2 are a software construct of CORE that indicates the function has been further decomposed.)

The concurrent structure (Fig. 3.11) allows multiple functions to work in parallel; thus, this structure is sometimes called "parallel." However, the concurrent structure should not be confused with the concepts of parallel in electric circuits or redundant systems. Essentially, control is activated on all lines *exiting* the first AND node, and control cannot be closed at the second AND node until all functions on each control line entering this second AND are completed. This control structure is almost always appropriate for the external systems diagram; the external systems typically act concurrently with each other and the system in which we are interested. The concurrent control structure is also common for the first-level functional decomposition of the system function.

A selection structure and a multiple-exit function achieve essentially the same purpose: the possibility of activating one of several functions. The multiple-exit function (see Fig. 3.12) achieves this by having a function placed at the fork of the selection process to make the selection explicit; this is the preferred approach to the selection structure.

When the selection function has been completed, one of the two or more emanating control lines is activated. Each control line can have zero, one, two, or more functions on it. Additional control

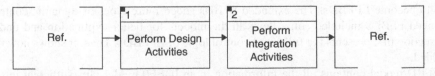

FIGURE 3.10 Series function flow block diagram. (Developed with CORE.)

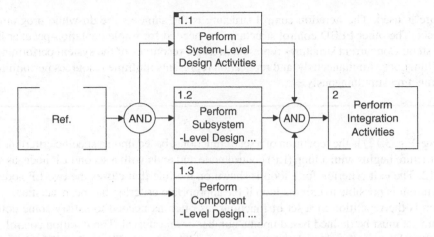

FIGURE 3.11 Concurrent control structure in an FFBD. (Developed with CORE.)

structures, such as concurrent, can be placed on any of these existing control lines. Once all the functions on the activated line have finished execution, control passes through the closing OR node. Each exit criterion for the control lines exiting the multiple-exit function appears has a label on the control line. (Note there is an exit criterion for every function with only one exit, but the exit criterion is not commonly shown on the exiting control line. The engineer may add a label for this purpose if desired.)

For the selection construct, which is an exclusive OR, the first OR node passes control to one of the exiting control lines in a manner that is unspecified on the diagram. This control line stays active until the set of functions on that control line is completed; control then passes through the second OR node. Figure 3.11 would be a selection construct if the AND nodes were OR nodes. Since the passing of control at the first OR node is not defined on the diagram, the authors strongly recommend the use of a multi-exit function instead of the selection control construct.

Additional control structures have been added to FFBDs: iteration, looping, and replication. See Sidebar 3.3 for a comparison of FFBD control constructs to structured programming.

SIDEBAR 3.3: STRUCTURED PROGRAMMING AND FFBD CONSTRUCTS

These constructs are quite analogous to those of structured programming, which began in the late 1950s and early 1960s with people such as Bohm, Dijkstra, Jacopini, and Warnier [De Marco, 1979]. Initially, the goal of structured programming was to define programming control structures that enhanced readability and improved testing. However, the goal evolved to define the control structures that would enable proving the correctness of an algorithm. While correctness proofs are still a goal, it was clear to these early investigators that program simplicity was critical. An intermediate goal to a correctness proof became the identification of the minimum set of logical constructs that would be sufficient to write any program. Bohm and Jacopini [1966] showed that only two constructs are necessary beyond the obvious series processing construct: "if-then-else" and "do-while." The if-then-else construct is the equivalent of the multi-exit function in FFBDs for situations in which a function does not need to be repeated. For repetitive activities that fit within if-then-else, the looping control structure is used. The iteration control structure is the same as the do-while programming construct. The other FFBD control structures are needed for implementation-peculiar issues of a system: Concurrent structures represent multiple resources of the system performing different functions simultaneously, and replication represents multiple resources performing the same function simultaneously.

Looping (Fig. 3.12) is the repetition of a set of functions, based upon a specific criterion. The loop control structure begins with a loop (LP) control node and ends with a second LP node, as shown in Figure 3.12. The exit criterion for a loop is shown on the line that closes the two LP nodes. In the loop structure it is possible to exit the loop if the appropriate criterion has been satisfied.

Iteration is the repetition of a set of functions, as often as needed to satisfy some domain set; this domain set must be defined based upon a number or an interval. The iteration control structure begins with an IT control node and ends with a second IT node, see Figure 3.12. The domain set for the iterative repetition is shown on the line closing the two IT nodes.

Finally, replication is the repetition of the same function concurrently using identical resources. This repetition is shown using the stacked paper icon; the reader can see an example of this in the section of Chapter 12 on behavior diagrams. This control structure is appropriate for certain physical designs and some functional architectures.

The Enhanced FFBD (EFFBD) overlays data, material, and energy inputs/outputs between functions onto the control structure of the functions (see Fig. 3.13). The EFFBD distinguishes between triggering and non-triggering inputs to functions. Adding duration times and resource constraints to functions makes the EFFBD executable as a discrete event model so that the system model can be validated both statically and dynamically. Static validation assures structural integrity, and dynamic validation assures performance integrity. Note that discrete event models can represent both continuous time and discrete-time systems. Duration times and resource constraints can be deterministic or non-deterministic. Monte Carlo simulation would be appropriate to characterize dynamic system performance when system operating conditions are non-deterministic.

The SysML activity diagram (see Fig. 3.14) is equivalent to the FFBD and EFFBD [Bock, 2006]. The EFFBD and activity diagram do show inputs and outputs, bringing those diagrams closer to the IDEF0 diagram. The activity diagram brings that diagram even closer to the IDEF0 in that the activities, equivalent to functions in the FFBD and EFFBD, can be placed in swim lanes representing the physical components (not shown in Fig. 3.14). Remember, IDEF0 has no way to capture the dynamic information that the EFFBD and activity diagram do.

3.7 STRUCTURAL MODELING OF THE SYSTEM'S COMPONENTS

Systems engineers have been using block diagrams since the beginning of systems engineering. However, there has been no standardization of how to construct these block diagrams and no uniform syntax and semantics. SysML has provided a much-needed syntax and semantics. A block is some element within the spectrum from the metasystem down to CI. Each element represents a set of resources (people, hardware, software, etc.) that can be used to perform one or more functions as inputs are transformed into outputs. The purpose of the block diagram is to display which blocks are connected to others based on either a hierarchical relationship or a peer-to-peer basis. *Block definition diagrams* represent hierarchical relationships such as how one block is composed of several other blocks. *Internal block diagrams* show which blocks within a higher-level block are connected to each other via interfaces.

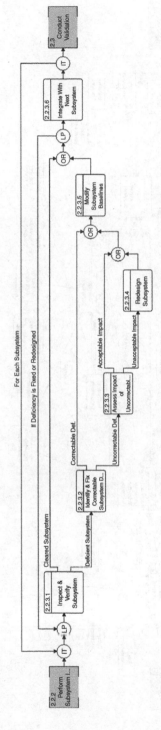

FIGURE 3.12 Selection and multiple-exit functions in an FFBD. (Developed with GENESYS.)

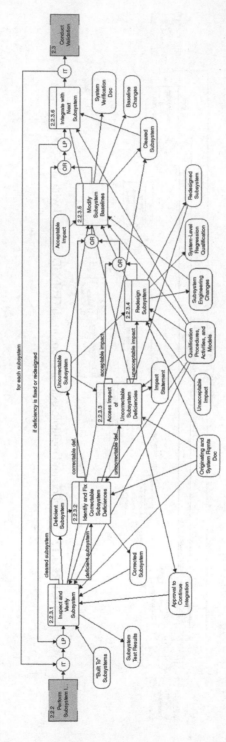

FIGURE 3.13 Selection, multiple-exit functions, and inputs/outputs between functions in an EFFBD. (Developed with GENESYS.)

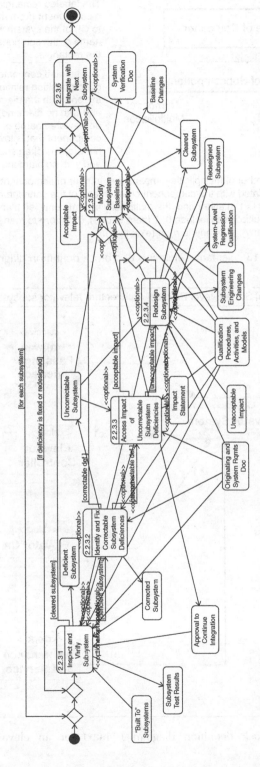

FIGURE 3.14 Selection, multiple-exit activities, and inputs/outputs between activities in an activity diagram. (Developed with GENESYS.)

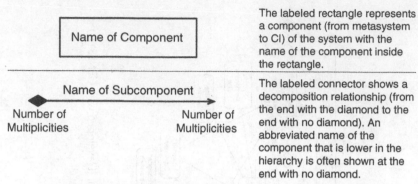

The labeled rectangle represents a component (from metasystem to CI) of the system with the name of the component inside the rectangle.

The labeled connector shows a decomposition relationship (from the end with the diamond to the end with no diamond). An abbreviated name of the component that is lower in the hierarchy is often shown at the end with no diamond.

Note: "Number of multiplicities" means the number of components that are associated with the component on each end of the connector. If the multiplicity is 1 at either end, the multiplicity is commonly left blank. Sample multiplicities include 0..1 (zero to one), 0..* (zero to many), 1..* (one to many), 1..n (one to n), n (exactly n).

FIGURE 3.15   Semantic elements of a block definition diagram.

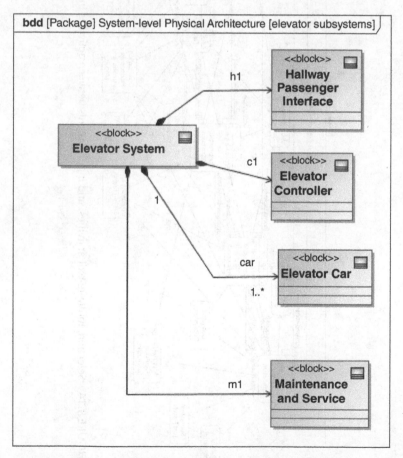

FIGURE 3.16 Exemplary block definition diagram (syntax) for an elevator. (Developed with MagicDraw.)

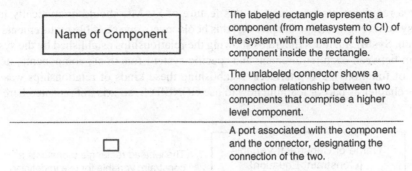

The labeled rectangle represents a component (from metasystem to CI) of the system with the name of the component inside the rectangle.

The unlabeled connector shows a connection relationship between two components that comprise a higher level component.

A port associated with the component and the connector, designating the connection of the two.

FIGURE 3.17 Semantic elements of the internal block diagram.

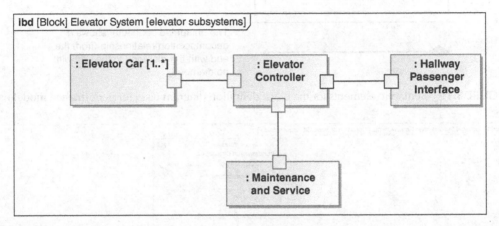

FIGURE 3.18 Exemplary internal block diagram for subsystems of an elevator system. (Developed with MagicDraw.)

The semantics for the block definition diagram include a labeled rectangle to define blocks, a labeled connector with a diamond on one end and an arrowhead on the other to show the hierarchical relationships. Figure 3.15 shows these two syntactic elements. Note, the full SysML semantics [Friedenthal and Moore, 2014] includes many other elements, but these two are the basic ones that are used in Chapter 8. Figure 3.16 shows the syntax of a block definition diagram for the elevator system and its subsystems that is discussed in Chapter 2.

The semantics of an internal block diagram (see Fig. 3.17) include a labeled rectangle for the specific blocks that compose the higher-level block that is the subject of the diagram, small unlabeled blocks on the boundary of the larger labeled blocks to define the connection between the block and the interface to another block, and unlabeled lines to show the interfaces or ports that connect blocks. Again, there are more elements of the semantics for an internal block diagram, but these will suffice for an introduction. Figure 3.18 shows an internal block diagram showing the interface connections among the subsystems of the elevator.

3.8 REQUIREMENTS MODELING

SysML also includes diagrams for requirements modeling. These diagrams show the requirements taxonomy being used by the systems engineering team. Far too many systems engineering teams do not have a requirements taxonomy, so this feature of SysML should dramatically improve the practice of systems engineering. Chapter 6 of this book covers one possible requirements taxonomy.

In addition, SysML includes diagrams showing the relationships established by the systems engineering team between each requirement and specific system functions, components, items (inputs and outputs of functions), and interfaces. Establishing these kinds of relationships was covered in the previous chapter as part of learning how to use GENESYS, so it is not repeated here.

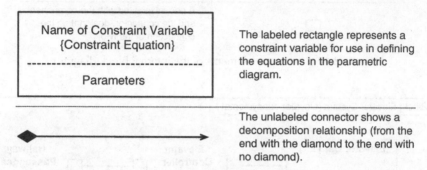

FIGURE 3.19 Semantic elements for the block definition diagram used for performance modeling.

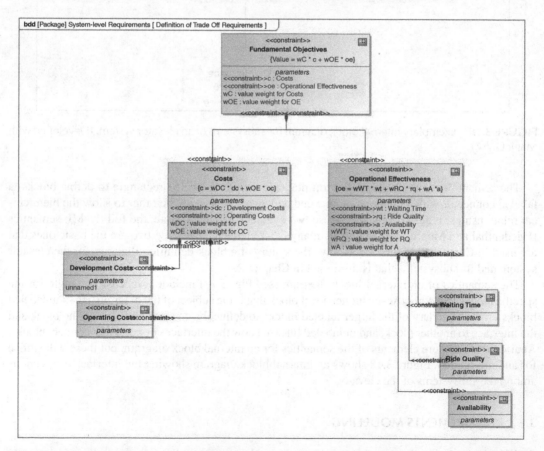

FIGURE 3.20 Exemplary block definition diagram for the fundamental objectives hierarchy of an elevator system. (Developed with MagicDraw.)

3.9 PERFORMANCE MODELING

SysML uses a combination of block definition and parametric diagrams to enable the systems engineer to define performance and trade-off models for use as part of the design process. The semantics of the block definition diagrams for performance modeling are not quite the same as those for block diagrams; see Figure 3.19. A rectangle, called a constraint block, is used to define each major variable for which an equation or constraint is defined. Besides the name of the variable appearing in the rectangle, the constraint equation appears inside the delimiters – { ... }. In addition, a list of parameters used in the equation with their mathematical abbreviations is shown in the rectangle below a separating line. The same sort of connecting line is used to show decomposition as in the block diagram case. Multiplicities are not needed. Figure 3.20 shows an example of a partial fundamental objectives hierarchy for a hypothetical elevator system.

The second SysML diagram used as part of specifying a performance model is called a parametric diagram. The parametric diagram contains roundtangles for the variables with equations and rectangles for the input variables associated with those equations. Regular lines are used to connect the concepts in the roundtangles and rectangles. Finally, a small rectangle is used to show connecting ports for the roundtangles. These connecting ports are associated with variables being used in the equation. Figure 3.21 shows the semantics of the parametric diagram.

3.10 SUMMARY

The role of qualitative modeling in the engineering of systems is essential. This chapter introduces modeling, purposes of models, and categories of models and discusses how engineers use models in the engineering of a system. Models are used to answer questions for which better answers are needed than currently exist; each modeling technique has its own language of symbols and conventions for combining symbols into higher-level concepts. A model is an abstraction of reality; models were characterized for the purposes of this book as mental, qualitative, quantitative, and physical. Each type of model has its advantages in terms of the types of questions that it answers best, as well as the development and operational costs for the model.

SysML's diagrams are introduced. The metasystem approaches of use case diagrams and sequence diagrams were described and illustrated for the elevator system that will be used throughout this book to illustrate the engineering of a system.

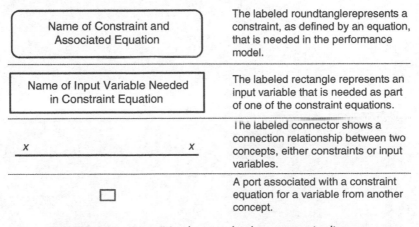

FIGURE 3.21 Semantic elements for the parametric diagram.

Next, IDEF0, a commonly used process modeling technique, is introduced and described in sufficient detail so that the reader should not only be able to read an IDEF0 model authored by someone else but will be able, with additional practice, to develop IDEF0 models on her or his own. This process modeling technique is introduced here because this book concentrates on the methods to be used in the engineering of systems, and some process modeling technique is needed to describe these methods. IDEF0 has the advantage of being a good communication tool as well as having a standardized syntax and semantics that do not vary by organization and discipline.

EFFBDs capture the dynamic execution of functions within the system. EFFBDs have a general set of control structures that overlay the functional decomposition in an IDEF0 model to capture the unique dynamics envisioned within the system. The EFFBDs also overlay the data, material, and energy inputs/triggers and outputs between functions.

Next, the block diagram semantics and syntax introduced by SysML are presented for both block definition diagrams and internal block diagrams. The former shows the decomposition of the physical architecture. The second shows the interface connections within a specific decomposition of a component.

Finally, the new concept of parametric diagrams to define the performance modeling being done within the engineering of the system is presented.

PROBLEMS

Reproduce the IDEF0 diagrams of the process for engineering a system in Appendix B using CORE. You must pay attention to the details of the content as well as the format. Both will be graded very carefully.

Create an FFBD diagram in CORE for each page of the IDEF0 model in Appendix B using CORE. Write a justification for the control logic of each diagram.

Describe at least three ways to estimate how much storage space would be needed if all the emails sent during a 24-hour period from all the people in the United States to anyone else in the United States were intercepted.

Chapter 4

Discrete Mathematics: Sets, Relations, and Functions

4.1 INTRODUCTION

Chapter 4 introduces material from the field of discrete mathematics. Much of this chapter will be review material (e.g., sets and functions) for most readers. The concepts of sets, relations, and functions are defined, discussed, and illustrated. A function, with which almost everyone is familiar, is shown to be a specialization of a relation, which in turn is a specialization of a set.

There are some key concepts introduced here that will be referred to in many of the succeeding chapters. For example, we will be discussing requirements and requirement documents in Chapter 6. Many system-level requirement documents are very large, larger than they need to be. These large system-level requirement documents can contain thousands and even tens of thousands of requirements. Examples might include:

- The system shall be able survive attacks from another computer system.
- The system shall be able survive buffer overflow attacks from another computer system.
- The system shall be able to survive stack-based buffer overflow attacks from another computer system.
- The system shall be able to survive stack-based buffer overflow attacks from an internal employee.
- The system shall be able to survive buffer overflow attacks against its operating system.
- The system shall be able to survive buffer overflow attacks against its application programs.
- The system shall be able to survive buffer overflow attacks originating in emails.
- The system shall be able to survive buffer overflow attacks while connected to websites on the Internet.
- And more of the same.

In Chapter 6, we will present an approach to writing such requirements and make the point that only one or a few of the above requirements should be in the system-level requirements document.

The Engineering Design of Systems: Models and Methods, Fourth Edition. Dennis M. Buede and William D. Miller
© 2024 John Wiley & Sons, Inc. Published 2024 by John Wiley & Sons, Inc.
Companion website: www.wiley.com/go/engineeringdesignofsystems4e

We will use the concept of a partition, introduced and defined here in Chapter 4, to make this case. A partition, based on the set theory introduced in this chapter, ensures that the requirements are not overlapping and are complete. Satisfying the nonoverlapping part will be relatively easy, but it is amazing how often it happens in practice. Achieving the completeness is a goal that is seldom, if ever, achieved. But there are approaches based on a partition that can help. Many requirements' documents contain duplicate, triplicate, and higher copies of requirements. Over time, some of these copies of requirements get changed while others do not, resulting in inconsistent requirements such as what happened on the Space Shuttle for operations in ambient temperatures, resulting in part in the explosion of the Challenger in 1986. Getting the concept of a partition of a set is key to many aspects of systems engineering.

In Chapter 7, we will discuss functions that systems perform in transforming their inputs into their outputs. When we have this discussion, you should remember the definition of mathematical function, which we cover here in Chapter 4. What you may not have learned previously is the concept of a mathematical relation, which is a weaker concept than that of a mathematical function. To perform mathematical analyses of our system's functional architecture, we will eventually need to be able to satisfy the mathematical definition of a function, not simply a relation, provided in this chapter. We will also need to recognize that we are dealing with relations when we are dealing with higher-level functions of a system. Ensuring that our functional decomposition is a partition that will arise again and again.

As part of the discussion of functional architectures in Chapter 7, we will be talking about decomposing higher-level functions into sets of lower-level functions. (Note the word *set* has been used again.) The mathematical concept of composition is defined in this chapter and discussed relative to hierarchical decomposition; mathematical composition will be shown to be a very limited representation of the functional modeling described in Chapter 7.

Two advanced concepts, power set and partial ordering, are introduced in this chapter. These concepts have great usefulness to the theoretical development of the engineering of systems, most of which is beyond the scope of this book but elements of which are discussed in Chapters 6, 7, and 9. The interested reader is referred to Mott et al. [1986] and Rosen [1995] for more details on set theory. Larsen and Buede [2002] provide a mathematical structure for performing early validation of requirements using many of the set theory concepts presented in this chapter.

Section 4.2 introduces the general concept of a set and then discusses special characteristics of sets, including operations on sets, the partition of a set, and the power set of a set. Section 4.3 defines relations in terms of sets. In particular, important characteristics of relations are defined. The partial ordering on a set is introduced and illustrated. Section 4.4 discusses functions and the composition of functions.

There are no models introduced in this chapter, but all this material is critical in understanding the development of models, as well as the power and limitations of models. Software engineers often make much more use of the discrete mathematics presented here than do the engineers of systems, but the material has the same richness and importance to engineers of systems and should be utilized to a fuller degree in the future. In addition, having a grasp of this material is essential to carrying on a conversation about architectures with many software engineers. I have seen systems engineers lose important and valid arguments to software engineers because the systems engineers were not equipped to understand what the software engineers were saying.

4.2 SETS

A *set* is a collection of well-defined objects, called elements or members. These *elements* or *members* are said to belong to the set. Sidebar 4.1 defines the mathematical symbols used in these and other definitions.

SIDEBAR 4.1: GLOSSARY OF MATHEMATICAL SYMBOLS

$\in$	is an element of
$\notin$	is not an element of
$\subseteq$	is a subset of
$\subset$	is a proper subset of
$\not\subset$	is not a subset of
$\supseteq$	is a superset of
$\supset$	is a proper superset of
$\cap$	intersection
$\cup$	union
$\rightarrow, \Rightarrow$	implies
$\Longleftrightarrow$	if and only if
$\neq$	is not equal to
Φ	the null set
$\underline{U}$	the universal set
$\overline{A}$	the complement of A
$\forall$	for all
$\exists$	there exists
$\ni$	such that
$\mid$	given that
$\sim, \neg$	not (negation)
$\wedge$	and
$\vee$	or

Examples of sets are

- An interval of numbers [7, 21]
- The students in SYST 520 at George Mason University during the spring semester of 1996
- The categories of inputs to elevator
- The possible states or outcomes that a particular input to the elevator can take
- The functions of an ATM (automated teller machine)

4.2.1 Writing Set Membership

A set is denoted by capital italic letter A, B, X, Y, with the exception of sets that are functions, which will be denoted by a lowercase italic letter. Members are also denoted by lowercase italic letters: a, b, x, y. The mathematical expression of set membership is

$$x \in A : x \text{ is an element of } A$$

$$x \notin A \text{ or } \neg(x \in A) : x \text{ is not an element of } A$$

4.2.2 Describing Members of a Set

There are at least five ways to describe the members of a set.

FIGURE 4.1   Set inclusion.

1. A is the set of elements, x, that satisfies the property (or predicate), $p(x)$. $A = \{x \mid p(x) \text{ is true}\}$ (braces are the common delimiter of a set's definition).

   The property $p(x)$ must *be well- defined*, that is, able to be determined by means of rules. One test of such a property is called the *clairvoyant's test – a* clairvoyant is able to predict the future or describe the past/present perfectly. Is the property or rule defined sufficiently well that the clairvoyant can answer the question? For example, the property "is tall" does not meet the clairvoyant's test, but the property "is taller than 6 feet 3 inches" does.

2. Complete enumeration is the listing of all the members of the set.

$$A_1 = \{0, 1, 2, 3, 4\}$$

$$A_2 = \{\text{student}_1, \text{student}_2, \dots, \text{student}_{31}\}$$

3. Use the characteristic function of the set

$$\mu_A(x) = \begin{cases} 1 & \text{for } x = 0, 1, 2, 3, 4 \\ 0 & \text{otherwise} \end{cases}$$

   where $\mu A(x)$ is the characteristic function of set A for elements, x, in the set, U, of all elements. For conventional (crisp, nonfuzzy) sets, $\mu A(x)$ may only take the values 0 for nonmembers or 1 for members.

4. Use recursive definition: $A = \{x_{i+1} = x_{i+1}, i = 0, 1, 2, 3; \text{ where } x_0 = 0\}$. Here, A is defined by a recursive formula.

5. Use one or more set operators such as union, intersection, and complement. These operations should be familiar to most readers and will be defined shortly.

4.2.3 Special Sets

U: the universal set or set of all possible members.

Φ: the null set, a set with no elements. Φ and $\{\Phi\}$ are not the same. Φ has no elements, while $\{\Phi\}$ has one). We can write $\Phi = \{x \in U \mid x \neq x\}$.

Singleton Set: a set with only one element.

Finite Set: a set with a finite number of distinct elements. *Infinite set*: a set with an infinite number of distinct elements.

For example: $A_1 = \{1, 2, 3, 4, \dots, 101\}$ is finite, $A_2 = \{1, 2, 3, 4, \dots\}$ is infinite, and $A_3 = \{x, \{1, 2\}, y, \{z\}\}$ may be finite or infinite. The finiteness of A_3 depends on whether x and y are finite or infinite. (Note $\{1,2\}$ and $\{z\}$ are sets, but each is only one element of A_3. Also note that z is *not* an element of A_3, but $\{z\}$ is).

Subsets or Set Inclusion: If A and B are two sets, and if every element of A is an element of B, then A is a *subset* of B, $A \subseteq B$. If A is a subset of B, and if B has at least one element that is *not* in A, then A is a *proper* subset of B, $A \subset B$. See Figure 4.1.

Equality of Sets: If A and B are sets, and A and B have precisely the same elements, then A and B are *equal*, $A = B$.

FIGURE 4.2 Absolute complement.

The following properties follow from the above definitions:

$A \subseteq A$; a set is a subset of itself.

$\Phi \subseteq A, A \subseteq U$. The null set is a subset of every set; every set is a subset of the universal set.

If $\Phi \neq A$, then $\Phi \subset A$. If a set is not the null set, then the null set is a proper subset of the set.

If $A \subseteq B$ and $B \subseteq A$, then $A = B$. If two sets are subsets of each other, then they are equal.

If $A \subseteq B$ and $B \subseteq C$, then $A \subseteq C$. Set inclusion is transitive, a property that we will formally define later.

4.2.4 Operations on Sets

The following operations are performed on sets:

Absolute Complement, $\overline{A}$: Let $A \subseteq U$. $\overline{A} = \{x \mid x \notin A\}$. (Note $\overline{\Phi} = U, \overline{U} = \Phi, \overline{\overline{A}} = A$.) See Figure 4.2.

Relative Complement of A with Respect to B, B − A: Let A and B be sets, $B - A = \{x \mid x \in B$ and $x \notin A\}$. The relative complement is also called *set difference*. See Figure 4.3.

Union of A and B, A ∪ B: $A \cup B = \{x \mid x \in A$ or $x \in B$ or both$\}$.

Intersection of A and B, A ∩ B: $A \cap B = \{x \mid x \in A$ and $x \in B\}$. (Note A and B are called *disjoint* if $A \cap B = \Phi$.) See Figure 4.4.

Boolean Sum (symmetrical difference), *A + B* or *A* Δ *B*:

$$A + B = \{x \mid x \in A \text{ or } x \in B, \text{but not both}\} = (A - B) \cup (B - A).$$

The following properties of the above set operations can be easily derived:

FIGURE 4.3 Relative complement.

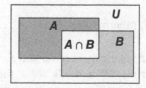

FIGURE 4.4 Set intersection.

1. $A \cup \Phi = A$, and $A \cap \Phi = \Phi$.
2. $A \cup U = U$, and $A \cap U = A$.
3. Idempotent: $A \cup A = A$, and $A \cap A = A$.
4. Associative:

$$(A \cup B) \cup C = A \cup (B \cup C)$$

$$(A \cap B) \cap C = A \cap (B \cap C).$$

5. Commutative: $A \cup B = B \cup A$, and $A \cap B = B \cap A$.
6. Distributive:

$$A \cup (B \cap C) = (A \cup B) \cap (A \cup C)$$

$$A \cap (B \cup C) = (A \cap B) \cup (A \cap C).$$

7. De Morgan's Laws: $\overline{(A \cup B)} = \overline{A} \cap \overline{B}$, and $\overline{(A \cap B)} = \overline{A} \cup \overline{B}$.

Example   Use De Morgan's laws to prove that the complement of $(\overline{A} \cap B) \cap (A \cup \overline{B}) \cap (A \cup C)$ is $(A \cup \overline{B}) \cup (\overline{A} \cap (B \cup \overline{C}))$.

Solution   Starting with $(\overline{A} \cap B) \cap (A \cup \overline{B}) \cap (A \cup C)$, note that $(A \cup \overline{B}) \cap (A \cup C)$ is the same as $A \cup (\overline{B} \cap C)$.

Step 1: Making this substitution, we want to find the complement $\overline{(\overline{A} \cap B) \cap (A \cup (\overline{B} \cap C))}$.

Step 2: By De Morgan's law, the complement of an intersection is the union of set complements. So, this can be written as $\overline{(\overline{A} \cap B)} \cup \overline{(A \cup (\overline{B} \cap C))}$.

Step 3: Again, the complement of an intersection is the union of the set complements. So, this can be written as $(A \cup \overline{B}) \cup \overline{(A \cup (\overline{B} \cap C))}$.

Step 4: Also by De Morgan's law, the complement of a union is the intersection of the set complements. So this can be written as $(A \cup \overline{B}) \cup (\overline{A} \cap \overline{(\overline{B} \cap C)})$.

Step 5: Again, the complement of an intersection is the union of the set complements. This yields $(A \cup \overline{B}) \cup (\overline{A} \cap (B \cup \overline{C}))$. QED

4.2.5 Partitions

A *partition* on a set A is a collection P of disjoint subsets of A whose union is A. For a collection B_i ($I = 1, 2, \ldots, n$) to be a partition P of A:

1. $B_i \subseteq A$ for $I = 1, 2, \ldots, n$.
2. $B_i \cap B_j = \Phi$ for $I \neq j$.
3. for any $x \in A$, $x \in B_i$ for some I; (alternatively $B_1 \cup B_2 \cup \ldots \cup B_n$).

The concept of a partition (Fig. 4.5) is the most basic and *far-reaching mathematical concept* to our development of systems engineering. We will talk about the importance of creating a partition of the system's requirements, a partition of the system's function, and a partition of the system's physical resources. This is just the beginning.

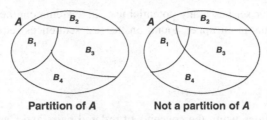

Partition of **A**   Not a partition of **A**

FIGURE 4.5   Set partition.

4.2.6   Power Set

The power set of a set A is denoted as $\mathbf{P}(A)$. The *power set* is the set of all sets that are subsets of A. Mathematically, the power set is the family (or set) of sets such that $X \subseteq A \Longleftrightarrow X \in \mathbf{P}(A)$, or $\mathbf{P}(A) = \{X \mid X \subseteq A\}$.

1. Let $A_0 = \Phi$, $\mathbf{P}(\Phi) = \{\Phi\}$, (where A_0 is a set with **zero** elements and $\mathbf{P}(A_0)$ has one element).
2. Let $A_1 = \{a\}$; $\mathbf{P}(A_1) = \{\Phi, A_1\} = \{\Phi, \{a\}\}$ (where A_1 is a set with **one** element and $\mathbf{P}(A_1)$ has two elements).
3. Let $A_2 = \{a, b\}$; $\mathbf{P}(A_2) = \{\Phi, \{a\}, \{b\}, \{a, b\}\}$ (where A_2 is a set with **two** elements and $\mathbf{P}(A_2)$ has four elements).

How many elements does the power set of a set of A_n have?

Theorem: If A_n is a set with n elements, then $\mathbf{P}(A_n)$ has 2^n elements.

Proof: We will use mathematical induction. For $n = 0, 1, 2, 3, \ldots$, let $S(n)$ be the statement: If A_n is a set with n elements, then $\mathbf{P}(A_n)$ has 2^n elements.

(i) First show that if A_0 has 0 elements, then $\mathbf{P}(A_0)$ has $2^0 = 1$ element.

$$A = \Phi, \mathbf{P}(A) = \{\Phi\}$$

(ii) Assume $S(k)$ is true and then show that $S(k+1)$ is true. Let A_{k+1} be a set with $k+1$ elements. Define B to be a proper subset of A_{k+1} with k of A_{k+1}'s elements:

$$A_{k+1} = \{a_1, a_2, \ldots, a_k, a_{k+1}\}$$
$$B = A_k = \{a_1, a_2, \ldots, a_k\}$$

So $A_{k+1} = \{a_{k+1}\} \cup B$.

Therefore, every subset of A_{k+1} either contains a_{k+1} or it does not.

1. If a subset does not contain a_{k+1}, then it is a subset of B, and we know there are 2^k subsets of B, by induction.
2. If a subset does contain a_{k+1}, then it is the union of a subset of B and a_{k+1}. There must be 2^k of these since there are 2^k subsets of B.

So there are $2^k + 2^k = 2^k (1+1) = 2^k \, 2 = 2^{k+1}$ subsets of A_{k+1} or 2^{k+1} elements of $\mathbf{P}(A_{k+1})$. QED

The concept of a power set has many potential uses in systems engineering. For example, the power set of system inputs is an upper bound on the test sequences required to test the system exhaustively.

4.3 RELATIONS

This section defines relations using the concepts of ordered pairs and Cartesian products. Important properties of relations are defined, followed by definitions of partial orderings and equivalence relations.

4.3.1 Ordered Pairs and Cartesian Products

An *ordered pair* is (x, y) if $x \in A$, $y \in B$. A *Cartesian product*, $A \times B$, is defined over two sets, A and B, such that $A \times B = \{(a, b) \mid a \in A \text{ and } b \in B\}$. That is, the Cartesian product of two sets is the set of all possible ordered pairs of those two sets. The following are examples of Cartesian products:

1. $A = \{1\}$, $B = \{2\}$: $A \times B = \{(1, 2)\}$ and $B \times A = \{(2, 1)\} \neq A \times B$.
2. $X = \{$students of SYST 520 during the spring semester of 1996$\} = \{S_1, S_2, \ldots, S_{31}\}$, $Y = \{A, B, C\}$: $X \times Y = \{(S_1, A), (S_1, B), (S_1, C), \ldots, (S_{31}, A), (S_{31}, B), (S_{31}, C)\}$

An ordered n-tuple is defined to be $A_1 \times A_2 \times \ldots \times A_n = \{(a_1, a_2, \ldots, a_n) \mid a_i \in A_i, i = 1, 2, \ldots, n\}$, where $(a_1, a_2, \ldots, a_n)$.

4.3.2 Unary and Binary Relations

A *unary relation* on a set A relates elements of A to itself and is a subset, R, of $A \times A$. R is usually described by a predicate that defines the relation. Examples are $\leq$, $=$, $>$, "taller than," and "older than." If a_1 and $a_2 \in A$, we write $(a_1, a_2) \in R$, which means that $a_1 \, R \, a_2$ or a_1 "is related to" a_2.

A binary *relation* is a relation R that relates elements of A to elements of B and is a subset of $A \times B$. The *domain* of R, written as "dom R," is defined as: dom $R = \{x \mid x \in A \text{ and } (x, y) \in R \text{ for some } y \in B\}$. The *range* of R, written as "ran R," is defined as: ran $R = \{y \mid y \in B \text{ and } (x, y) \in R \text{ for some } x \in A\}$. Again $(a_1, b_1) \in R \Longleftrightarrow a_1 \, R \, b_1$.

Example   Let R be the relation from $A = \{1, 3, 5, 7\}$ to $B = \{1, 3, 5\}$, which is defined by "x is less than y." Write R as a set of ordered pairs.

Solution

$$R = \{(x, y) \mid x \in A, y \in B, x < y\}$$

$$R = \{(1, 3), (1, 5), (3, 5)\}$$

Recall the relations within and between systems engineering classes that were discussed in Chapter 2. The hierarchy of requirements was defined by the relation "incorporates" in moving from the top of the requirements hierarchy to the bottom; "incorporated in" was the relation that moved from bottom to top. The relation "is decomposed by" moved from the top of the functional decomposition to the bottom; "decomposes" moves in the opposite direction. The physical hierarchy of a system and its components used the relation "is built from" in moving from top to bottom and "is built in" for moving from bottom to top.

Binary relations included the tracing from requirements to functions or the system, the performance of functions by the system and its components, and inputs and outputs of items for functions. The relation "is traced to" was used for the binary relations of input/output stakeholders' requirements being mapped to functions and for system-wide/technology requirements being mapped to the system. The binary relation for the system and components being related to functions used the relation "pertains." The relations "inputs" and "outputs" addressed functions being related to items.

To discuss the properties of unary relations, some additional information is needed concerning the possible ways to prove an implication. An implication is an "If …, then …" statement, which is commonly written as "If p is true, then q is true" or "$p \rightarrow q$." There are eight common methods for proving implications of this form.

1. *Trivial Proof*: Show that q is true independently of the truth of p.
2. *Vacuous Proof*: By mathematical convention, whenever p is false, $p \rightarrow q$ is true. The vacuous proof involves showing that p is false. *This method is key to understanding the full implications of the properties of unary relations that are discussed below.*
3. *Direct Proof*: Assume that p is true and use arguments based upon other known facts and logic to show that q must be true.
4. *Indirect Proof*: Use direct proof of the contrapositive of $p \rightarrow q$. The contrapositive of a true implication is known to be true; the contrapositive of $p \rightarrow q$ is $\sim q \rightarrow \sim p$ (or q is false implies p is false). Here, we assume q is false and prove via logic and known facts that p must be false.
5. *Contradiction-Based Proof*: De Morgan's laws can be used to show that $p \rightarrow q$ is equivalent to $\sim(p \wedge (\sim q))$, that is, the statement "p is true and q is false" is false. Proof by contradiction starts by assuming that $(p \wedge (\sim q))$ is true and then proving, based on this assumption, that some known truth must be false. If the only weak link in the argument is the assumption of $(p \wedge (\sim q))$, then this assumption must be wrong.
6. *Proof by Cases*: If p can be written in the form of p_1 or p_2 or … or p_n ($p_1 \vee p_2 \vee \ldots \vee p_n$), then $p \rightarrow q$ can be proven by proving $p_1 \rightarrow q, p_2 \rightarrow q, \ldots, p_n \rightarrow q$ as separate arguments.
7. *Proof by Elimination of Cases is an Extension of the Method Above*: Recall from the second method that $p \rightarrow q$ is equivalent to $[(p \wedge q) \vee (\sim p)]$, that is ($p$ and q are true) or (p is false). Now p can be partitioned into a set of cases as done in 6 and attacked one at a time.
8. *Conditional Proof*: If we are to prove $p \rightarrow (q \rightarrow r)$, we can prove the equivalent $(p \wedge q) \rightarrow r$.

4.3.3 Properties of Unary Relations on *A*

The seven properties discussed here are reflexive, irreflexive, symmetric, antisymmetric, asymmetric, transitive, and intransitive.

1. *Reflexive*: $x \, R \, x$ for all $x \in A$, e.g., equality, $\leq, \geq$.
2. *Irreflexive*: $x \, \not R \, x$ for all $x \in A$, for example, greater than, is the father of.
3. *Symmetric*: If $x \, R \, y$, then $y \, R \, x \, \forall x, y \in A$, for example, equality, is spouse of. Note, if $x \, \not R \, y$ for all x and y in A, then the relation is symmetric by a vacuous proof.
4. *Antisymmetric*: If $x \, R \, y$ and $y \, R \, x$, then $x = y \, \forall x, y \in A$, for example, equality, $\leq, \geq$. Note, if there is no situation in which "$x \, R \, y$ and $y \, R \, x$" is true, then the relation is antisymmetric by vacuous proof.
5. *Asymmetric*: If $x \, R \, y$, then $y \, \not R \, x \, \forall x, y \in A$, e.g., $<, >$.

6. *Transitive*: If $x R y$ and $y R z$, then $x R z \; \forall x, y, z \in A$, for example, $\leq, \geq, =, >$. This property is the most difficult to grasp. If there is no situation in which "$x R y$ and $y R z$," then the relation is transitive by vacuous proof.

7. *Intransitive*: If for some $x, y, z \in A$, it is true that $x R y$, $y R z$, but $x \not R z$, the relation is considered intransitive.

Example Let L be the set of lines in the Euclidean plane and let R be the relation on L defined by "x is parallel to y." Is R a reflexive relation? Why? Is R a symmetric relation? Why? Is R a transitive relation?

Solution

1. This question reduces to whether a line is parallel to itself. If the definition of parallel is having no points in common (everywhere equidistant), then a line cannot be parallel to itself because the two lines have every point in common. So, R is not a reflexive relation.

2. R is a symmetric relation. Consider each $x \in L$. x will have an infinite number of $y \in L$ which satisfies the parallel relationship. Each such y is in turn parallel to x. Thus, $(x, y) \in R$ for all x and y that are parallel, and $(y, x) \in R$, so the relation is symmetric.

3. R is a transitive relation. Again, consider $(x, y) \in R$ and $(y, z) \in R$; x will be parallel to z, so $x R z$ and R is transitive for all $x, y, z \in L$.

Example Let F be the set of functions in the functional decomposition for a system. Let R be the relation on F defined by "is decomposed by." Is R a reflexive relation? Why? Is R a symmetric relation? Why? Is R a transitive relation?

Solution

1. R is not a reflexive relation because a function does not decompose itself.

2. R is not a symmetric relation because if f_1 decomposes f_0, then f_0 cannot decompose f_1.

3. R is not a transitive relation. The function f_0 is decomposed by f_1, f_2, and f_3, and f_1 is decomposed by f_{11}, f_{12}, and f_{13}. However, f_0 is not decomposed by f_{11}, f_{12}, or f_{13}.

4.3.4 Partial Ordering

A relation R on A is a *partial ordering* if R is reflexive, antisymmetric, and transitive. Examples of partial orderings are $\geq$ or $\leq$ on the real number line, or $\supseteq$ or $\subseteq$ on $\mathbf{P}(A)$. Examples of nonpartial orderings are $<$ or $>$ on the real number line, $\subset$ or $\supset$ on $\mathbf{P}(A)$. (Both of these are asymmetric and antisymmetric.)

4.3.5 Equivalence Relations

A relation R on a set A is an *equivalence relation* if R is reflexive, symmetric, and transitive. An example of an equivalence relation is equality.

4.4 FUNCTIONS

This section defines functions and discusses the composition of functions.

4.4.1 Definitions

Let A and B be two nonempty sets. We write a function f as $f : A \rightarrow B$ and say that f maps every element of A (the domain) to one and only one element of B (the range). If $(a, b) \in f$, then element b is the *image* of element a under f. Note that a function can map elements of A onto itself, $f : A \rightarrow A$. A *function f* from A to B is a *relation* such that

(a) dom $f = A$
   (i) f is defined for each element of A, $a \in A$.
   (ii) $\exists$ (a, b) where $b \in B$ for each element of A, $a \in A$.
(b) if $(a, b) \in f$ and $(a, c) \in f$, then $b = c$; that is, f is single-valued, or no element of A is related to two elements of B.

A function is called *one-to-one* or *injective* if $(a, b) \in f$ and $(c, b) \in f$ implies $a = c$. That is, no two elements of A can be mapped into the same element of B by f.

A function $f : A \rightarrow B$ is *onto* or *surjective* if and only if the range of $f = B$, that is, f is defined for every $b \in B$.

If a function is both one-to-one and onto (or *bijective*), then the relation f^{-1} is single-valued and maps every element of B onto some element of A; f^{-1} is therefore a function, called the *inverse* function.

Example If $A = \{1, 2, 3, 4\}$ and $B = \{a, b, c, d\}$, determine if the following functions are one-to-one or onto.

(a) $f = \{(1, a), (2, a), (3, b), (4, d)\}$
(b) $g = \{(1, d), (2, b), (3, a), (4, a)\}$
(c) $h = \{(1, d), (2, b), (3, a), (4, c)\}$

Solution
(a) f is NOT one-to-one since $f^{-1}(a) = \{1, 2\}$.
    f is NOT onto since $f^{-1}(c) = \Phi$.
(b) g is NOT one-to-one since $g^{-1}(a) = \{3, 4\}$.
    g is NOT onto since $g^{-1}(c) = \Phi$.
(c) h is *one-to-one* since all elements of B correspond to unique elements in A.
    h is *onto* since every element of B has some pre-image in A.

So, we have progressed mathematically from sets to relations to functions.

Functions $\subseteq$ Relations $\subseteq$ Sets, or a function is a relation is a set.

As systems engineers, we will focus on functional architectures. We will represent the functions of the system as relations or functions in graph-like structures. The underlying theory is set theory.

4.4.2 Composition

Let R be a relation from A to B, and S be a relation from B to C. (a, c) is an element of the *composition* of R and S (denoted $R \cdot S$ or $R\ S$), if and only if there is an element $b \in B$ such that $a\ R\ b$ and $b\ S\ c$. That is, a and c must be linked together by b; a is mapped to b and b is mapped to c. (Note that some authors write the composition of R and S as $S \cdot R$ so be careful.)

The composition of functions is defined in the same way as the composition of relations.

Example   Assume R and S are relations from A to A. If $A = \{1, 2, 3, 4\}$, $R = \{(1, 2), (2, 3), (3, 4), (4, 2)\}$, and $S = \{(1, 3), (2, 4), (4, 2), (4, 3)\}$, then compute $R \cdot S$, $S \cdot R$, and $R \cdot R$.

Solution

$R \cdot S = \{(1, 4), (3, 2), (3, 3), (4, 4)\}$.

$(1, 2)$ from R is composed with $(2, 4)$ from S (this is written $(1, 2) \cdot (2, 4)$) and yields $(1, 4)$.

$(1, 2)$ from R cannot be composed with any of the other elements of S because they do not begin with a 2.

$(3, 4) \cdot (4, 2) = (3, 2)$.

$(3, 4) \cdot (4, 3) = (3, 3)$.

$(4, 2) \cdot (2, 4) = (4, 4)$.

$S \cdot R = \{(1, 4), (2, 2), (4, 3), (4, 4)\}$, which is not equal to $R \cdot S$.

$R \cdot R = \{(1, 3), (2, 4), (3, 2), (4, 3)\}$.

As systems engineers, we will employ functional decomposition to develop the functional architecture. Composition is the mathematical property from which decomposition derives its name. However, as discussed in Chapter 7, composition is only applicable to functional decomposition in limited situations.

4.5   SUMMARY

This chapter began with the introduction of a set, the foundation of a branch of mathematics called discrete mathematics. A great deal of terminology was introduced to define special sets such as the universal and null sets and operations on sets.

During the discussion of sets, the concept of partition was defined. The partition is perhaps the most important mathematical concept introduced in this chapter for application in this book. A partition is a subdivision of a set into subsets, which contain no common members, and yet the union of the subsets contains every element of the original set. In future chapters, requirements will be partitioned, functional decompositions will be defined to be partitions, and the physical decomposition will be defined to be a partition.

The power set of a set is the set of all subsets of that set. This notion of a power set is not exploited fully in this book but will become key to the future development and application of mathematics to the engineering of systems.

The next major section of this chapter dealt with relations and the key properties associated with relations. A relation is a set of ordered pairs; the elements of the ordered pairs come from one or two sets. If the functions of a system are not fully defined in terms of inputs, then these system functions are, in fact, mathematical relations.

Functions are relations that satisfy certain properties; a function maps every element of the domain of the function to some element of the range but does not map any element of its domain to more than one element of the range. One-to-one and onto properties of functions were also discussed. Finally, the composition of functions was defined.

PROBLEMS

4.1 Define the students enrolled in this class during this semester as a set, S.

   a. Specify a partition of S into two subsets.

b. Specify a partition of S into three subsets.

c. Specify a partition of S into five subsets.

4.2 Let $A_1 = \{1, 3, 5, 7, 9, 11\}$, $A_2 = \{-2, 6, 9, 11\}$, $A_3 = \{-2, 4, 6, 9, 11\}$. Show that:

a. $A_1 + A_2 = (A_1 - A_2) \cup (A_2 - A_1)$

b. $A_1 \cup (A_2 \cap A_3) = (A_1 \cup A_2) \cap (A_1 \cup A_3)$

4.3 Prove that the following relations are true in general:

a. $A_1 + A_2 = (A_1 - A_2) \cup (A_2 - A_1)$

b. $A_1 \cup (A_2 \cap A_3) = (A_1 \cup A_2) \cap (A_1 \cup A_3)$

4.4 Let R be a relation from A to B and defined "x is at least twice as big as y." Write R as a set of ordered pairs for

a. $A = \{1, 3, 5, 7\}$ and $B = \{2, 3, 4, 6\}$

b. $A = \{0, 1\}$ and $B = \{0, 1\}$

c. $A = \{1, 2, 3, 4, 5, 6, 7\}$ and $B = \{3, 6\}$

4.5 Let R be relation from A to B where "x is greater than or equal to y squared."

Then define R as a set of ordered pairs for the following:

a. $A = \{1, 2, 3, 4, 5\}$, $B = \{1, 2, 3, 4, 5\}$

b. $A = \{25\}$, $B = \{5, 6, 7\}$

4.6 There are three families defined by the sets A, B, and C; each family has a dad, mom, and three kids:

$$A = \{\text{Dad, Mom, Doris, Bill, Tom}\}$$

$$B = \{\text{Dad, Mom, Doris, Daisy, Debbie}\}$$

$$C = \{\text{Dad, Mom, Bill, Bob, Biff}\}$$

Consider the relations "is the spouse of," "is the brother of," and "is the blood relative of." (Hints: I am not the brother of myself. Two people are blood relatives if they share the blood of a common ancestor, who may or may not be part of sets A, B, or C. I am the blood relative of myself. Biff is a male.)

Identify which of these relations satisfy which of the seven properties of unary relations for each of the three sets by placing a yes or no in the empty cells of the following table.

	Reflexive	Irreflexive	Symmetric	Anti-symmetric	Asymmetric	Transitive	Intransitive
"is the spouse of" on A							
"is the brother of" on A							
"is the blood relative of" on A							
"is the spouse of" on B							
"is the brother of" on B							
"is the blood relative of" on B							
"is the spouse of" on C							
"is the brother of" on C							
"is the blood relative of" on C							

4.7 Let R *be* a relation from A to B and S be a relation from B to C.

   a. Find $R \cdot S$ for $A = \{1, 3, 5, 7\}$, $B = \{1, 2, 4, 5, 7\}$, $C = \{1, 2, 3, 4, 5, 6\}$, $R = \{(1,2), (3,4), (5,2), (7,4)\}$, and $S = \{(1,2), (2,4), (4,3), (7,5)\}$.

   b. Are any of these relations R, S, $R \cdot S$ functions? One-to-one functions? One-to-one and onto functions?

4.8 If $A_1 = \{1, 2, 3, 4\}$ and $A_2 = \{1, 4, 9, 25\}$, determine if the following functions that map A_1 onto A_2 are one-to-one, onto, or both one-to-one and onto.

   a. $f_1 = \{(1, 1), (2, 4), (3, 4), (4, 25)\}$

   b. $f_2 = \{(1, 1), (2, 4), (3, 25), (4, 25)\}$

   c. $f_3 = \{(1, 1), (2, 4), (3, 9), (4, 25)\}$

4.9 Develop two relations R (from A to B) and S (from B to C) that have to do with people. Show the result of $R \cdot S$.

4.10 Let R and S be relations from $A \to A$, where $A = \{1, 2, 3, 4\}$ and:

$$R = \{(1, 1), (2, 2), (3, 3), (1, 2), (2, 3), (1, 3), (2, 1), (3, 1), (3, 2)\}$$

$$S = \{(2, 3), (1, 2), (2, 1), (3, 1), (1, 3)\}$$

   a. Find if these relations are symmetric, reflexive, and transitive.

   b. Find $R \cdot S$, $S \cdot R$, and $R \cdot R$.

4.11 Let A be a set of three colors: {red, blue, green}. What are the elements of the power set of A?

4.12 Let *SIBLINGS* = {Andrea, Bobby, Catherine, David, Eric}. Find the elements of the power set of *SIBLINGS*, $\mathbf{P}(SIBLINGS)$.

4.13 Show that the $\mathbf{P}\{$Andrea, Bobby$\}$ is a subset of the $\mathbf{P}(SIBLINGS)$ from Problem 4.12.

4.14 Prove that for any two sets A and B, $(\mathbf{P}(A) \cap \mathbf{P}(B)) = \mathbf{P}(A \cap B)$.

4.15 Find two sets A and B that show $(\mathbf{P}(A) \cup \mathbf{P}(B)) \neq \mathbf{P}(A \cup B)$.

4.16 Prove that for any two sets A and B, $(\mathbf{P}(A) \cup \mathbf{P}(B)) \subseteq \mathbf{P}(A \cup B)$.

4.17 Prove that the seven properties of set operations in Section 4.2.4 are true.

Chapter 5

Graphs and Directed Graphs (Digraphs)

5.1 INTRODUCTION

This chapter introduces the mathematics of graph theory, the formal representation of a relation (or function) among elements of a set or a pair of sets. The concept of a relation discussed in this chapter is the same concept introduced in Chapter 4. A graph in mathematics is a set of nodes and a set of edges between pairs of those nodes; the edges are ordered or nonordered pairs, or a relation, that defines the pairs of nodes for which the relation being examined is valid. As an example, the people working as systems engineers on a project could be the members of a set. One relation defined over this set could be "works for." Another relation could be "respects." The edges can either be undirected or directed; directed edges depict a relation that requires the nodes to be ordered, while an undirected edge defines a relation in which no ordering of the edges is implied. The "works for" and "respects" relations would be examples of ordered relations. An example of an undirected relation would be "sits next to."

A graph enables us to visualize a relation over a set, which makes the characteristics of relations such as transitivity and symmetry easier to understand. The reader will hopefully comprehend the power of visualizing mathematical concepts, as enabled by mathematical graph theory, by the end of reading this chapter.

There is a great deal of terminology associated with graph theory; most of the basics are introduced in this chapter. Notions such as paths and cycles are key to understanding the more complex and powerful concepts of graph theory. There are many degrees of connectedness that apply to a graph; understanding these types of connectedness enables the engineer to understand the basic properties that can be defined for the graph representing some aspect of his or her system. The concepts of adjacency and reachability are the first steps to understanding the ability of an allocated architecture of a system to execute properly.

In addition to aiding in the visualization of relations, graph theory is the basis of many modeling languages. However, there are many more modeling languages such as IDEF0, which look like graphs, but which have no underlying mathematics. The material presented in this chapter is necessary but not sufficient to be able to detect if a modeling language with graphical representations has a mathematical basis or not. For example, understanding the seven properties of unary relations

The Engineering Design of Systems: Models and Methods, Fourth Edition. Dennis M. Buede and William D. Miller
© 2024 John Wiley & Sons, Inc. Published 2024 by John Wiley & Sons, Inc.
Companion website: www.wiley.com/go/engineeringdesignofsystems4e

presented in this chapter will enable the reader to detect key assumptions such as transitivity being made or assumed by a modeling language.

Similarly, understanding the difference between a partial order and a total order will give the reader an appreciation of the restrictions and power of a modeling language. A specific example of the use of some of the key concepts in this chapter relates to total and partial orders of elements of a set based upon the relation defined over the set. When a relation induces a total order, the elements of the set over which the relation is defined can be numbered from 1 to n. However, the concept of a partial order suggests that there is more than one possible order from 1 to n of the set's elements that is consistent with the relation. There are a number of applications of a partial order in systems engineering. For example, the set of functions being executed by the system's components can often be executed in more than one sequence. Understanding the many partial orders of functional execution is key to developing test plans to verify the system's performance characteristics. The interested reader is referred to Goodaire and Parmenter [1998], Harary [1972], and Harary et al. [1965] for more details on graph theory. Shin and Levis [2003] provide a performance prediction model based upon a creative application of Petri nets, which is a graph theoretic modeling language based on set theory.

Another specific example of the use of concepts from this chapter relates to the power of hierarchies in the engineering of systems; hierarchies for requirements, functions, and physical components are discussed in Chapter 2. In graph theory, a hierarchy is represented as a directed tree. This chapter introduces the terminology associated with trees in graph theory.

The state-of-the-art practice in the engineering of systems is to use a number of graphical concepts that have various amounts of grounding in mathematics as communication mechanisms. The challenge for the future is to develop additional modeling techniques that have significantly more grounding in mathematics while maintaining the quality of the communication among the stakeholders and the engineers in the various disciplines. The software engineering community has been moving in this direction for at least 15 years. The systems engineering community has just started this trek with SysML.

5.2 TERMINOLOGY

A *graph*, G, is a pair of sets, $V(G)$ and $E(G)$. $V(G) = \{n_1, n_2, \ldots, n_N\}$ is the set of vertices or nodes. $E(G) = \{e_{ij}\} \subseteq (V(G) \times V(G))$ is a relation that defines the set of edges that are unordered, not necessarily distinct pairs of nodes. $V(G)$ is a finite, nonempty set; $E(G)$ may be empty and is a subset of the Cartesian product of $V(G)$ with itself.

Due to the undirected nature of the edges in a graph, the edges represent symmetric relations such as "____ is next to ____," "____ is the sibling of ____," "____ is married to ____." Due to the symmetry, the order in which the nodes are placed does not matter.

The following Konigsberg bridge problem is one of the earliest known graph theory problems (see Sidebar 5.1). Euler's graph of the Konigsberg bridge problem is known as a *multigraph*, in which two or more edges connecting the same nodes are possible. This graph is also known as a *simple* graph because there are no loops. A *loop* is an edge connecting a node to itself, e_{ii}.

A *directed graph* or *digraph*, G, is a pair of sets, $V(G)$ and $E(G)$; $V(G) = \{n_1, n_2, \ldots, n_N\}$ is the set of vertices or nodes. $V(G)$ is again a finite, nonempty set; $E(G) = \{e_{ij}\}$ is a subset of $V \times V$ or *ordered* pairs of nodes; e_{ij} is said to be from n_i to n_j. Again, $E(G)$ may be empty.

The edges in a digraph represent *antisymmetric* or asymmetric relations. Examples are "____ is a parent of ____" and "____ is higher than ____." Here, the order in which the nodes are placed in the blanks does matter. Examples include Markov chains and Program Evaluation Review Technique (PERT) charts.

Figure 5.1 shows a sample digraph for the relation "is the parent of." Nodes that are connected by a directed edge are often discussed in terms of parent and child. The node at the tail of the edge is often called the *parent*, and the node at the arrow of the edge is called the *child*.

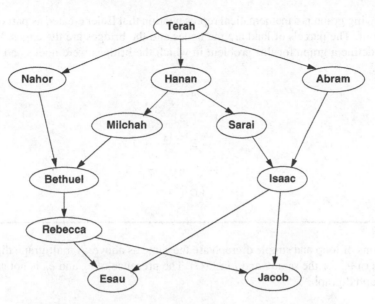

FIGURE 5.1 Sample-directed graph for "is the parent of."

SIDEBAR 5.1: THE KONIGSBERG BRIDGE PROBLEM

In the 1700s, the inhabitants of Konigsberg in eastern Prussia were entertained by a puzzle involving seven bridges over the Pregel River. The puzzle posed by mathematicians was whether it was possible to start at any one of the four distinct parcels of land (*A*, *B*, *C*, or *D*) and find a tour that crossed every bridge once and only once in such a way that the tourer ends up at the same parcel of land from which the tour began. L. Euler, the Swiss mathematician, proved that such a tour could not be done, and in 1736 gave precise conditions for when such a tour could be defined for any system of interconnected bridges.

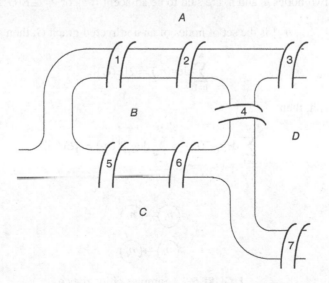

The following graph is a mathematical representation that Euler created as part of his mathematical proof. The parcels of land are the nodes and the bridges are the edges. Would it be possible to define a graph for this problem in which the bridges were nodes and the parcels were edges?

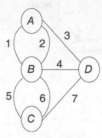

The definitions of loop and simple digraph are the same as above. A multigraph digraph requires multiple copies of e_{ij} for the same I and j in $E(G)$. The presence of e_{ij} and e_{ji} is not sufficient for G to be a multigraph digraph.

Cardinality of a set $A = |A| =$ the number of elements of A. Note, the cardinality of φ is 0. If A has n elements, then $\mathbf{P}(A)$ has cardinality of 2^n.

Order of $G = |V(G)| =$ the number of nodes of G.

Size of $G = |E(G)| =$ the number of edges of G.

The incidence of edges (Fig. 5.2) is defined as (a) e_{ij} is *incident* on n_i and n_j in a graph and (b) e_{ij} is *incident* from n_i to n_j in a digraph.

Degree of node $n_i =$ the number of edges connected to n_i in a graph, $\deg(n_i)$.

Out-degree of node $n_i =$ the number of edges incident from (or exiting) n_i in a digraph, $\deg_G{}^-(n_i)$.

In-degree of node $n_i =$ the number of edges incident to (or entering) n_i in a digraph, $\deg_G{}^+(n_i)$.

Adjacency – two nodes n_i and n_j are said to be adjacent if e_{ij} or $e_{ji} \in E(G)$.

If $V = \{n_1, n_2, \ldots, n_N\}$ is the set of nodes of an undirected graph G, then

$$\sum_{i=1}^{N} \deg(n_i) = 2|E(G)|.$$

If G is a digraph, then

$$\sum_{i=1}^{N} \deg_G{}^-(n_{\mathrm{i}}) = \sum_{i=1}^{N} \deg_G{}^+(n_{\mathrm{i}}) = |E(G)|.$$

FIGURE 5.2 Samples of incidence.

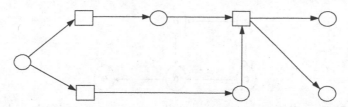

FIGURE 5.3 Sample bipartite graph. The box nodes are in *A* and the circles in *B*.

Edge labeling of a graph or digraph *G* is a function *f*: $E(G) \rightarrow D$, where *D* is a domain of labels.
Node labeling of a graph or digraph *G* is a function *f*: $V(G) \rightarrow D$, where *D* is a domain of labels.

Recall from Chapter 3 that IDEF0 (Integrated Definition for Function Modeling) uses edge and node labeling.

Bipartite graph is a graph (digraph) whose set of nodes can be partitioned into two sets *A* and *B* such that no edge connects a node in *A* to another node in *A* and, similarly, no edge connects a node in *B* to another node in *B*. See Figure 5.3. Is the family tree in Figure 5.1 a bipartite graph?

5.3 PATHS AND CYCLES

A *walk* in a digraph is a sequence of one or more nodes $\{n_0, n_1, \ldots, n_k\}$ and zero or more edges $\{e_{01}, e_{12}, \ldots, e_{k-1,k}\}$. See Figure 5.4. A walk may revisit the same node more than once. A walk is *closed* if its initial and end vertices are the same; otherwise, it is *open*. A walk is nontrivial if it has one or more edges.

A *path* is a walk in which each node is distinct (i.e., there are no repeats), except possibly the end nodes. See Figure 5.4. Note, since the nodes cannot repeat, the edges cannot repeat.

A *trail* is a walk in which each edge is distinct. Note, the same node may be revisited more than once. A closed trail is a *circuit*.

A *circuit* is a nontrivial walk with no repeated edges and whose endpoints are the same. Figure 5.5 has a circuit: *a, b, c, d, e, c, a.*

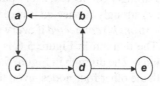

FIGURE 5.4 Digraph with a walk (*d–b–a–c–d–e*), closed walk, path, and a cycle (*a–c–d–b*).

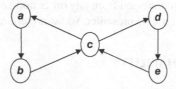

FIGURE 5.5 Digraph with two cycles (*a–b–c* and *c–d–e*) and a circuit (*c–a–b–c–d–e–c*).

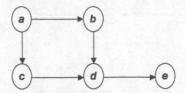

FIGURE 5.6 Digraph with a semipath (*b–a–c–d–e*) and semicycle (*d–b–a–c*).

A *cycle* is a circuit in which all the nodes are distinct except the first and last. See Figures 5.4 and 5.5. The nodes *a, c, d, b* in Figure 5.4 are a cycle. This cycle could be defined as (*d, b, a, c*) or (*b, a, c, d*) or (*c, d, b, a*) as well, but there is only a single cycle in this graph.

A *nondirected walk* (or *semiwalk*) in a digraph is a sequence of one or more nodes $\{n_0, n_1, \ldots, n_k\}$ and zero or more edges $\{e_{10}$ or e_{01}, e_{21} or $e_{12}, \ldots, e_{k,k-1}$ or $e_{k-1,k}\}$. A semiwalk can travel the wrong way on a directed edge.

A *semipath* (or *chain*) is a semiwalk in which each node is distinct, again with the possible exception of the end nodes. See Figure 5.6.

A semicircuit is a nontrivial semiwalk in which the first and last nodes are the same and no edges are repeated.

A semicycle is semicircuit in which the only repeated nodes are the first and last. See Figure 5.6.

A digraph is *acyclic* if there exists no subgraph that is a cycle.

By now, most readers are probably wondering how these definitions are going to be useful. The vocabulary provided by these definitions is very useful in describing when a graph has the seven unary characteristics (e.g., reflexivity and transitivity) from Section 4.3.3. In addition, there are other concepts that will be introduced in this chapter that have general applicability to the engineering of a system, for which this vocabulary will also be useful.

5.4 CONNECTEDNESS

Another vocabulary that proves very useful is connectedness. A *pair of nodes* in a digraph is *weakly connected* if there is a semipath between them, for example, nodes *b* and *c* in Figure 5.6. The nodes are *unilaterally connected* if there is a path between them, for example, all the pairs of nodes in Figure 5.6 except *b* and *c*. Finally, the nodes are *strongly connected* if there is a path in both directions. No pair of the nodes in Figure 5.6 is strongly connected; every pair of nodes in Figure 5.5 is strongly connected. Note, a pair of nodes that is strongly connected is also weakly and unilaterally connected.

A *digraph is weakly* (*unilaterally, strongly*) *connected* if every pair of nodes in the graph is weakly (unilaterally, strongly) connected. The digraph in Figure 5.6 is weakly connected because of the weak connection between nodes *b* and c. The digraph in Figure 5.4 is unilaterally connected because node *e* is unilaterally connected with the other four nodes, even though each of the other four nodes is strongly connected to each of the other three. The digraph in Figure 5.5 is strongly connected. The digraphs in Figures 5.1 and 5.3 are weakly connected.

A pair of nodes is *disconnected* if there is no path or semipath between them. A digraph is disconnected if one of its nodes is disconnected from any other node of the graph. A graph is connected if it is not disconnected. All the digraphs presented so far are connected.

5.5 ADJACENCY AND REACHABILITY*

The adjacency matrix of a graph *G*, *A*(*G*), provides a mathematical representation of which nodes in a digraph are adjacent to each other. Recall that a relation from *N*(*G*) to *N*(*G*) is defined by the edges

*Advanced Material.

of G, $E(G)$. So, in fact, $A(G)$ is a description of the relation $E(G)$ from $N(G)$ to $N(G)$.

$$A(G) = [a_{ij}]$$

is an $N \times N$ Boolean matrix where N is the order (number of nodes) of G.

$$a_{ij} = \begin{cases} 1 & \text{if } e_{ij} \in E(G) \\ 0 & \text{if } e_{ij} \notin E(G) \end{cases}$$

Note, a Boolean matrix is one whose elements are 0 or 1. The row sums of $A(G)$ give the out-degrees of the associated node; the column sums give the in-degrees. If G is not a digraph but a graph, $A(G)$ will be a symmetric matrix.

A node n_j of G is said to be reachable from node n_i of G if there exists a path from n_i to n_j in G. The reachability matrix, $R(G)$, is a Boolean matrix that indicates which nodes can be reached from which other nodes.

$$R(G) = [r_{ij}]$$

is an $N \times N$ Boolean matrix where N is the order of G. To compute $R(G)$, we first compute A, A^2, A^3, ... , $A^{|E(G)|}$.

$$r_{ij} = \begin{cases} 1 & \text{if } i = j \\ 1 & \text{if } a_{ij}^{(k)} > 0 \text{ for some } A^k \\ 0 & \text{otherwise} \end{cases}$$

Node n_j is reachable from node n_i if $r_{ij} = 1$. $R(G)$ is also called the *transitive, reflexive closure* of $E(G)$ because $R(G)$ is defined to be a reflexive relation that adds the edges necessary to make $E(G)$ a transitive relation. $R(G)$ is sometimes denoted $R^*(G)$.

The *transitive closure*, $R^+(G)$, is defined to be $R^+(G) = [r^+_{ij}]$, where

$$r^+_{ij} = \begin{cases} 1 & \text{if } a_{ij}^{(k)} > 0 \text{ for some } A^k \\ 0 & \text{otherwise} \end{cases}$$

Note, in this case, the reflexivity of the transitive closure is determined by the reflexivity of $E(G)$.

The *distance* between two nodes is the smallest number of edges between the nodes on any path connecting the two nodes. The distance matrix, $D(G)$, reflects these numbers.

$$D(G) = [d_{ij}]$$

is an $N \times N$ matrix where N is the order of G.

$$d_{ij} = \begin{cases} 0 & \text{if } i = j \\ k & \text{if } n_j \text{ is reachable from } n_i, k \text{ is the exponent of the first } A^k \text{ in which } a_{ij}^{(k)} > 0 \\ \infty & \text{if there is no path from } n_i \text{ to } n_j \end{cases}$$

5.6 UNARY RELATIONS AND DIGRAPHS

Now directed graphs will be used to visualize the seven properties of unary relations that were introduced in Chapter 4.

Reflexivity: $\forall x, x \, R \, x$. That is, all nodes must have loops. The top of Figure 5.7 shows a reflexive relation.

Irreflexivity: $\forall x, x \not{R} x$. That is, no nodes can have loops. The relations shown in the digraphs of Figures 5.1 and 5.3 through 5.6 are irreflexive. The bottom of Figure 5.7 shows an irreflexive relation.

Note digraphs can depict relations that are neither reflexive nor irreflexive when some of the nodes have loops and others do not.

Symmetry: $\forall x, y$, if $x R y$, then $y R x$. That is, there must be a cycle between any two nodes that are adjacent to each other. There is no limitation about arcs besides this. The relations shown in the digraphs of Figures 5.4–5.6 are not symmetric. The relation in the digraph shown in Figure 5.8 is symmetric.

Antisymmetry: $\forall x, y$, if $x R y$ and $y R x$, then $x = y$. That is, there cannot be a cycle between any two nodes that are adjacent to each other. Again, there is no limitation about arcs besides this one, so cycles containing three or more nodes can exist. Any node can have a loop. The digraphs in Figures 5.1 and 5.3–5.6 show antisymmetric relations; the relation in the digraph shown in Figure 5.8 is not.

Asymmetry: $\forall x, y$, if $x R y$, then $y \not{R} x$. That is, there can be no cycle between any two nodes, and there can be no loops. Asymmetric relations must be irreflexive. Again, cycles among three or more nodes are allowed. The relations in the digraphs shown in Figures 5.1 and 5.3–5.6 are asymmetric; the digraph in Figure 5.8 shows a relation that is not.

Note a relation that is irreflexive but in which no node is adjacent to any other node (completely disconnected) is symmetric, antisymmetric, and asymmetric due to the vacuous proof in Chapter 4.

Transitivity: $\forall x, y, z$, if $x R y$ and $y R z$, then $x R z$. This condition only applies to triplets of nodes and requires that there be a semicycle among the three nodes in the triplet. (Note the first and third node in a triplet can be the same, in which case there must be cycle between the two nodes and loops at each node.) A relation to which this condition, or left-hand side, is

A Sample Reflexive Relation

A Sample Irreflexive Relation

FIGURE 5.7 Reflexive and irreflexive relations.

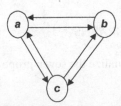

FIGURE 5.8 Digraph of a symmetric relation.

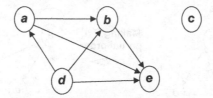

FIGURE 5.9 Digraph of a transitive relation.

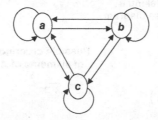

FIGURE 5.10 Transitive version of the digraph in Figure 5.8.

not applicable (i.e., the "if condition" is never satisfied) will be transitive. Figure 5.9 shows a transitive relation:

dRa and *aRb dRb*,
aRb and *bRe aRe*,
dRb and *bRe dRe*,
dRa and *aRe dRe*.

Intransitivity: For some *x*, *y*, *z*, if *x R̸ z*, then *x R y* and *y R z*. Relations are either transitive or intransitive. Cycles may exist in transitive relations, but note that a transitive relation with cycles that contains three or more nodes means that there must be a cycle between every pair of nodes that is part of the cycle, resulting in a symmetric relation with loops for the subset of nodes in the cycle. The relation in Figure 5.8 is symmetric but not transitive because *aRb* and *bRa*, but *a* is not related to *a*; the same applies for nodes *b* and *c*. Figure 5.10 shows the transitive version of the relation of Figure 5.8; the loops are added at each node.

It should be obvious that it is easier to use a directed graph to visualize the properties of unary relations than the mathematical expressions discussed in Chapter 4. Likewise, graphical techniques for visualizing functional relationships together with inputs and outputs are much more comprehensible than purely written or tabular methods for most people. "A picture is worth 1000 words."

5.7 ORDERING RELATIONS*

Relation *R* is a *partial order* on set *A* when *R* is reflexive, antisymmetric, and transitive on the set *A*. In this case, *A* is called a *partially ordered set*, or *POSET*, written as [*A*; *R*]. *Therefore, a relation that is a partial order cannot have any cycles*. As discussed in the previous section, a relation that is transitive and has cycles must have pairs of nodes that are symmetric. If any pair of nodes in a relation is symmetric, then the relation cannot be antisymmetric.

Two elements a_1 and a_2 in *A* are said to be *comparable* under *R* if either $a_1 \ R \ a_2$ or $a_2 \ R \ a_1$. Otherwise, the elements are incomparable. If every pair of elements is comparable, then [*A*; *R*] is *totally ordered*.

*Advanced Material.

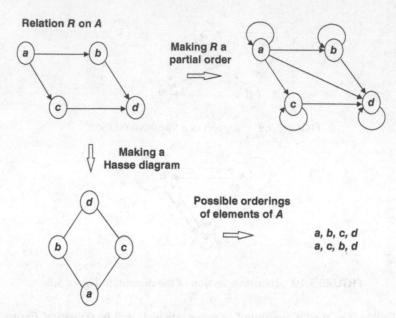

FIGURE 5.11   Partial order on a set A, Hasse diagram, and partial orderings of A.

A *Hasse diagram* is an undirected graph of the relations between the elements of a partially ordered set. See Figure 5.11. Each element of A is represented as a node. Reflexivity is not represented in the Hasse diagram, thereby eliminating all loops from the graph. Edges that are required by the transitivity property are also omitted; that is, any edge that depicts a shorter path to another node than some other combination of edges is deleted. To draw a Hasse diagram, we place the nodes on a piece of paper such that a_i is below a_j if $a_i R a_j$. We connect a_i to a_j with an undirected edge if and only if $a_i R a_j$ and there is no a_k such that $a_i R a_k$ and $a_k R a_j$. Figure 5.12 provides a second example of a Hasse diagram and the resulting partial orderings of A.

If there is only one node at the top of the Hasse diagram and only one node at the bottom, then the POSET is called a *lattice*. That is, with the transitivity property in force, there must be one and only one element, the upper bound or α of A, such that $\alpha R a_i \, \forall I$, and a second element, the lower bound or ζ of A, such that $a_i R \zeta \, \forall i$.

5.8   ISOMORPHISMS*

Two graphs, $G_1 = (V_1, E_1)$ and $G_2 = (V_2, E_2)$, are *isomorphic* if there exists a one-to-one and onto function, f, such that $f: V_1 \rightarrow V_2$ and f preserves adjacency. That is, $E_2 = \{(f(v), f(w)) \mid (v, w) \in E_1\}$. Note that "___ is isomorphic to ___" is an equivalence relation.

An isomorphism f from G_1 to G_2 is not necessarily unique. Some necessary properties for G_1 and G_2 to be isomorphic are (1) $|V(G_1)| = |V(G_2)|$, (2) $|E(G_1)| = |E(G_2)|$, and (3) if $n_1 \in V(G_1)$, then $\deg_{G_1}^{+}(n_1) = \deg_{G_1}^{+}(f(n_1))$ and $\deg_{G_1}^{-}(n_1) = \deg_{G_1}^{-}(f(n_1))$.

5.9   TREES

A *tree* is a graph G with no loops in which there is a unique, simple (no loops), nondirected path (or semipath in the case of a digraph) between each pair of nodes. Figure 5.13 shows a graph that is a tree.

*Advanced Material.

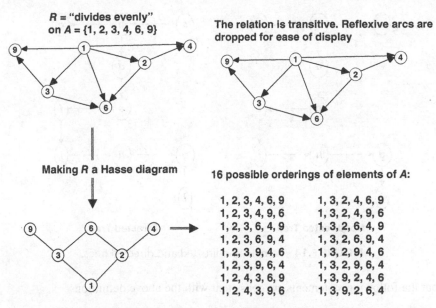

R = "divides evenly"
on A = {1, 2, 3, 4, 6, 9}

The relation is transitive. Reflexive arcs are
dropped for ease of display

Making R a Hasse diagram

16 possible orderings of elements of A:

1, 2, 3, 4, 6, 9	1, 3, 2, 4, 6, 9
1, 2, 3, 4, 9, 6	1, 3, 2, 4, 9, 6
1, 2, 3, 6, 4, 9	1, 3, 2, 6, 4, 9
1, 2, 3, 6, 9, 4	1, 3, 2, 6, 9, 4
1, 2, 3, 9, 4, 6	1, 3, 2, 9, 4, 6
1, 2, 3, 9, 6, 4	1, 3, 2, 9, 6, 4
1, 2, 4, 3, 6, 9	1, 3, 9, 2, 4, 6
1, 2, 4, 3, 9, 6	1, 3, 9, 2, 6, 4

FIGURE 5.12   Second Hasse diagram example.

A *rooted tree* is a tree in which there is a designated "root" node. In a graph, the root node must have a degree of 1. In Figure 5.13, nodes *a*, *c*, and *j* could be root nodes. In a directed tree, the root node must have no parents, or an in-degree of 0. In Figure 5.14, in the left digraph, nodes *a* and *c* could be root nodes; in the right digraph, only node *a* can be root node.

A *directed tree* is a rooted tree in which there is a (directed) path from the root to every other node. Note that the tree in Figure 5.13 is not a directed tree because the graph is not a digraph. The right-hand digraph in Figure 5.14 is a directed tree in which node *a* is the root. The left-hand graph is a tree because there exists a semipath from every node to every other node; that is, the graph is weakly connected. The graph is not a directed tree because there is not a path from any root (*a* or *c*) to every other node.

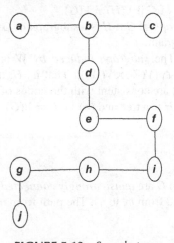

FIGURE 5.13   Sample tree.

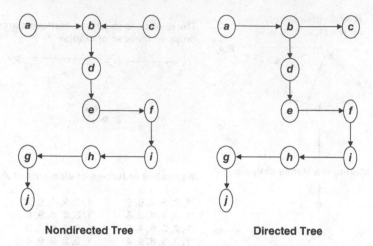

Nondirected Tree   Directed Tree

FIGURE 5.14   Sample nondirected and directed trees.

Note that the following statements are consistent with the above definitions:

(1) A simple nondirected graph G is a tree if and only if G is connected and contains no cycles.

(2) A tree with n nodes has exactly $n - 1$ edges.

(3) A graph G is a tree if and only if G has no cycles and $|E(G)| = |V(G)| - 1$.

A directed tree is a graphical representation of a partition, the fundamental construct of our requirements, functional, and physical decompositions.

5.9.1   Spanning Trees*

A graph H is a *subgraph* of a graph G if $V(H) \subseteq V(G)$ and $E(H) \subseteq [E(G) \cap (V(H) \times V(H))]$. That is, the nodes in the subgraph must be a subset of the nodes in the graph, and the edges in the subgraph must be a subset of those in the graph, with the added stipulation that all of the edges are connected to two nodes, one on each end of the edge.

Graph H is a *proper subgraph* of G if $V(H) \neq V(G)$.

A graph H is a *spanning subgraph* of G if H is a subgraph of G and $V(H) = V(G)$. So, a spanning subgraph cannot be a proper subgraph.

Let W be a subgraph of G. The *subgraph induced by* W is the subgraph H of G in which $V(H) = V(W)$ and $E(H) = [E(G) \cap (V(W) \times V(W))]$. That is, H, the subgraph of G induced by W, contains all of the edges of G that are consistent with the nodes of W. A subgraph H of a graph G is called a *spanning tree* of G if (a) H is a tree and (b) $V(H) = V(G)$. A spanning tree that is a directed tree is a *directed spanning tree*.

5.9.2   Directed Trees

Two nodes, n_1 and n_2, in a digraph G are *quasi-strongly connected* if there exists a node n_3 such that there is a path(s) from n_3 to n_1 and from n_3 to n_2. The path from n_3 to n_2 can pass through n_1.

---

*Advanced Material.

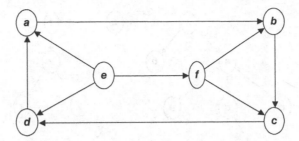

FIGURE 5.15   Quasi-strongly connected digraph.

Digraph G is a *quasi-strongly connected digraph* if and only if there is at least one node, r, in G such that there exists a path from r to all the remaining nodes of G. See Figure 5.15.

Let G be a digraph with $|V(G)| > 1$. Then the following statements are equivalent:

(1) G is a directed tree.
(2) There is a node r in G such that there exists a unique path from r to every node in G.
(3) G is quasi-strongly connected, and G – (any edge) is not quasi-strongly connected.
(4) G is quasi-strongly connected and contains a node r such that the in-degree of r is 0 and the in-degree of every other node in G is 1.

Note, that a and d are quasi-strongly connected via e; and b and c are quasi-strongly connected via both e and f; a and b are quasi-strongly connected via e by considering the path e–a–b, as well as the paths from e to a and from e to f to b. A directed tree can only be formed with e as the root.

The height of a directed tree is the length of the longest path. The height of the directed tree in Figure 5.14 is 8. A directed tree has levels. Level 0 is associated with the root of the directed tree. The first level of the directed tree contains all nodes adjacent to the root, or the children of the root. The second level contains the children of all nodes in level 1, and so on. See Figure 5.16. Note that a directed tree need not be symmetric, that is, reach the same level along every path.

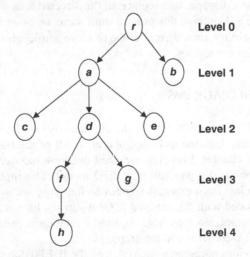

FIGURE 5.16   Levels of a directed tree.

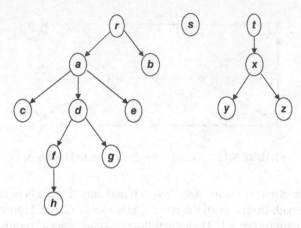

FIGURE 5.17   Sample directed forest.

5.9.3   Forest

A *directed forest* is a collection of directed trees. See Figure 5.17. Forests are important in systems engineering as we practice concurrent engineering. Recall from Chapter 1 that we must be concerned not only with the system that will be used during the operational phase but also with the development, manufacturing, training, deployment, refinement, and retirement systems. The concurrent requirements form a requirements forest.

5.10   FINDING CYCLES AND SEMICYCLES IN A GRAPH

In very large digraphs, it will not always be apparent that there are cycles or semicycles. To find the cycles, remove all barren nodes (nodes without children) and border nodes (nodes without parents). Continue this process until there are no remaining barren or border nodes. If there are any nodes remaining, then there are one or more cycles, and the remaining nodes are part of at least one of the cycles.

To find the semicycles in a digraph, first replace all the directed arcs with nondirected arcs. Then remove all nodes of degree 1. Continue this process until there are no remaining nodes of degree 1. If there are any nodes remaining, then there are one or more semicycles, and the remaining nodes are part of at least one of the semicycles.

5.11   REVISITING IDEF0 DIAGRAMS

At a superficial level, IDEF0 diagrams resemble the digraphs that we have been discussing. On any IDEF0 page, there are nodes, depicted as boxes, and arcs. All of the boxes and edges are labeled as discussed earlier in this chapter. However, we need not look too deep to see some major discrepancies between digraphs and a page of an IDEF0 model. The inputs, controls, outputs, and mechanisms (ICOMs) coming from external sources to the page are not nodes but labels on the edges. These edges, associated with the external ICOMs, do not have a node at one of their ends; this never happened in a digraph since an edge depicted a relation between two elements of a set A and all of the elements of A were shown in the graph.

As mentioned in the previous paragraph, each edge on the IDEF0 diagram is labeled. While there can be labels on the edges in digraphs, all the digraphs presented in this chapter had none. In a digraph, each edge represents the fact that a single relation exists between each pair of connected nodes, *aRb*.

Each node in the IDEF0 diagram is called a function and is named consistently with our understanding of a function, namely a transformation. Yet, digraphs represent a specific relation, which may be a mathematical function if certain conditions are satisfied (see Chapter 4). The relation, or function, in a digraph is represented by the edges, not the nodes.

At an even deeper level, each label on the edge of an IDEF0 arrow represents a set of possible items that can become an input, control, or output of the relevant function. All of the possible inputs and controls entering a function must then be represented by n-tuple of the Cartesian product across all input and control arrows entering that function. Similarly, the Cartesian product represents all possible outputs of a function across all output arrows exiting a function. So, there are, in fact, many important differences between a digraph and a page of an IDEF0 diagram.

Several people have attempted to transform an IDEF0 model into a bipartite graph. The first step is to turn the arc labels into nodes of a second type, say circles. The IDEF0 diagram (without mechanisms) in the top of Figure 5.18 is converted into a bipartite graph in the bottom of Figure 5.18. Each label is replaced by a circular node. Each external label is connected by the edge entering or

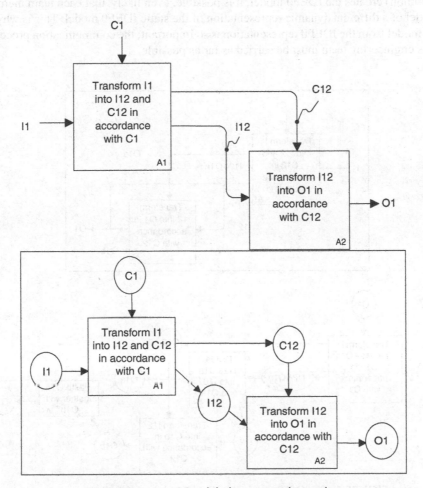

FIGURE 5.18 ICOM labels converted to nodes.

leaving the appropriate function. The new nodes for I12 and C12 are now connected by two edges; one going into the new node and one coming out of the new node. We have now satisfied the basic requirements of a bipartite graph; there are two types of nodes, and no edge connects two nodes of the same type. There are, in essence, two types of edges; those that connect boxes to circles (outputs of the function in the box) and those that connect circles to boxes (inputs to the function in the box).

However, there are two remaining problems. First, IDEF0 differentiates between arcs entering a function from the top and left. There is no provision for such differentiation in digraphs. Other process modeling techniques in Chapter 13 do not differentiate between inputs and controls; it is necessary to drop this distinction between inputs and controls, as is done in Petri nets, which is the only graph-theoretic modeling tool discussed in Chapter 13.

Second, there is a problem with branches and joins. There is no analogous construct in graph theory. To solve this problem, a function must be inserted at each branch to accomplish a divide or copy and at each join to accomplish a paste. See Figure 5.19.

With all these workarounds, IDEF0 remains a static snapshot of a dynamic process. There are potentially infinite dynamic models that can be created from each IDEF0 model. The information that separates the proper dynamic model from the rest of the possible dynamic models is not in the IDEF0 model but remains in the mental model of the creator of the IDEF0 model. If a team (which is most common) creates the IDEF0 model, it is possible, even likely, that each team member has a mental model of a different dynamic representation of the static IDEF0 model. This is why creating a dynamic model from the IDEF0 representation is so important; the communication process among the systems engineering team must be carried as far as possible.

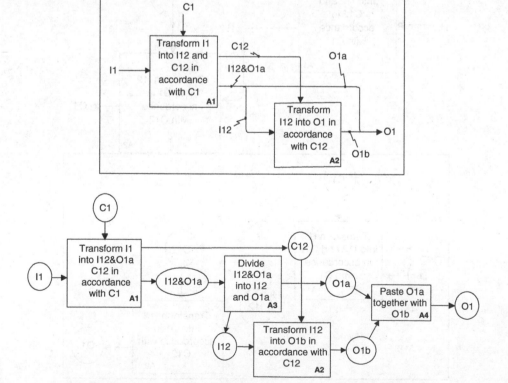

FIGURE 5.19 IDEF0 page with divide and paste functions added.

5.12 SUMMARY

A graph consists of a set of nodes and a set of edges. The edges define a relation over the set of nodes. The relation can require an order of the nodes in which case the edges are directed; directed graphs are the most applied in the engineering of systems. Bipartite graphs are a special form of a directed graph in which there are two types of nodes, and the edges cannot connect nodes that are the same type.

Sequences of nodes in a graph can be defined by the terms walk, path, trail, circuit, and cycle. Graphs can be connected or disconnected; there are variations of connectedness, ranging from weakly to strongly. Nodes that are not adjacent to each other in a graph can be reachable via a path in the graph. This notion of reachability can be critical if attaining some output requires the execution of a set of functions, but the set of functions is not part of a reachable set.

The properties of reflexivity, irreflexivity, symmetry, antisymmetry, asymmetry, transitivity, and intransitivity were defined in Chapter 4 and then redefined in terms of graphs in this chapter. Visualizing these relations provides a much greater understanding of their meaning and ability to detect their absence or presence in a graph.

Partial orders of the elements of a set were defined as alternative orders of the nodes based upon the relation defined over the nodes. The Hasse diagram was defined and illustrated for finding the partial order on the set and then enumerating the possible partial orders.

Trees and several variations of trees were introduced as a special form of a graph. A directed tree describes the notion of a hierarchical decomposition. Hierarchies of requirements, functions, and components were discussed in Chapter 2 and will be revisited in Chapters 6–11. These hierarchies must be partitions (as defined in Chapter 4) and can be represented as directed trees.

Finally, the IDEF0 process modeling technique was revisited and discussed in terms of mathematical graph theory. The reasons why an IDEF0 model is not a directed graph were discussed, as well as the difficulty associated with turning an IDEF0 model into a graph.

PROBLEMS

5.1 For the following graph, G_1:

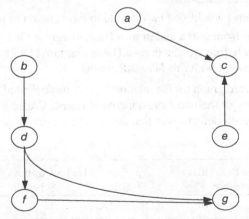

   a. Find $|V(G_1)|$ and $|E(G_1)|$.
   b. Write the relation depicted by G_1 as a set of ordered pairs.
   c. Define the adjacency matrix of G_1.

   d. What is the out-degree of each node of G_1? What is the in-degree of each node of G_1?

   e. Could G_1 be a bipartite graph? If no, why? If yes, what is the partition into two subsets of nodes that makes this a bipartite graph?

   f. Is the relation depicted here reflexive? irreflexive? symmetric? antisymmetric? asymmetric? transitive? intransitive?

   g. What arcs (if any) would you have to add to this relation to make it transitive?

5.2 For the following graph, G_2:

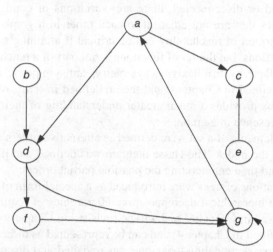

   a. Write the relation depicted by G_2 as a set of ordered pairs.

   b. Define the adjacency matrix of G_2.

   c. Could G_2 be a bipartite graph? If no, why? If yes, what is the partition into two subsets of nodes that makes this a bipartite graph?

   d. *Is there a cycle in G_2? How many?

   e. *Is there a semicycle in G_2? Which nodes are included?

   f. Is the relation depicted here reflexive? irreflexive? symmetric? antisymmetric? asymmetric? transitive? intransitive?

   g. What arcs (if any) would you have to add to this relation to make it transitive?

   h. *Delete the arc from g to a and draw a Hasse diagram for G_2. Why must we delete the arc from g to a before we can draw a Hasse diagram? Define at least 10 different node orderings consistent with this Hasse diagram.

5.3 a. Develop a directed graph for the relation "_____ has defeated _____." using the following won/lost records of the two 1993 superbowl teams. Create a single node for each team and an arc for each defeat. Note that this will be a multigraph.

	Buffalo Bills (BB) Schedule			Dallas Cowboys (DC) Schedule		
BB	38	NEP	14	DC	16	WR	35
		BB	13	DC	10		
BB	13	MD	22	DC	17	PC	10
BB	17	NYG	14	DC	36	GBP	14

	Buffalo Bills (BB) Schedule				Dallas Cowboys (DC) Schedule		
BB	38	NEP	14	DC	16	WR	35
BB	35	HO	7	DC	27	IC	3
BB	19	NYJ	10	DC	26	SFF	17
BB	24	WR	10	DC	23	PE	10
BB	0	PS	23	DC	20	PC	15
BB	13	NEP	10	DC	31	NYG	9
BB	23	IC	9	DC	14	AF	27
BB	7	KCC	23	DC	14	MD	16
BB	24	LAR	25	DC	23	PE	17
BB	10	PE	7	DC	37	MV	20
BB	47	MD	34	DC	28	NYJ	7
BB	16	NYJ	14	DC	38	WR	3
BB	30	IC	10	DC	16	NYG	13
BB	29	LAR	23	DC	27	GBP	17
BB	30	KCC	13	DC	38	SFF	21
SUPER BOWL		BB	13	DC	30		

b. Is this directed graph reflexive? irreflexive? transitive? asymmetric?

c. *There will be cycles in the graph created in part (a). Break these cycles by eliminating arcs in favor of the two Super Bowl teams; that is, if there is a cycle between a Super Bowl team and another team, eliminate the arc showing that the Super Bowl team was defeated by the other team. Assume the resulting relation is a partial order and draw a Hasse diagram of the relation.

5.4 For the following adjacency matrix:

	a	b	c	d	e	f	g	h	i
a	0	1	0	0	0	0	0	0	0
b	0	0	1	0	1	0	0	0	0
c	0	0	0	1	0	0	0	0	0
d	1	0	0	0	0	0	0	0	0
e	0	0	0	0	0	0	0	0	1
f	0	0	0	0	0	0	0	0	1
g	0	0	0	0	0	1	0	0	0
h	0	0	0	0	0	0	1	0	0
i	0	0	0	0	0	0	0	1	0

a. Draw the graphical representation, G_4, that is defined by the adjacency matrix.

b. Find $|V(G_4)|$ and $|E(G_4)|$.

c. Write the relation depicted by G_4 as a set of ordered pairs.

d. What is the out-degree of each node of G_4? What is the in-degree each node of G_4?

e. Could G_4 be a bipartite graph? If no, why? If yes, what is the partition into two subsets of the nodes that makes G_4 a bipartite graph?

f. Which of the seven properties (reflexive, irreflexive, transitive, intransitive, symmetric, asymmetric, antisymmetric) does this relation satisfy?

5.5 *Drop the arc from b to c in Figure 5.15 and draw a Hasse diagram for the resulting graph. How many orderings of the nodes in the digraph are consistent with this Hasse diagram?

5.6 There are three families defined by the sets A, B, and C; each family has a dad, mom, and three kids:

$$A = \{\text{Dad, Mom, Doris, Bill, Tom}\}$$

$$B = \{\text{Dad, Mom, Doris, Daisy, Debbie}\}$$

$$C = \{\text{Dad, Mom, Bill, Bob, Biff}\}$$

Consider the relations "is the spouse of," "is the brother of," and "is the blood relative of." (Hints: I am not the brother of myself. Two people are blood relatives if they share the blood of a common ancestor, who may or may not be part of sets A, B, or C. I am the blood relative of myself.)

Create a digraph for each of the three relations on each of the three sets. Identify which of these relations satisfy which of the seven properties of unary relations for each of the three sets by placing a yes or no in the empty cells of the following table.

	Reflexive	Irreflexive	Symmetric	Antisymmetric	Asymmetric	Transitive	Intransitive
"is the spouse of" on A							
"is the brother of" on A							
"is the blood relative of" on A							
"is the spouse of" on B							
"is the brother of" on B							
"is the blood relative of" on B							
"is the spouse of" on C							
"is the brother of" on C							
"is the blood relative of" on C							

5.7 A city street snapshot is shown in the figure. Note there are streets with arcs on them indicating one-way streets. The streets with double-headed arcs are two-way streets. There are 11 intersections, labeled 1 through 11.

a. Draw a directed graph that represents this street system. (Hint: use a node to represent street intersections.)

b. Is this digraph quasi-strongly connected? If not, what is the minimum number of arcs that must be added and what nodes must they connect to make it quasi-strongly connected? If yes, why?

c. If you think the digraph in part (a) is quasi-strongly connected, draw a directed spanning tree for it. If you do not think the digraph in part (a) is quasi-strongly connected, add arcs so that it is and then draw a directed spanning tree for it.

d. What is the height of the tree that you have drawn?

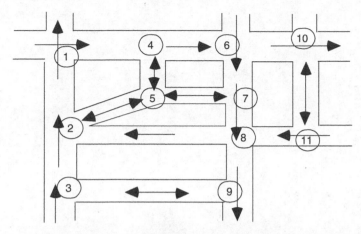

5.8 For the set of all possible relations, create a partition using combinations of the properties symmetric, antisymmetric, and asymmetric where each subset in the partition cannot be empty. As an example, a partition of all relations using the properties reflexive and irreflexive would be: (reflexive relations), (irreflexive relations), (relations that are neither reflexive nor irreflexive). Note, the subset of relations that are both reflexive and irreflexive is left out because this combination is impossible.

5.9 Consider an IDEF0 model in which the function A0 has two inputs (I_1 and I_2), three controls (C_1, C_2, and C_3), and three outputs (O_1, O_2, and O_3). The IDEF0 function, A0, can be considered a relation that maps elements of $\Delta = (I_1 \times I_2 \times C_1 \times C_2 \times C_3)$ into elements of $\Pi = (O_1 \times O_2 \times O_3)$. The five-tuple for inputs and controls to A0 and the three-tuple for outputs are used because each input, control, and output represents a set of possible inputs, controls, or outputs, respectively. The n-tuples define all possible combinations of inputs and outputs, respectively. Under what restrictions is A0 a function? Why?

Design and Integration

Chapter 6

Requirements and Defining the Design Problem

6.1 INTRODUCTION

Requirements are the cornerstone of the systems engineering process: Stakeholders' requirements provide operational statements by the stakeholders concerning their needs; derived requirements enable the engineers of systems to partition the design problem into components that can be worked in parallel while maintaining design control through the requirements partition and the interfaces between the components; derived requirements enable the verification of the configuration items and components during the qualification activity during development; and stakeholders' requirements provide the means for validating the system's design during qualification.

Requirements do not just show up on the systems engineer's desk. Obtaining "good" requirements is critical to the successful engineering of a system [Blum, 1992, pp. 68–81; Davis, 2005, pp. 3–39]. The systems engineer must work hard with the stakeholders of the system to develop the requirements. Fortunately, there is a tried-and-true method with some valuable modeling techniques that can be used in this effort.

There are few references that provide a coherent view of the systems engineering process for developing stakeholders' requirements for a system, including a definition of how these requirements might be usefully characterized to aid the generation process. Grady [1993] provides an excellent discussion of what requirements are, how requirements should be written one at a time and in documents, and how requirements should be allocated. Faulk et al. [1992] describe a software engineering method for real-time requirements that has many of the characteristics that are important. Crowe et al. [1996] adapt the method of Faulk et al. [1992] to software-intensive systems. This chapter (an expansion of Buede [1997]) defines such a process that is consistent with most systems engineering practices.

The chapter begins by discussing what the requirements are. Definitions that are key to putting a system in its context with external and enabling systems and the environment are provided next. Section 6.4 defines the process or method by which requirements are developed. The various categories of requirements found in the literature of systems engineering are then discussed, followed by the partition of requirements that will be used in this book. The proposed outline for a stakeholders'

The Engineering Design of Systems: Models and Methods, Fourth Edition. Dennis M. Buede and William D. Miller
© 2024 John Wiley & Sons, Inc. Published 2024 by John Wiley & Sons, Inc.
Companion website: www.wiley.com/go/engineeringdesignofsystems4e

requirements document that addresses all phases of the system's life cycle is provided in Section 6.7. The literature on requirements has proposed several characteristics that define either a sound individual requirement or a set of sound requirements; these characteristics of sound requirements are given in Section 6.8. The convention for writing requirements is discussed in Section 6.9.

Sections 6.10–6.13 describe in detail the portions of the process for developing requirements: defining the operational concept for each phase of the system's life cycle, creating an external systems model for each phase of the life cycle, establishing an objectives hierarchy for each phase of the life cycle, and conducting prototyping and usability testing to analyze the potential requirements in each phase of the life cycle. Section 6.14 provides a detailed discussion of the four segments of the requirements partition for each phase of the life cycle: the input/output requirements, the system-wide and technology requirements, the trade-off requirements, and the qualification requirements. Finally, the issue of managing requirements during the development of a system is discussed.

The focus of this chapter is the method for defining requirements for a system and all the enabling systems associated with each phase of the system's life cycle. There are seven activities associated with this method: developing the operational concept; defining the system boundary; developing an objectives hierarchy; developing, analyzing, and refining the requirements (including prototyping and usability testing); ensuring requirements feasibility; defining the qualification system requirements; and obtaining approval of the requirements.

Several models are introduced to support the process of defining requirements. A qualitative model, an input/output trace, defines a scenario that is part of the system's operational concept. An application of IDEF0 (Integrated Definition for Function Modeling) modeling is described for defining the process of a system's interaction with other (external and enabling) systems; this external system model defines all of the inputs and outputs associated with the system. A hierarchical decomposition of the objectives for a system is another example of a qualitative model used in this requirements definition process.

The exit criterion for this initial activity in the engineering of a system is the approval of the requirements document by the stakeholders. Often, the engineers of a system are focused on obtaining this approval as quickly as possible, often without defining all the requirements suggested in this chapter. The trade-off and qualification requirements are missing from most requirements documents. The contention of this chapter is that the real exit criterion of the requirements definition process is the approval by the stakeholders of the acceptance plan for the system. If the acceptance plan is affirmed, then all of the other portions of the requirements document are presumed to be defined in acceptable detail.

6.2   REQUIREMENTS

Many authors have defined the term *requirement*. The list below provides several definitions that highlight key concepts (the italics are the author's).

Sailor [1990]: identifiable capabilities expressed as *performance measurables* of functions that the system must possess to meet the mission objectives.

MIL-STD 499B [Military Standard, 1993a]: identifies the *accomplishment levels* needed to achieve specific objectives.

Chambers and Manos [1992]: the attributes of the final design that must be a part of any *acceptable solution to the design problem*.

Grady [1993]: an *essential attribute* for a system or an element of a system, *coupled by a relation statement with value and units information for the attribute*.

Davis [2005]: an *externally observable characteristic of a desired* system.

The requirements for a system set up standards and measurement tools for judging the success of the system design. These requirements should be viewed hierarchically. At the top are mission-level requirements that establish how the stakeholders will benefit by introducing the system in question into the super (or meta)-system of the system. These *mission requirements* relate to the objectives of the stakeholders that are defined in the context of the metasystem, not the system itself. For example, Boeing identified two primary **mission** requirements when starting on the Boeing 777 commercial aircraft: trip cost per seat and total trip cost. Each airline company that purchases a 777 is the metasystem that most influence an aircraft company during the development phase.

Stakeholders' requirements are developed next in the context of these mission requirements and should focus on the boundary of the system. If the stakeholders' requirements are defined internally to the system, the risk of having design statements embedded in the requirements goes up substantially. A major emphasis of this chapter is that the stakeholders' requirements should be as design-independent as possible. However, the stakeholder requirements are always in the context of the ecosystem around the system to be developed. That ecosystem is dynamic over time, not static, calling for detailed non-deterministic modeling of the ecosystem over the life cycle of the system being developed. Goode and Machol [1957] describe this as the "design of the exterior system." In particular, Goode and Machol call out probability as the basic tool of exterior system design. Hall [1962] addresses ecosystem modeling as part of the problem definition as environmental research, integrating the physical, technical, economic, business, and social environments. Boeing's stakeholders' requirements for the 777 included such topics as lift able weight of the aircraft at specified conditions, the empty weight of the aircraft, the drag force on the aircraft for certain specified flight conditions, and the fuel consumption of the aircraft at certain specified flight conditions.

As discussed in Chapter 1, *system requirements* are a translation (or derivation) of the stakeholders' requirements into engineering terminology. Once this translation occurs, the derivation process of requirements continues. Recall from Chapter 1 that the goal of the design process is to create a system specification that can be developed into specifications for the system's components, which are then segmented into specifications to the level of the system configuration items (CIs). As a result, the design process creates two hierarchies of requirements, as shown in Figure 6.1.

The stakeholders' requirements are produced in conjunction with the stakeholders of the system, based upon the operational needs of these stakeholders. Some systems engineers believe the systems engineering process begins when the Stakeholders' Requirements Document (StkhldrsRD) arrives; however, the position taken here and supported by Pragmatic Principle 1 [DeFoe, 1993] of the International Council on Systems Engineering (INCOSE), is that the systems engineers must be involved with the stakeholders to have any hope of producing a useful StkhldrsRD, note italicized

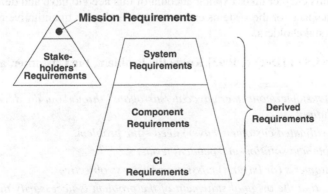

FIGURE 6.1 Requirements hierarchies.

items. In fact, the process described in this chapter is focused on methods and models for developing a valid and complete StkhldrsRD.

The Systems Requirements Document (SysRD), which is derived from the StkhldrsRD, is a translation from the language of stakeholders to the language of engineers. The system's requirements are traced directly from the stakeholders' requirements.

Note the term *stakeholder* is used in the above discussion in place of the more common term *user*. This is to emphasize the fact that there are usually multiple categories of users of a system: owner and/or bill payer, developer, producer or manufacturer, tester, deployer, trainer, operator, user, victim, maintainer, sustainer, product improver, and decommissioner. Each stakeholder has a significantly different perspective of the system and the system's requirements. If one perspective is singled out as the only appropriate one, the developers of the system will miss key information, and the system will be viewed negatively or as a failure from the other perspectives.

The systems engineering process for creating a system design is decision-rich. That is, the systems engineer is searching via a great deal of analysis and experience to find a very good (optimum is usually not possible to determine) solution that satisfies all of the mandatory requirements of the stakeholders and delivers as much performance as possible within the guidelines of cost and schedule.

This *search process* involves making many decisions about the system's physical character (or resources) and allocations of functions to resources that are usually only revisited if necessary. This search process occurs as the top-down onion-peeling process of systems engineering occurs. Figure 6.1 shows derived requirements at the component level (which may be several layers of the onion) and the CI (or bottom) level. Chapters 7–10 will describe this process of architecture development and the creation of appropriate derived requirements supported by analysis and judgment. To continue the story of the Boeing 777, Boeing created requirements for a major subsystem of the 777 – the engine. These derived requirements for the engine included the weight of the engine (derived from the weight of the empty aircraft), the thrust of the engine at specified conditions (derived from the lift able weight of the aircraft), the drag of the engine at specified conditions (derived from the drag of the aircraft), and the fuel consumption of the engine at specified conditions (derived from the fuel consumption of the aircraft).

A major impediment to this design process being successful is the over-constraint of the solution space by the stakeholders' requirements. The systems engineers' job is to work with the stakeholders to define the stakeholders' requirements to ensure that there is significant design freedom within these requirements and that many feasible designs exist. Stakeholders and (all too often) engineers are willing to constrain the requirements space very tightly without fully understanding or appreciating the potential value of the design options that they are eliminating. The stakeholders' requirements process defined in this chapter takes explicit account of this need to have and define a large tradable region in the design space for the systems engineers to search with quantitative techniques utilizing the priorities of the stakeholders.

Pragmatic Principle 1 [DeFoe, 1993] Know the Problem, the Customer, and the Consumer

1. *Become the "customer/consumer advocate/surrogate" throughout the development and fielding of the solution.*
2. *Begin with a validated customer (buyer) need – the problem.*
3. *State the problem in solution-independent terms.*
4. *Know the customer's (or buyer's) mission or business objectives.*
5. *Don't assume that the original statement of the problem is necessarily the best, or even the right one.*

6. *When confronted with the customer's need, consider what smaller objective(s) is/are key to satisfying the need, and from what larger purpose or mission the need drives; that is, find at the beginning the right level of problem to solve.*

7. *Determine customer priorities (performance, cost, schedule, risk, etc.).*

8. *Probe the customer for new product ideas, product problems/shortfalls, and identification of problem fixes.*

9. *Work with the customer to identify the consumer (user) groups that will be affected by the system.*

10. *Use a systematic method for identifying the needs and solution preferences of each customer group.*

11. *Don't depend on written specifications and statements of work. Face-to-face sessions with the different customer/consumer groups are necessary.*

12. *State as much of each need in quantified terms as possible. However, important needs for which no accurate or quantified measure exists still must be explicitly addressed.*

13. *Clarify each need by identifying the power and limitations of current and projected technology relative to the customer's larger purpose, the environment, and ways of doing business.*

6.3 DEFINITIONS

Before discussing the process for developing stakeholders' requirements, the definitions presented in Chapter 2 are reviewed.

A *system* is a set of components (subsystems, segments) acting together to achieve a set of common objectives via the accomplishment of a set of tasks.

A *system task, function, or activity* is a set of functions that must be performed to achieve a specific objective. (Function will be used to cover this class.)

A *human-designed system* is (1) a specially defined set of segments (hardware, software, physical entities, humans, facilities) acting as planned, (2) via a set of interfaces, which are designed to connect the components, (3) to achieve a common mission or fundamental objective (i.e., a set of specially defined objectives), (4) subject to a set of constraints, (5) through the accomplishment of a predetermined set of functions.

The external *systems* [Levis, 1993] of a system are a set of entities that interact with the system via the system's external interfaces. Note in Figure 6.2, the external systems can impact the

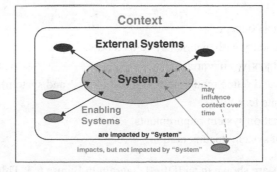

FIGURE 6.2 Depiction of the system, external systems, enabling systems, and context.

system and the system does impact the external systems. The system's inputs may flow from these external systems or from the context, but all of the system's outputs flow to these external systems. The external systems, many or all of which may be legacy (existing) systems, play a major role in establishing the stakeholders' requirements.

The context [Levis, 1993] of a system is a set of entities that can impact the system but cannot be impacted by the system. The entities in the system's context are responsible for some of the system's requirements. See Figure 6.2. Wieringa [1995] uses the phrase "universe of discourse" to label the context and external systems as that part of the world about which the system registers data and controls its behavior.

6.4 STAKEHOLDERS' REQUIREMENTS DEVELOPMENT: DEFINING THE DESIGN PROBLEM

Developing a good and complete set of requirements is very difficult. First, we have to figure out what topics we should be writing requirements about. These topics for the system-level requirements should all be at the same level of granularity, a level of granularity that is consistent with the system-level and not higher (the metasystem) or lower (subsystems). To facilitate defining these topics, we will introduce the concepts of an operational concept, external systems diagram, and objectives hierarchy.

After we determine what the topics of the requirements conversation are going to be, we can start writing specific requirements. Now, we have to determine what we want to say in that requirement. What is the threshold we are going to set for the minimum level of acceptable achievement? Here, we will talk about prototyping, analysis, elicitation, and usability testing.

Next, the requirements should be analyzed to determine that at least one feasible solution exists. A common problem is that we have defined thousands of requirements, and together, they are so constraining that there is no solution with enough performance at a low enough cost and a quick enough schedule. Often, it is very difficult to determine that there is a feasible solution, so this step is skipped. Typically, the selected design proves to be insufficient for 5–20 requirements, meaning it was not a feasible solution. Late in the design process systems engineers are confronted with the problem of should we search for a new design or accept the fact the current design cannot meet all of the requirements.

The last step before approval should be defining qualification or test requirements that are appropriate for the level of requirements being defined. When defining system-level requirements, these qualification requirements should address how will system-level verification and validation be done.

So the seven functions of this stakeholders' requirements development process are:

1. Develop operational concept,
2. Define system boundary with external systems model,
3. Develop system objectives hierarchy,
4. Develop, analyze, and refine requirements (stakeholders' and system),
5. Ensure requirements feasibility,
6. Define the qualification system requirements,
7. Obtain approval of system documentation.

These seven functions are shown in an IDEF0 diagram in Figure 6.3. This diagram is taken from the IDEF0 model of the process for engineering a system in Appendix B (see Wiley website for

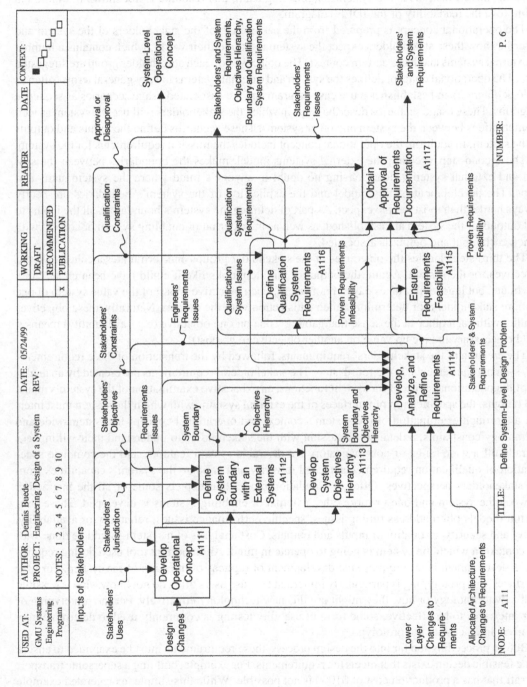

FIGURE 6.3 IDEF0 diagram of the system-level design process.

this appendix). To define this process fully, the first three functions must be defined in meaningful terms to justify their presence and provide explicit inputs to the fourth function. The last three functions are important but follow-on from the development of the StkhldrsRD. The resource that performs these functions is the systems engineering team; this resource is not shown in Figure 6.3 to improve the readability of the IDEF0 diagram.

The operational concept is prepared from the perspective of the stakeholders of the system and describes how these stakeholders expect the system to fit into their world, which contains a number of external systems and has a certain context. The objectives of each stakeholder group are suggested here. The operational concept defines the system and external systems in very general terms (often as a block diagram) and establishes a use case diagram and the associated usage scenarios as sequence diagrams. These usage scenarios describe ways in which the stakeholders will use the system as well as interactions between the system and other systems. These scenarios define the inputs and outputs of the system. In addition, the operational concept includes the mission requirements for the system.

The second step, creating the external systems model, makes the boundaries between the system and external systems clear, leaving no doubt in anyone's mind where the system starts and stops. The development of this model and the explication of the system's boundaries are nearly always harder than most people expect. As part of defining the system's boundaries, all the inputs to and outputs of the system are established, as well as the external or enabling system or context with which each input and output is associated.

The third step clarifies the objectives of the stakeholder groups and formulates a coherent set of objectives for the system. Again, the output of this step looks like it could have been created in a few hours, but it generally takes days, if not weeks. Each objective is part of the value system of one or more stakeholders for determining their satisfaction with the system. Naturally, these objectives conflict with each other in the sense that gaining value on one objective (e.g., availability) means it will be necessary to give up value on another objective (e.g., cost).

The creation of the stakeholders' requirements, followed by the translation of these requirements into system requirements, is the fourth step. The stakeholders' requirements are created by an analysis of the operational concept for system functions, an exhaustive examination of the system's inputs and outputs, the specification of interfaces of the external systems with which the system must interact, a thorough examination of the system's context and operational concept for system-wide and technology constraints, a detailed discussion with the stakeholders to understand their willingness to trade-off a wide range of non-mandatory but desirable system features, and the complete specification of qualification requirements needed to verify and validate the system's capabilities from the stakeholders perspectives. Often, a simulation model that depicts some or all the interaction between the system and one or more other external or enabling systems is developed. These simulation models often address timing issues, specific performance issues, reliability or availability, safety and security, or quality of inputs and outputs. Cost analyses of a system should be done with the context in which the system is going to operate in mind. An important tool used during requirements development is prototyping, the development of replicas of the parts of the system. For user interfaces, this prototyping is particularly important because users often do not know what is possible with new technology or how they might use this new technology effectively. For the prototyping of user interfaces to be effective, some form of usability testing is commonly used to determine how the users interact with the prototype.

Before proceeding too far into the design process, these requirements must be examined to ensure that a feasible design exists that meets the requirements. For example, building a supersonic transport aircraft that has a production cost of $1000 is not possible. While this simple, exaggerated example illustrates the problem, in practice, the development of hundreds, or even thousands, of requirements makes the test for feasibility quite difficult.

The sixth step is the development of requirements for the qualification system needed to verify and validate the resulting system. This involves the development of input/output requirements for the qualification system, as well as system-wide requirements. Trade-off requirements are also needed for the qualification system. Finally, the qualification system must also be qualified.

Finally, the stakeholders must approve the requirements documents. This approval process works best when the stakeholders are actively involved in and understand the previous steps.

Before defining and discussing requirements, noting that requirements must be developed for each phase of the system's life cycle is important. The life cycle phases used in this book are:

1. Development (design and integration),
2. Manufacturing or production,
3. Deployment,
4. Training,
5. Operations, maintenance, and support,
6. Refinement,
7. Retirement.

There is a strong correlation between the stakeholders and the life cycle phases. These seven functions should be applied to each stakeholder group and phase of the system's life cycle. Note that some of these phases may not be relevant for some systems. Most of the discussion from here on out will focus on the operations, maintenance, and support phases, but keep in mind that all phases of the life cycle should be addressed. Table 6.1 discusses who is involved in this requirements generation process and what their roles are.

6.5 REQUIREMENTS CATEGORIES

Many authors have categorized requirements. Here are some of the often-discussed categories:

1. *Specification Level: Stakeholders', Derived, Implied and Emergent.* Stakeholders' requirements, derived from operational needs, are those top-level statements defined in language that is understandable to the stakeholders, leaving substantial room for design flexibility. Stakeholders' requirements should define the essence of the stakeholders' needs sufficiently clearly for the stakeholders to be completely satisfied with whatever system results from the systems engineering process. Derived requirements are those requirements defined by the systems engineering team for the segments, subsystems, and CIs in engineering terms during the design process. Derived requirements are needed to complete the design to sufficient detail for the specification to be delivered to the design teams responsible for the physical CIs of the system. *Implied* requirements are those requirements not specifically identified in the StkhldrsRD, but that can be inferred based upon information in the StkhldrsRD. *Emergent* requirements are those requirements that are not even hinted at in the StkhldrsRD but whose presence is made known by stakeholders later in the systems engineering process. These last two sets of requirements are to be avoided, if possible, by a sound and systematic stakeholders' requirements development process.

2. *Performance Requirements Versus Constraints. Performance requirements* are defined on some index that establishes a range of acceptable performance from a minimum acceptable threshold to a design goal. Constraints simply rule out certain possible designs; for example, the system must be painted a specific shade of green. A performance requirement defines

TABLE 6.1 Roles and Responsibilities during Requirements Generation

Who has the right to have a stakeholders' requirement?	Any individual/organization with a need involved in the development (design and qualification), production, deployment, training, operation, maintenance, support, refinement, decommissioning of, **and payment** for the system.
What does one call a person who has a requirement?	Customer or stakeholder
Who must respond to the requirer(s) having a requirement and how?	System's **requirements team**, a collection of *stakeholders* and *systems engineers*. Response is acceptance, request for clarification, or rejection.
By what criteria does the Systems Requirements Team respond?	This team establishes the external systems diagram and fundamental objectives hierarchy of the system, and then determines if the requirement fits within the scope of the system's boundary and fundamental objective. Stakeholders' requirements also have to be assessed for the proper level of abstraction. A requirement should not be too strategic (mission-oriented) or means (or solution) oriented.
How does one know that the requirement is "right"?	There is no right or wrong, only acceptable or unacceptable at this time. Over time, some of the stakeholders' requirements will change.
How are these requirements conveyed to the people who get involved once a requirer has enunciated a requirement?	The system's requirements team documents the collection of stakeholders' requirements. This stakeholders' requirements document (StkhldrsRD) is distributed to the stakeholders and systems engineers. Included in this document is a discussion of the operational concept of the system and the external systems and context associated with the system, that is, how each stakeholder expects to interact with the system. By reviewing the stakeholders' requirements document, each stakeholder can see how the requirement s/he suggested fits into the envisioned operation of the system and can judge whether this vision makes sense from her/his perspective.
What does the Systems Requirements Team do next?	The system's stakeholders' requirements team remains active throughout the system's life cycle. During design, there will be many occasions when the system's stakeholders' requirements must be reviewed and modified. These occasions will diminish in frequency once the system is deployed, but the requirements process is still critical as requirements changes and system modifications are envisioned, agreed to, developed, and fielded.

a desired direction of performance; for an elevator system (which is used throughout this book as an example), a performance requirement might be to "minimize passengers' waiting time during peak periods." For any performance requirement, there must also be a minimum acceptable performance constraint or threshold associated with the index, beyond which designs with such poor performance are not feasible (e.g., average passengers' waiting time during peak periods shall be less than 35 seconds). Often, there is also a maximum threshold or goal on the performance index that states the stakeholders do not noticeably value performance beyond this point (e.g., average passengers' waiting time during peak periods need not be less than 27 seconds).

3. *Application – System Versus Program.* System requirements relate to characteristics of the system's performance (in the broadest sense). Program requirements relate to the first life cycle phase of the systems engineering process and usually address the treatment of the cost and schedule for this phase. Program requirements relate either to the programmatic tasks

that must be performed, programmatic trade-offs among cost and schedule, and programmatic products associated with the systems engineering process (e.g., the Up and Down Elevator Corporation shall own full rights to the design data of the elevator).

4. *Functional, Interface, or System-wide Requirements.* Functional requirements relate to specific functions (at any level of abstraction) that the system must perform while transforming inputs into outputs. As a result, a functional requirement is a requirement that can be associated with one or more of the system's outputs. Interface requirements are usually constraints that define the reception of inputs and transmission of outputs between the system and the system's ecosystem. System-wide requirements (often called "-ilities") are characteristics of the entire system; examples include availability, reliability, maintainability, durability, supportability, safety, trainability, testability, extensibility (growth potential), and affordability (e.g., operating cost).

6.6 REQUIREMENTS PARTITION

There is great value in having a structure for various types of requirements. If the requirements are listed in random order in a requirements document, it is nearly impossible to be sure that a given requirement is not addressed multiple times in that single requirements document. It is also difficult to find a specific requirement in a large document. There are other benefits of a requirements structure, especially if the structure is a partition. A partition is a structure that has subcategories that are mutually exclusive, meaning a requirement can only be put in one category. A partition also needs to be exhaustive, meaning every requirement has some category that is appropriate for it. By creating such a partition, it is easy to review the partition to ensure that there are as many requirements in that category as expected and that every requirement in the category is appropriate for that category.

The partition that is introduced here has both a vertical spectrum and a horizontal spectrum. The vertical spectrum was introduced in Figure 6.1, which shows two vertical levels of requirements written for the stakeholders and three or more levels of derived requirements written for the engineers. The horizontal spectrum addresses the life cycle as well as categories of requirements within each phase of the life cycle. The life cycle steps or phases include development, production, operations, etc.; recall Figure 1.1. The categories of requirements within each phase of the life cycle are discussed next.

Wymore [1993] identifies six types of system design requirements: input/output, technology and system-wide, performance trade-off, cost trade-off, cost–performance trade-off, and test. These six types of requirements are condensed into four categories: input/output, technology and system-wide, trade-off, and qualification (test). From a concurrent engineering perspective, each requirements category should be used to address the relevant system (e.g., development system and manufacturing system) in each phase of the system's life cycle (development, production, deployment, training, operation and maintenance, refinement, and retirement). Table 6.2 provides examples of various types of requirements; these examples have been collected from a wide variety of sources.

(1) *Input/output requirements* include sets of acceptable inputs and outputs, trajectories of inputs to and outputs from the system, interface constraints imposed by the external and enabling systems, and eligibility functions that match system inputs with system outputs for the life cycle phase of interest. Clearly, there are several requirements in this category during the operations phase of the life cycle. However, the system may have inputs and outputs in all portions of the system's life cycle (e.g., training stimulations, standardized internal interfaces for product improvement); if so, the requirements for these activities would be found in this category in the appropriate life cycle phase. This category is partitioned into four subsets: (1) inputs, (2)

TABLE 6.2   Exemplary Requirement Dimensions

Requirements Category	Exemplary Requirement Dimensions
Input or Output Performance	Quality of output
	Accuracy (or precision)
	Correctness (or confidence, error rate)
	Security (or perishability, survivability)
	Quantity of output
	Intensity, size, or distance
	Number per unit time (throughput, velocity)
	Coverage (area or volume served by outputs)
	Timing of outputs
	Response time (timeliness, time to create an output)
	Update frequency
	Availability
Undesired or Unexpected Inputs	Unexpected or undesired inputs and appropriate response
	Bounds on expected inputs and appropriate response
Interface Constraint	Required format of an input or output as defined by the interface
	Timing constraints associated with an interface
	Physical form or fit of an interface
Suitability or Quality Issues of the System	Usability
	Weight of the system
	Form (volume) and fit (dimensions) of the system
	Survivability of the system
	Availability, reliability, and maintainability of the system
	Supportability of the system
	Safety of the system
	Security
	Trainability of the system
	Testability of the system
	Extensibility (expected changes/growth potential) of the system
Costs for Various Life Cycle Phases	Affordability (or operating and maintenance cost) of the system
	Development cost
	Production cost (manufacturability) of the system
	Deployment and training costs of the system
	Decommissioning cost of the system
Schedule for Various Life Cycle Phases	Development period
	Manufacturing time for each unit
	Training time to reach proficiency by category of user
	Deployment period
	Durability (or operational life) of the system

outputs, (3) external interface requirements, and (4) functional requirements. Input requirements state what inputs the system must receive and any performance or constraint aspects of each. Output requirements state what outputs the system must produce and any performance aspects. Table 6.2 provides an extensive list of possible performance issues for the outputs of any system, segmented by quality, quantity, and timeliness. External interface requirements deal with limitations placed upon the receipt of inputs and transmission of outputs by the interfaces of the external systems; see Table 6.2. Functional requirements can be endless unless organized; the functional requirements proposed here are the two to seven functions that are the first-level decomposition of the system's function.

The very strong position being taken here is that the input and output requirements are the key to defining the needs of the stakeholders in terms that they can understand. Stakeholders in each phase of the system's life cycle can relate to quantity, quality, and timing aspects of the outputs delivered by the system under question and the ability to deal with quantity, quality, and timing of inputs. The engineers of the system develop the system's functions during the design process. This development of a functional architecture (see Chapter 7) is a very valuable means for dealing with the complexity of the engineering problem. However, the stakeholders should not care a whit about the functions being performed by the system as long as they are happy with the characteristics of the inputs being consumed and the outputs being produced by the system. The concept of having a major section of Stakeholders' Requirements Document devoted to the functions of the system is misguided and guaranteed not to elicit the needs of the stakeholders.

(2) *Technology and system-wide requirements* consist of constraints and performance index thresholds (e.g., the length of the operational life for the system, the cost of the system in various life cycle phases, and the system's availability) that are placed upon the physical resources of the system. Many of the requirements from each phase of the system's life cycle are found in this category because these requirements specifically relate to the physical manifestation of the system. This category can be partitioned into four subsets: (1) technology, (2) suitability and quality issues, (3) cost for the relevant system (e.g., development cost and operational cost), and (4) schedule for the relevant life cycle phase (e.g., development time period and operational life of the system).

(3) *Trade-off requirements* are algorithms for comparing any two alternate designs on the aggregation of cost and performance objectives. These algorithms can be divided into (1) performance trade-offs, (2) cost trade-offs, and (3) cost–performance trade-offs. The performance trade-off algorithm defines how the relative performance of any two alternate designs can be compared in terms of the system's performance objectives. These performance objectives are defined within the input/output and non-cost system-wide requirements. The performance trade-off algorithm specifically defines how the performance parameters are to be compared to each other. Note schedule requirements are embedded within the performance requirements for simplicity of discussion. The cost trade-off algorithm defines how the relative cost of any two alternate designs can be compared across all cost parameters (life cycle phases) of interest to the stakeholders. Note that dollars spent at different times may not be comparable by present value computations when there are different bill payers at different times. Finally, the cost–performance trade-offs define how performance objectives should be traded with cost objectives.

These trade-off algorithms could be based upon many different mathematical logics; indeed, many have been proposed. The strong position taken in this book is that these trade-off algorithms must be based upon the value preferences of the stakeholders. Decision analysis provides a normative basis for these preference judgments and algorithms, as described in detail in Chapter 14. For applications of these decision analysis techniques (value curves and swing weights), see Buede and Bresnick [2007], Buede and Choisser [1992], Daniels et al. [2001], Ross et al. [2004], Thurston and Carnahan [1993], and Walton and Hastings [2004].

The ideal approach for quantifying the trade-off preferences of the stakeholders would be to obtain these preferences as statements of "willingness-to-pay" (in terms of money for development effort) for enhanced performance and decreased cost in each of the other life cycle phases. To make these statements of "willingness-to-pay" operationally meaningful, the appropriate contractual arrangements must be established that would permit the transfer of payments based upon the stated payment preferences. In addition, a warranty system must be established that requires the developers to stand behind their developmental phase claims of performance attainment during the remaining phases of the system's life cycle. For example, if a performance claim made during the development phase is not achieved during the operational phase, the developer would have to make a warranty payment to the stakeholders. Although this entire approach is known and obviously will work, the approach has never been used to the author's knowledge. For an example of the opposite of this happening, see Sidebar 15.3 (Chapter 15). In fact, users are quite cynical about the performance claims made by developers during the development phase.

(4) *System qualification requirements* address the need to qualify the system as being designed right, the right system, and an acceptable system. There are four primary elements:

   a. *Observance*: to state which qualification data for each input/output and system-wide requirement will be obtained by (1) demonstration, (2) analysis and simulation, (3) inspection, or (4) instrumented test, and the verification requirements on the system requirements and validation requirements based on the stakeholder requirements. These requirements are included in the Stakeholders' Requirements Document because the stakeholders need to consider the risks in terms of time, money, and facilities associated with different levels of these requirements.

   b. *Verification Plan*: to state how the qualification data will be used to determine that the real system conforms to the design that was developed. These should be characterized as verification requirements. As part of this plan, the engineering and exercising of the verification system also needs to be defined with requirements.

   c. *Validation Plan*: to state how the qualification data will be used to determine that the real system complies with the stakeholders' performance, cost, and trade-off requirements. These should be characterized as validation requirements. As part of this plan, the engineering and exercising of the validation system also needs to be defined with requirements.

   d. *Acceptance Plan*: to state how the qualification data will be used to determine that the real system is acceptable to the stakeholders. These should be characterized as acceptance requirements.

Note that the qualification requirements associated with the first objective define the basis for the requirements for the suite of qualification systems (e.g., simulations, instrumented test equipment) needed for the system under development. Having technology/system-wide requirements that limit the flexibility to develop new test equipment is common.

This requirements partition provides a solid basis and set of guidelines for guaranteeing that the system's requirements are complete, consistent, unique, comparable, and modifiable. (These terms will be defined a little later.) Success is not certain with this basis and guidelines but is greatly enhanced over current industry practice.

Figure 6.4 traces the origins of the performance requirements to the objectives hierarchy by showing that the objectives hierarchy defines the performance parameters defined to have a performance range rather than a single point. These performance parameters can fall within the categories of input, output, "-ilities," cost, and schedule requirements. The thresholds and goals for these tradable requirements are defined as part of the input, output, "-ilities," cost, and schedule requirements. The algorithms that define the tradable space over these performance parameters are documented in the performance, cost, and cost–performance trade-off requirements. The performance, cost, and cost–performance trade-off requirements combine to define the iso-value lines in the tradable space; these iso-value lines will be the basis for all design trade-offs.

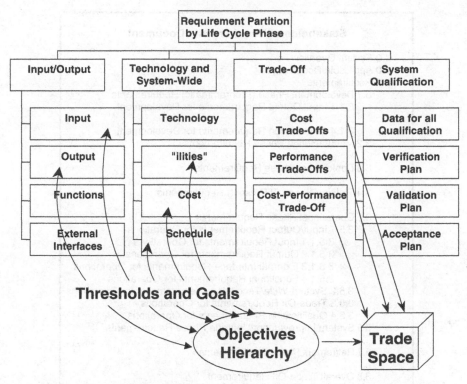

FIGURE 6.4   Objectives hierarchy, requirements partition, and trade space.

If every set of requirements contained the information defined by Wymore [1993], there would be far fewer problems in system development efforts. Very few requirements documents contain performance, cost, and cost–performance trade-off requirements as defined by Wymore. These elements should be defined in the stakeholders' requirements document from the stakeholders' perspective; otherwise, the systems engineers must guess at the ultimate trade-offs of the stakeholders; the ability of engineers to do a complete and effective job of guessing iso-value trade-offs is questionable at best.

6.7   STAKEHOLDERS' REQUIREMENTS DOCUMENT (StkhldrsRD)

The format for a StkhldrsRD (Fig. 6.5) should include sections for a brief overview of the system, references to relevant documents from which the stakeholders' requirements have been traced, and the requirements. The requirements should be organized by the life cycle phase. Within each life cycle phase, requirements from the four segments of the above taxonomy should be developed. The life cycle phases are being called out explicitly to highlight the criticality of the concurrent engineering nature of the design problem. The designs of the life cycle systems needed to obtain an operational system are not that straightforward. Requirements in one phase of the life cycle will often have a major impact on the design of a system in another phase. For example, a requirement that the manufacturing system be operational by a specified date precludes many interesting designs of the operational system. This interaction of requirements and design options across life cycle phases is a major contributing factor to failure in the real world; in addition, this interaction makes the concept of formulating the design problem as an optimization problem nonsensical to practitioners.

Stakeholders' Requirements Document

1.0 System Overview
2.0 Applicable Documents
3.0 Requirements
 3.1 Development Phase (Programmatic) Requirements
  3.1.1 Input/Output Requirements for Development
  ...
  3.1.4 Qualification Requirements for Development
 3.2 Manufacturing Phase Requirements
 ...
 3.3 Deployment Phase Requirements
 ...
 3.4 Training Phase (If Present) Requirements
 ...
 3.5 Operational Phase Requirements
  3.5.1 Input/Output Requirements for Operations
   3.5.1.1 Input Requirements for Operations
   3.5.1.2 Output Requirements for Operations
   3.5.1.3 External Interface Requirements for Operations
   3.5.1.4 Functional Requirements for Operations
  3.5.2 System-Wide/Technology Requirements for Operations
  3.5.3 Trade-Off Requirements for Operations
  3.5.4 Qualification Requirements for Operations
 3.6 System Improvement/Upgrade Phase Requirements
 ...
 3.7 Retirement Phase Requirements
 ...
 3.8 Overall Trade-Off Requirement

Appendix A. Operational Concepts by Phase
Appendix B. External System Diagrams by Phase

FIGURE 6.5   Outline of stakeholders' requirements document.

Rather, the segregation of requirements by life cycle phase is meant to aid in attaining the desired attributes (e.g., complete, consistent) of requirements discussed in Table 6.3 of the next section.

Given the organization of the StkhldrsRD shown in Figure 6.5, an overall trade-off requirement (Section 3.8 of the StkhldrsRD) that addresses comparisons across life cycle phases is needed to enable coherent evaluations of design options.

6.8 CHARACTERISTICS OF SOUND REQUIREMENTS

A number of authors [Frantz, 1993; Davis, 1993; Mar, 1994] have developed various numbers of attributes for requirements. The literature is not in total agreement about the meaning of these attributes. Table 6.3 is the result of a detailed examination of the literature. The characteristics are divided into those that are related to individual requirements and those relevant to groups of requirements.

In any systems engineering effort, as many correct requirements must be developed as possible; these correct requirements should be verifiable. In addition, as many incorrect requirements should he eliminated as possible. In summary, the requirements document should contain a complete, consistent, comparable, design-independent, modifiable, and attainable statement of the design problem.

TABLE 6.3   Attributes of Requirements

Individual Requirement Attributes

(1) *Unambiguous* – every requirement has only one interpretation

(2) *Understandable* – the interpretation of each requirement is clear to those selected to review the requirement

(3) *Correct* – the requirement states something required of the system, as judged by the stakeholders

(4) *Concise* – no unnecessary information is included in the requirement

(5) *Traced* – each stakeholders' requirement is traced to some document or statement of the stakeholders

(6) *Traceable* – each derived requirement must be traceable to a higher-level requirement via some unique name or number

(7) *Design independent* – each requirement does not specify a particular solution or a portion of a particular solution

(8) *Verifiable* – a finite, cost-effective process can be defined to check that the requirement has been attained

Attributes of the Set of Requirements

(9) *Unique* – requirement(s) is(are) not overlapping or redundant with other requirements

(10) *Complete* – (1) everything the system is required to do throughout the system's life cycle is included, (2) responses to all possible (realizable) inputs throughout the system's life cycle are defined, (3) the document is defined clearly and self-contained, and (4) there are no "to be defined" (TBD) or to be reviewed (TBR) statements; completeness is a desired property but cannot be proven at the time of requirements development, or perhaps ever

(11) *Consistent* – (1) internal – no two subsets of requirements conflict and (2) external – no subset of requirements conflicts with external documents from which the requirements are traced

(12) *Comparable* – the relative necessity of the requirements is included

(13) *Modifiable* – changes to the requirements can be made easily, consistently (free of redundancy) and completely

(14) *Attainable* – solutions exist within performance, cost and schedule constraints

(15) *Organized* – grouped according to a hierarchical set of concepts, such as life cycle and categories.

6.9   WRITING REQUIREMENTS

Certain procedures have been developed [Grady, 1993; Hooks, 1994] for writing requirements – these procedures guide requirements writers toward the achievement of the above attributes. First, a set of terms has been developed. Specifically, a statement of a requirement includes the use of the word "shall" to indicate the limiting nature of a requirement; statements of fact use "will"; and goals use "should." The requirements statement shall include a subject (the relevant life cycle system), the word "shall," a relation statement (e.g., less than or equal to), and the minimum acceptable threshold with units. Data clarifying the terms in the requirement can also be added. Examples of appropriate grammar are:

> The system shall provide the customer a receipt at the end of each transaction. The receipt shall contain Bank Name, Account Number, Date and Time of Day, Type of Transaction, Account Balance at the end of the Transaction, and Automatic Teller Location Code Number.

> The system shall stop the flow of liquid hydrogen in 0.5 seconds or less. The liquid stopping time is measured from the time the control signal for stopping is received until the flow through reaches zero.

It is important to avoid compound predicates and negative predicates:

The system shall fit …, weigh …, cost … (this causes traceability problems).

The system shall not … (attempt to turn this into a positive statement of what the system shall do).

Similarly, the "and/or" colloquialism is inappropriate because "and/or" provides the designer with a choice; be specific about whether you mean "and" or "or." The requirement should not start with an "If …" statement. Conditions under which the requirement is true should be placed at the end of the requirement.

Ambiguous terms are a plague on requirements. Common verbs that are not specific enough include "maximize" and "minimize" because the system is seldom operating in an environment in which optimization is possible. "Accommodate" is another example of a vague verb. Adjectives are a major source of ambiguity; examples include "adaptable," "adequate," "easy," "flexible," "rapid," "robust," "sufficient," "supportable," and "user-friendly." Requirements should start with the system of interest, be followed by a verb phrase starting with the word "shall," be followed by an object that describes an input, output, etc., and end (if necessary) with conditions under which the previous was true. Examples include:

The development system shall receive inputs from stakeholders. (Input requirement)

The manufacturing system shall have a scrappage rate that is less than $x\%$. The design goal is $0.7x\%$. (Output requirement)

The deployment system shall accept boxes of x ft^3 or less. The design goal is $0.5\ x$ ft^3. (Input requirement)

The training system shall complete training in x hours per student or less. The design goal is $0.9x$ hours. (Output requirement)

The operational system shall have an operational life of x years or more. The design goal is $2x$ years. (System-wide schedule requirement)

The refinement system shall be compatible with the following new technologies (x, y, z) for the central processing unit. (Input requirement)

The retirement system shall retire units for less than $\$x$ each. (System-wide cost requirement)

6.10 OPERATIONAL CONCEPT

An *operational concept* [Lano, 1990a] is a vision for what the system is (in general terms), a statement of mission requirements, and a description of how the system will be used. Hooks and Farry [2001] describe the operational concept as a "day in the life of your product." This operational concept is an opportunity to create a vision that is shared among all the stakeholders for the really major interactions of people and things with the system of interest. The shared vision is from the perspective of the system's stakeholders, addressing how the system will be developed, produced, deployed, trained, operated, maintained, refined, and retired to overcome some operational problems and achieve the stakeholders' operational needs and objectives. The development of the operational concept serves the purpose of obtaining consensus in the written language of the stakeholders about what needs the system will satisfy and the ways in which the system will be used. Remember that there is a system for each phase of the system's life cycle and that an operational concept is needed for each of the systems. By describing how the system will be used, the operational concept provides substantial (but incomplete) information about the system's interaction with other systems and the context of the system.

The conceptual design identifies alternate concepts and performs the decision analysis to select a preferred concept for development, see Table 1.3. The alternatives, relatively few in number, span the range of key drivers within the projected technology capabilities, beginning with and continuing through the operational life of the system, e.g., minimum time or minimum energy to complete a mission. Section 2.3 on key systems engineering concepts identifies constructs equally applicable to conceptual design as to development: operational concept, external systems diagram, objectives hierarchy, requirements, functions, items, components, interfaces, verification, validation, and acceptance. Conceptual design requires a level of abstraction and fidelity for each of the alternatives, much less than that for the development of the selected concept. Chapter 14 on Decision Analysis for Design Trades is equally applicable for conceptual trades, having an objectives hierarchy for each alternate concept that embodies stakeholder (risk/reward) preferences for values, measures, and weights. Similarly, each alternate concept has a functional, physical, and allocated architecture that can be verified (usually by analysis and simulation) and validated. There are several possible outcomes regarding acceptance: one or more alternatives are selected for development, an alternative is selected subject to modification, the decision to go through a further conceptual design with additional alternatives or hybrid alternatives combining aspects of existing alternatives, or a decision that no alternative is selected, and development does not proceed.

Figure 6.6 shows the three primary choices that were considered by the National Aeronautics and Space Administration (NASA) engineers in determining an operational concept for landing on the moon during the 1960s [Brooks et al., 1979; Murray and Cox, 1989]. The NASA engineers called these concepts modes and started out favoring the direct ascent from Earth to the moon and back to Earth.

However, calculations concerning the thrust required for this concept quickly proved that the concept was infeasible. As a result, the second and third concepts (Earth rendezvous and lunar

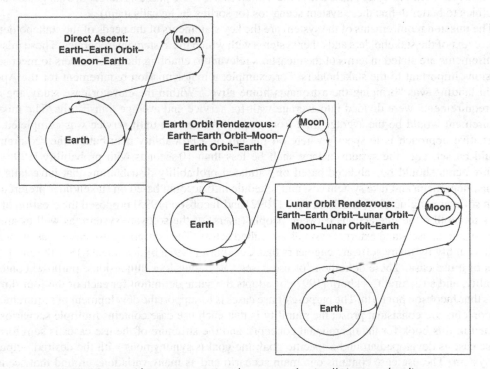

FIGURE 6.6   Alternate operational concepts for Apollo's moon landing.

rendezvous) were defined and explored in detail. Werner von Braun had previously developed the concept of staged rockets for lifting payloads into Earth orbit; with staged rockets, the weight that is no longer relevant can he shed. The same concept applied to Earth and lunar rendezvous. Many teams conducted calculations and simulations of these two concepts over several years, focusing primarily on cost (using energy as a surrogate) and safety. The results estimated that the lunar orbit rendezvous concept was almost $1.5 billion cheaper and had a 6- to 8-month shorter timeline for landing on the moon. There was some controversy at the end about which was safest; many engineers felt they were about equal with respect to safety, each having different strengths and weaknesses.

The operational concept includes a collection of scenarios as described in a use case diagram (see Fig. 3.1). One or more scenarios are needed for each group of stakeholders in each relevant phase of the system's life cycle. The use case diagram is used to provide a "big picture" of how the individual scenarios relate to each other in defining how the system is to be employed. Each scenario addresses one way that a particular stakeholder(s) will want to use, deploy, and fix the system; the scenario defines how the system will respond to inputs from other systems in order to produce a desired output. Included in each scenario are the relevant inputs to and outputs from the system and the other systems that are responsible for those inputs and outputs. The scenario should not describe how the system is processing inputs to produce outputs. Rather, each scenario should focus on the exchange of inputs and outputs by the system with other systems. It is critical that this shared vision be consistent with the collection of scenarios comprising the operational concept.

Hunger [1995] uses the phrase "mission analysis" for the development of the operational concept. The collection of scenarios in the operational concept includes sortie missions (or scenarios) and life missions, both from the perspective of the stakeholders. Sortie missions are scenarios that describe how the system will be used during the operational phase, capturing the reasons the system has for existing. The life missions address the nonoperational, life cycle aspects of the system, resulting in scenarios for each life cycle phase and some that cross life cycle phases. Hunger has suggested using timelines to better define these system scenarios (or sorties, as he calls them).

The mission requirements of the system are the key statements of the needs of the stakeholders in the context of the stakeholders and other systems with which the system interoperates. These mission requirements are stated in terms of the measures relevant to enabling the stakeholders to meet some missions important to the stakeholders. For example, a major mission requirement for the Apollo moon landing was "bringing the astronauts home alive." Within the elevator case study, the output requirements were divided into average wait for service and average transit time. The mission requirement would be the average time from request for service until service was completed. An alternative approach is to specify where on the tail of a probability distribution the requirement should be set, e.g., the system latency shall be less than 10 seconds with probability 0.99; such requirements should be validated based on validated probability distributions that have tails that fit the phenomena and take system cost and schedule into account based on stakeholder preferences.

In software engineering, Jacobson et al. [1992] and Jacobson [1995] proposed the creation of use cases to capture the interactions between people (users) of the software system, as well as among other systems; users and external systems are called actors. The concept of use cases was embraced so thoroughly by many software engineers that Cockburn [1997a,b] documents 18 different definitions of a use case. These definitions of use cases vary along four dimensions: purpose, contents, plurality, and structure. Cockburn [1997a,b] adopts the same definition for each of the four dimensions that Jacobson put forth. The purpose of use cases is to support the development of requirements; the contents are consistent prose; the plurality is that each use case contains multiple scenarios (as defined in this book for the operational concept); and the structure of the use cases is semiformal. A use case is developed around a specific goal; the goal is synonymous with the desired output of the system. The use case contains one main scenario and as many variations around that scenario as are meaningful. For our elevator system, variations may relate to the types of people using the elevator system, for example, blind people, deaf people, small children, and people in wheelchairs. So far, a collection of use cases is very consistent with a collection of scenarios as defined for the operational concept. However, several authors [Jacobson, 1995; Eriksson and Penker, 1998] illustrate the use case with statements of functions that the system and actors are performing rather than the flow of information and physical entities between the system and the actors. As stressed so far in this chapter, the focus during the development should be on defining requirements related to the inputs and outputs of the system and not on the functions of the system and functional requirements. There is quite a bit of confusion and sloppiness in discussions of use cases on this issue; several of the authors [Cockburn, 1997a,1997b; Eriksson and Penker, 1998] are really clear that the system should be treated as a black box with no visibility into functions, yet the functions show up in the discussion and diagrams documenting the use cases [see Jacobson, 1995; Eriksson and Penker, 1998].

The emphasis in this book has been on defining all aspects of the life cycle system. Consistent with Hunger's [1995] concept for sortie and life missions, the engineers for a system should develop scenarios for the system of interest in every phase of the life cycle. There should be scenarios and mission requirements for the development, manufacturing, training, deployment, refinement, and retirement phases unless one or more of these phases is not relevant.

To generate these scenarios, start with the key stakeholder, the operator/user, and generate a few simple scenarios. Then, scenario generation is expanded to other stakeholders while staying simple. Finally, complexity is added to all scenarios for each stakeholder, explicitly addressing atypical weather situations, failure modes of external systems that are relevant, and identifying key failure modes, constraints, standards, and external system interfaces that the system should address *in every phase of the life cycle*. In all scenarios, the focus should be on *what* the stakeholders and external systems do and *not on how* the systems accomplish their tasks. The system of interest should be viewed as a black box; that is, the system's internals are blacked out, leaving only the inputs and outputs to the system. Table 6.4 shows sample operational concept scenarios for an elevator.

There are some common operating scenarios for nearly every system:

- Initialization of the system
- Normal steady-state operation in standard operating modes of the system for all possible contexts (environments) in which the system may be placed (e.g., extreme cold, outer space)
- Extremes of operations due to high and low peaks of the external systems in each standard operating mode in each context
- Standard maintenance modes of the system
- Standard resupply modes of the system
- Reaction to failure modes of other systems
- Failure modes due to internal problems provide as much graceful degradation of the metasystem as possible
- Shutdown of the system
- Termination (phase out) of the system

The total number of scenarios for a common (relatively simple) system would be 25–50.

The SysML modeling technique called a sequence diagram (formerly called an input/output trace in the first edition of this book) can be used to make the description of each scenario as explicit as possible. A sequence diagram (see Fig. 6.7) has a timeline associated with each major actor (our system and other systems) in the scenario. The systems involved are listed across the top of the diagram, with the timelines running vertically down the page under each of the systems. Time moves from top to bottom in an input/output trace; the system of concern is highlighted with a bold label and heavier line. Interactions involving the movement of data, energy, or matter from the originating

TABLE 6.4 Sample Operational Concept Scenarios for an Elevator

(1) Passengers (including mobility, visually and hearing challenged) request up service, receive feedback that their request was accepted, receive input that the elevator car is approaching and then that an entry opportunity is available, enter the elevator car, request floor, receive feedback that their request was accepted, receive feedback that door is closing, receive feedback about what floor at which elevator is stopping, receive feedback that an exit opportunity is available, and exit elevator with no physical impediments.

(2) Passengers receive transportation in the elevator system when a fire breaks out in the building; the building alarm system sends the signal to the elevator system to stop elevator cars at the nearest floor, provide exit opportunities, and sound a fire alarm. Passengers leave elevator cars. Elevator cars are reactivated by special access available to maintenance personnel after the building is re-opened.

(3) Passengers are entering (exiting) an elevator car when doors start to shut; passengers can stop doors from shutting and continue to enter (exit).

(4) Elevator car stops functioning. Passengers in the elevator car push an emergency alarm that notifies building personnel to come and help them. Passengers use a phone system in the elevator car to call a centralized service center and report the problem to the people who answer. Elevator maintenance personnel arrive and create an exit opportunity.

(5) Too many passengers enter an elevator car, and the weight of passengers in the elevator car exceeds a preset safety limit; the elevator car signals a capacity problem and provides a prolonged exit opportunity until some passengers exit the car.

(6) Maintain a comfortable environment in the elevator by sensing the temperature in the elevator car that is based upon heat loss/gain of the passengers and the building and then supplying the necessary heat loss/gain to keep the passengers comfortable.

(7) A maintenance person needs to repair an individual car; the maintenance person places the elevator system in "partial maintenance" mode so that the other cars can continue to pick up passengers while the car(s) in question is (are) being diagnosed, repaired, and tested. After completion, the maintenance person places the elevator system in "full operation" mode.

(8) Electric power is transferred to the elevator from the building.

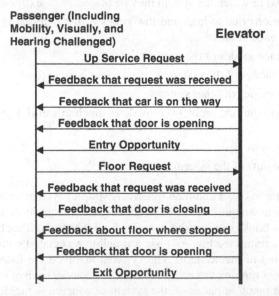

FIGURE 6.7 Sequence diagram of first elevator scenario.

system to the receiving system are represented as horizontal arcs. A label is shown just above each arc to describe the item being conveyed. Double-headed arcs are permissible to represent dialog in a compact manner. Having two or more arcs in quick succession is also common to illustrate that the same item is being transmitted from one system to multiple systems or that multiple systems are potentially transmitting the same item to one system. Figure 6.7 shows the first of these scenarios documented as an input/output trace diagram. See the elevator case study on the companion web site for more examples.

The purpose of these sequence diagrams is to be more explicit than written text can be about the systems involved, with a specific focus on the time-based interaction of systems and the transmission of data and items. Compare the sequence diagram in Figure 6.7 to the first scenario in Table 6.4. These sequence diagrams are not meant to be exact representations of dynamic interaction. An interval timescale is not being represented; rather, time is ordinal – any arc that is above another happens earlier, but there is no indication as to how large the time interval is.

The *shared vision, mission requirements,* and *the use case diagram with sequence diagrams for the scenarios* define the system's mission and provide the first hints as to the boundary of the system. The external and enabling systems are defined in the scenarios, also defining the inputs and outputs of the system. The system's inputs and outputs cross this boundary, defining the input/output requirements of the system and the external interfaces. The mission requirements suggest the fundamental objectives (objectives hierarchy of the stakeholders). This objectives hierarchy becomes the basis of the system's performance requirements. Finally, the first-level decomposition of the system's function can be suggested by examining the operational concept. Thus, the operational concept also leads to the functional requirements.

Recall that multiple systems are being developed concurrently, one for each phase of the life cycle and a qualification system for each of those systems. Each of these systems should have an operational concept.

The American Institute of Aeronautics and Astronautics (AIAA) and the Institute of Electrical and Electronics Engineers (IEEE) have standard documents for the Concept of Operations and Operational Concepts for the interested reader.

6.11 ECOSYSTEM MODEL

The single, largest issue in defining a new system is where to draw the system's boundaries, see Figure 6.2. Everything within the boundaries of the system is open to change. Subject to the requirements, nothing outside of the boundaries can be changed, leading to many of the system's constraint requirements. The *external systems model* is the model of the interaction of the system with other (external and enabling) systems and environments in the relevant contexts, thus providing a definition of the system's boundary in terms of the system's inputs and outputs.

Who is responsible for drawing these boundaries? All the stakeholders have a say in drawing these boundaries. However, there are substantial cost and schedule implications, so the procurer of the system typically has a major input. Nonetheless, all the stakeholders should be prepared to discuss the impact upon them of various boundary-drawing options. The systems engineer is responsible for guiding this boundary-drawing process to a conclusion that the stakeholders understand and accept. The systems engineer uses these boundaries to establish and maintain control of the system's interfaces.

The system's boundaries need to be drawn early in the systems engineering process because so much else in the design phase is dependent upon them. As is discussed next, the fundamental objectives or measures of effectiveness of the system need to be focused just beyond the external interfaces of the system. The operational concept relies upon knowing where the boundaries are for each stakeholder. The interface requirements capture the implications of the boundaries on the system design.

Many graphical modeling techniques (e.g., IDEF0, N^2 charts, data flow diagrams, EFFBDs) can be used to define the system boundary see Davis [1990]. See the companion website for a discussion of these techniques. IDEF0 is used in this chapter to illustrate external systems diagrams *in* terms of the elevator. The boundary for the elevator is defined to exclude the passenger, the maintenance personnel, and the building.

First, the purpose and viewpoint are defined:

Purpose: Explicitly define the system's boundary and needed interfaces

Viewpoint: Systems Engineering Team

Next, the mechanisms or external systems are established, followed by the functions of these systems. The system and external system come directly from the input/output traces of the scenarios in the operational concept:

Mechanism (System/External System)	System Function
1. Elevator – the system	Provide elevator services
2. Passengers	Request and use elevator services
3. Maintenance personnel	Maintain elevator operations
4. Building	Provide structural support

Now, the inputs, controls, and outputs of these functions are developed to finish the external system diagram. Recall that as part of this analysis of the elevator boundaries, the focus is on the context or environment of the elevator, and these key variables are shown in the diagram as controls; see Figure 6.8.

The above discussion has focused on an external systems diagram for the operational phase of the system, in which the system interacts with the system's users and other systems. External systems diagrams can and should be developed for every phase of the system's life cycle.

In addition to the usual syntax and semantics requirements of IDEF0 diagrams, an external systems diagram introduces several new constraints for the diagram to be valid. First, all of the *outputs of the system's function* (the elevator in this case) have to go to at least one of the external systems' functions on the page and *cannot exit the diagram*. If the output did exit the page, there would be an external system that was not included in the diagram, invalidating the purpose of the effort. Similarly, *each of the external systems must receive at least one output of our system*; otherwise, the system should be part of the context. In some cases, part of the context could be shown on the external systems diagram to emphasize the importance of a particular input to the system.

6.12   OBJECTIVES HIERARCHY FOR PERFORMANCE REQUIREMENTS

Traditionally, systems engineers have used the terms measure of effectiveness (MOE) and measure of performance (MOP), sometimes called a figure of merit (FOM). A *measure of effectiveness* describes how well a system carries out a task or set of tasks within a specific context; an MOE is measured outside the system for a defined environment and state of the context variables and is used to define mission requirements. Note that the further outside the system that the MOE measurement

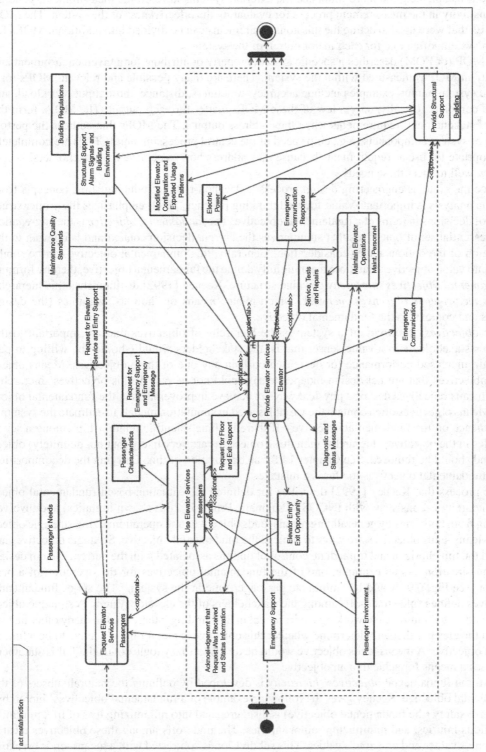

FIGURE 6.8 External systems' model diagram view for operational use of an elevator.

153

process is established, the more influence the external systems have on the measurement, yielding less sensitivity in the measurement process for evaluating the effectiveness of the system. The MOE or MOEs that were used to define the mission requirements can be divided into additional MOEs for a given system, often one for each major output of the system.

An MOP (or FOM) describes a specific system property or attribute for a given environment and context; an MOP is measured within the system. There are many possible and relevant MOPs for a specific system output; examples include accuracy, timeliness, distance, throughput, workload, and time to complete. Usually, only a few of these MOPs matter for each output. The MOPs form the basis of stakeholders' requirements when they address outputs. The MOPs that address the performance of system components (e.g., chip speed of the central processing unit [CPU]) are completely inappropriate for use as requirements because they address how to achieve the stakeholders' needs, not how well to meet these needs.

Since the systems engineering design process is decision-rich, introducing some concepts from decision analysis is important. Value-focused thinking [Keeney, 1992] emphasizes the proper structuring of decisions in terms of a fundamental objective. The *fundamental objective* is the aggregation of the essential set of objectives that summarizes the current decision context and is relevant to the evaluation of the options under consideration. Generally, this fundamental objective can be subdivided into value objectives that more meaningfully define the fundamental objective, thereby forming a *fundamental objectives' hierarchy* or value structure. Keeney [1992] distinguishes this hierarchy from a *means–ends objectives' network*, which relates means or "how to" variables (the design options and context) to the fundamental objective.

The *objectives' hierarchy* of a system is the hierarchy of objectives that are important to the system's stakeholders in a value sense; that is, the stakeholders would (should) be willing to pay to obtain increased performance (or decreased cost) in any one of these objectives. Means objectives (objectives that are not design independent) should not be part of this objectives' hierarchy. These means objectives describe physical ways to achieve improvements in the fundamental objectives. Means objectives often contain the variables used in simulation models to estimate the system's performance on the fundamental objectives. If there is some scientific relationship among a set of variables in the objectives' hierarchy, then these objectives are very likely (but not definitely) objectives and should be removed. See Chapter 14 for an exploration of this. Carrying the decomposition of the fundamental objectives too far is a mistake.

The process that Keeney [1992] describes for defining this situation-based fundamental objectives hierarchy is consistent with INCOSE Pragmatic Principle 2 (as shown in italics) and involves working from both ends by generalizing means–ends objectives and operationalizing strategic objectives. Means–ends objectives are ways to achieve the fundamental objective. Strategic objectives are beyond the time horizon and immediate control of options associated with the current system design decision situation. As an example, one of the fundamental objectives for the operation of a new elevator (see Fig. 6.9) would be "minimize passenger time in the system." The set of fundamental objectives defines value trade-offs among the stakeholders of the elevator system. A strategic objective would be to "improve the working environment in the building"; there are too many other factors beyond the elevator that will determine whether this objective is met for the objective to be a fundamental objective. A means–ends objective would be to "use a fuzzy logic controller"; this statement addresses a means for achieving an objective.

Next, the fundamental *objectives' hierarchy* is developed by defining the natural subsets of the fundamental objective. Keeney gives the following example of a fundamental objectives' hierarchy: maximize safety (the fundamental objective) is disaggregated into minimizing loss of life, minimizing serious injuries, and minimizing minor injuries. The trade-offs among these objectives clearly entail one's values and only one's values. This subdivision is contrasted with a means–ends breakout of maximizing safety that starts with minimizing accidents and maximizing the use of safety features

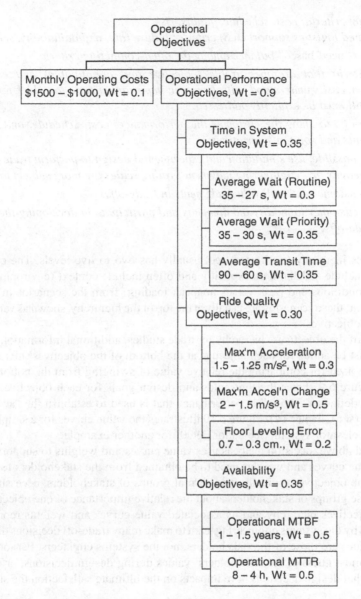

FIGURE 6.9 Fundamental objectives' hierarchy for operational phase of elevator.

on vehicles, both of which are means oriented and involve outcomes for which value trade-offs are difficult. Figure 6.9 provides the fundamental objectives hierarchy for the operation of the elevator.

Pragmatic Principle 2 [DeFoe, 1993] Use Effectiveness Criteria Based on Needs to Make System Decisions

1. *Select criteria that have demonstrable links to customer/consumer needs and system requirements*
   a. *Operational criteria: mission success, technical performance*

    b. *Program criteria: cost, schedule, quality, risk*

    c. *Integrated logistics support (ILS) criteria: failure rate, maintainability, serviceability*

2. *Maintain a "need-based" balance among the often-conflicting criteria.*

3. *Select criteria that are measurable (objective and quantifiable) and express them in well-known, easily understood units. However, important criteria for which no measure seems to exist, still must be explicitly addressed.*

4. *Use trade-offs to show the customer the performance, cost, schedule, and risk impacts of requirements and solutions variations.*

5. *Whenever possible, use simulation and experimental design to perform trade-offs as methods that rely heavily on "engineering judgment" rating scales are more subject to bias and error.*

6. *Have the customer make all value judgments in trade-offs.*

7. *Allow the customer to modify requirements and participate in developing the solution based on the trade-offs.*

The objectives hierarchy (a directed tree) usually has two to five levels. The objectives in the hierarchy may include stakeholders explicitly and often include context (environmental) variables (e.g., weather conditions and peak versus nonpeak loading) from the scenarios in the operational concept. If present, these scenarios are usually at the top of the hierarchy, shown as varying conditions for defining the objectives.

To make use of the objectives' hierarchy for trade studies, additional information must be added; value curves must be added for each objective at the bottom of the objectives' hierarchy, and value weights must be used for comparing the relative value of swinging from the bottom of each value scale to top. Figure 6.9 shows the thresholds and design goals for each objective; each threshold and design goal defines a "swing" in performance that is used to establish the "swing" weights in the value model (see Chapter 14). Figure 6.10 illustrates the value curves for a simplified objectives hierarchy for an elevator system. See Sailor [1990] for another example.

As mentioned above, decision analysis uses value curves and weights to support trade-off decisions. These value curves and weights need to be obtained from the stakeholders for two important reasons. First, the objectives typically span several groups of stakeholders, necessitating an agreement among these groups of stakeholders about the relative importance of one objective with others. Second, this objectives hierarchy and its associated value curves and weights represent the value structure needed by the systems engineering team to make many trade-off decisions during the design process. The values are those of the stakeholders, not the systems engineers. Far too often, the systems engineers must guess at the stakeholders' values during design decisions, or even worse, are not even aware that design decisions have impacts on the ultimate satisfaction the stakeholders will experience.

The *objectives' hierarchy* is typically used throughout the systems engineering design process as the *cornerstone of all the trade studies* that compare one design alternative with another. In doing trade studies, the evaluation should reveal which of several design alternatives is preferred; each design alternative will commonly have one advantage over the others, such as operational cost, reliability, accuracy of outputs, and the like. Since there is a system and associated qualification system for each phase of the life cycle, there should also be an objectives' hierarchy for each of these systems.

This decision analysis approach has been used for many military acquisitions, two of which are covered in Buede and Bresnick [2007] in which the objectives hierarchy, value curves, and weights were developed with government users and included in the request for proposal (RFP) to industry; Chapter 14 provides a discussion of one of these two acquisitions. This explicit, quantitative approach received very positive responses from the industry design teams. Watson and Buede [1987] describe

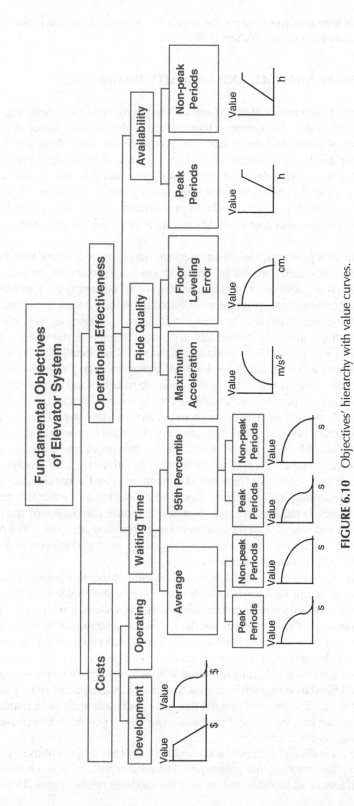

FIGURE 6.10 Objectives' hierarchy with value curves.

the analytic methodology that was used for these efforts. Other applications include Sailor [1990], Thurston and Carnahan [1993], and Walters [1994].

6.13   PROTOTYPING, ANALYSES, AND USABILITY TESTING

Prototyping can apply to any aspect of the system and is synonymous with modeling. A *prototype* is a physical model of the system that ignores certain aspects of the system, glosses over other aspects, and is representative of a subset of the system. The prototype can range from a subscale model of the system to a paper display (storyboard) of the user interface of the system. Prototyping became strongly associated with software development in the 1980s, and it is this context that will be the focus of this section. Most discussions of prototyping focus on the development of the prototype and assume that the answers for requirements and design alternatives magically appear. However, in the real world, the prototype has to undergo usability testing in order for this information to be gathered reliably.

The development of a prototype for a user interface ranges from a throwaway prototype to an evolutionary prototype [Connell and Shafer, 1989]. Throwaway prototypes are just what the name implies: prototypes that are developed for the main purpose of educating the users about the possibilities and extracting requirements from the users based upon their needs. Evolutionary prototypes are built for these educational and requirements development purposes as well, but with the idea that the prototype will eventually be turned into a working version of the system. The evolutionary prototype initially will only address a portion of the total functionality of the system, and that new functionality will be added on as the development and operational phases evolve together. Both concepts of prototyping have proven effective and continue today. In fact, software products for the rapid development of prototypes are now a business area in their own right.

In Chapters 9 and 15, we will introduce many types of analyses that should be conducted as part of the process for engineering systems. These analyses range from performance analyses to predict how far or well the system might be able to travel or see, timing analyses to determine how fast the system can respond or how many outputs the system can deliver per unit time, "ility" analyses to determine how available or safe the system is. There are also many cost and schedule analyses conducted. During the requirements phase, these analyses should be conducted on the metasystem to determine what difference it makes in the performance, cost and schedule parameters of the metasystem as the performance, cost and schedule of the system being engineered are varied. The results of these analyses provide very important information for the setting of minimum acceptable and desired marks in the system's requirements' statements.

Coupled with these analyses are many forms of elicitation of the viewpoints of the stakeholders. These elicitation sessions can be interviews with one or a few stakeholders, facilitated group sessions, observations of stakeholder performance on the current system or with prototypes of the new system, and questionnaires. Questionnaires are the last resort when no other approach is available since questionnaires produce lots of random responses from stakeholders who are too busy or too confused to do better. Valuable information is usually only achieved through human interactions. Individual interviews are best at soliciting information from quiet people who might be silent during group sessions. Facilitated meetings are best used to surface disagreements and try to find common ground or reasons for the differences of opinion that trace back to context and external system interactions. Observations are best for stressful periods during which people do things that they may not consciously recall during discussions.

Usability testing is obtaining samples of users and eliciting the reactions of these users about their needs and desires as they interact with prototypes. The prototypes can be as crude as written samples of screen interfaces or as sophisticated as working modules of the system. Usability [Bias and

TABLE 6.5   Metrics for Measuring Usability Elements.

Usability Element	Metrics
Learnability	Time to master a defined efficiency level, e.g., 50 words per minute
	Time to master a defined skill, e.g., cut and paste
Efficiency	Time for a frequent user to complete a defined task
	Rate of producing a defined set of products for a frequent user
Memorability	Time for a casual user to complete a defined task
	Time for a casual user to achieve previously achieved rate of production
Error Rate	Number of errors of a specific type in a given period for a given task
Satisfaction	Stress level associated with use
	Fun level associated with use

Mayhew, 1994; Nielsen, 1993; Wiklund, 1994] is a discipline associated with human-computer interaction that became very sophisticated in the 1980s and 1990s.

The performance elements of usability are ease of learning (learnability), ease of use (efficiency), ease of remembering (memorability), error rate, and subjectively pleasing (satisfaction). Table 6.5 provides a sample of common metrics for each of these elements. Each of these metrics has to be measured in the context of specific types of users and specific tasks. The tasks come from the scenarios in the operational concept. For the error rate element, categorizing errors into categories such as minor, major, and catastrophic is important. Care must be taken to separate random errors from those caused by the system. If necessary, the baseline capabilities of the users must be measured in order to define a baseline error rate for categories of users. Satisfaction typically has to be measured by subjective, categorical questions; see Nielsen [1993].

Users can be categorized along three dimensions: domain knowledge, computer experience, and system use experience. Segments of users along these three dimensions should be developed for testing purposes. When a sample of users is developed for usability testing, the population of actual system users must be considered, not the population of people in society.

Many guidelines have been developed for user interfaces. There is insufficient room to even summarize these guidelines here, but they should be consulted while developing requirements for user interfaces [see Brown 1988; Chapanis 1996; Marshall et al., 1987; Mayhew, 1992; Reason, 1990; Shneiderman, 1992].

6.14   DEFINING THE STAKEHOLDERS' REQUIREMENTS

The framework for defining requirements on the basis of the operational concept, the external system model, and the objectives hierarchy is presented here in detail. Recall that there are four requirement categories: input/output, system-wide and technology, trade-off, and qualification. The addendum, which can be downloaded from the companion website, provides a detailed example of these requirements for the life cycle phases of an elevator.

6.14.1   Input/Output Requirements

Input/output requirements are defined based on the inputs, controls, and outputs of the system identified while bounding the system with the external systems diagram. This external systems diagram is the primary tool used to support the development of input/output requirements.

The systems engineering team must examine each input, control, and output in detail to discover every requirement associated with each of these items. One or more input requirements are written for each input and control; similarly, one or more output requirements are written for each output. For example, the potential passengers of the elevator have certain characteristics that impact the provision of information about the floor location of the elevator. The requirements could state that audible feedback is needed, but this would be wrong. Rather, the requirements should dictate that feedback be provided to all relevant passengers, letting the engineers design a system to do this.

See Table 6.2 for examples of requirements that may be associated with inputs or outputs. Note, there will be some controls such as policies and procedures that were included because each function requires at least one control. These controls are not really data elements that the system receives, and therefore there need not be any input requirements established for them.

The environment (e.g., weather and elements that are outside the control of the system) or "context" is typically defined as part of the scenarios of the operational concept. This context should be addressed in the requirements. The questions typically addressed are:

1. What elements of the environment matter?
2. How much variation in the environmental elements must be planned for? At what priority?
3. How well can these variations be forecasted (predicted)? Can these forecasts be part of the system?
4. Can the environment be controlled by the system or external system? Must the system protect itself from the environment?

In addition to input and output requirements, there are external interface constraints and functions that should be used to decompose the system's function. Interface constraints address the physical aspects of the interface to which the system has to connect to obtain the inputs and disseminate the outputs. Examples include the standard connector type for electrical and mechanical connections. The characteristics of the power or data that come across the interface should be part of the input or output requirement.

Finally, the functional requirements are not meant to be a long list of specific, detailed functions the system has to perform to produce outputs needed by the system. Rather, the functional requirements should be the two to six functions that partition the system function in such a way that all the inputs to the system can be transformed into all of the outputs that have been identified as part of the external systems diagram.

Several examples of input/output requirements are:

The elevator shall receive "calls" from all floors of the building. (Input requirement)

The elevator shall indicate to a prospective passenger that he/she has successfully called the elevator. (Output requirement)

The elevator shall use a standard phone line from the building for emergency calls. (External interface requirement)

6.14.2 System-Wide and Technology Requirements

The system-wide and technology requirements relate to the system as a whole and not to specific inputs or outputs. These system-wide and technology requirements are not represented in the external systems diagram and are not addressed in a substantial way in the operational concept. Yet every system should have several system-wide and technology requirements that are key to the system's success. Recall that the four major categories are technology, suitability, cost, and schedule.

A typical category of requirements relates to regulations or laws that pertain to the system. Consider the following requirements:

The elevator system shall comply with the Americans with Disabilities Act.

This requirement is considered a system-wide requirement because the requirement, like all system-wide requirements, requires knowledge of the whole system to determine whether the requirement has been met. This is a deceptive requirement, though, because the requirement relates directly to an external system of the elevator, the passengers, and the ability of a special class of passengers to use the system. This requirement defines input and output restrictions with which the elevator must comply. For this reason, this requirement could be placed in both the input and output sections of the input/output requirements category. However, there are major disadvantages, as discussed before, in having one requirement in multiple places of the requirements document. For this reason, placing such a regulation in the system-wide requirements category of suitability is wise.

Technology requirements are the ones that engineers would prefer not to have because they really do constrain engineering creativity and should result from the other requirements if they are justifiable. These requirements are usually justified based on interoperability or compatibility with an existing product line, which ultimately should be reflected in cost savings. Examples are:

The elevator system's software shall be written in C++.
The elevator system's CPU shall be Pentium 4.

Table 6.2 provides a list of common *suitability* issues and topics that address quality concerns of a system and are system-wide in scope. There are technical engineering definitions that are expressed mathematically behind each of these suitability issues. In fact, many systems engineers make a career by specializing in one or several of these suitability areas. The detailed discussion of these suitability issues is critical for understanding the engineering of systems but is beyond the scope of this book, which is to provide a set of methods and models for getting to the definition of requirements for these issues and developing a design that meets such requirements. Conducting analyses of system concepts or designs related to suitability issues is discussed in more detail in Blanchard and Fabrycky [1998] and Pohl [2007].

Besides the technology and suitability requirements, cost and schedule requirements are also part of this segment of the requirements' partition. A *cost requirement* deals with the payment of money during the appropriate life cycle phase for the system in question to be useful. A *schedule requirement* deals with a timing issue for the relevant system for the phase of the life cycle in question. There is nearly always a cost and a schedule requirement for every phase of the system's life cycle. Table 6.2 provides examples of some of these.

The objectives hierarchy should address every system-wide requirement that is critical enough to be considered a performance requirement. These typically include the cost and schedule requirements as well as several suitability requirements.

6.14.3 Trade-Off Requirements

Trade-off requirements in the form of value curves and value weights were described above during the discussion of the objectives' hierarchy. Chapter 14 provides much more detail into the theory and elicitation techniques that can be used to obtain this requirements information. This set of requirements relies solely on the value judgments of each segment of the stakeholders. These value judgments must be obtained in a reliable manner from a reasonable sample of representatives of each segment of the stakeholders. For some segments, such as the bill payer, determining who should provide the value judgments is easy. For other systems that will be used by thousands or millions of people, talking to everyone is not feasible. Care must be taken to define a sufficiently large and representative sample of these users.

6.14.4   Qualification Requirements

The four elements of the qualification requirements for a system in any life cycle phase are: (1) *observance*: how the estimates (qualification data) for each input/output and system-wide requirement will be obtained, that is, test, analysis and simulation, inspection, or demonstration as well as the verification and validation requirements; (2) *verification plan*: how the qualification data will be used to determine that the real system conforms to the design that was developed; (3) *validation plan*: how the qualification data will be used to determine that the real system complies with the stakeholders' requirements; and (4) *acceptance plan*: how the qualification data will be used to determine that the real system is acceptable to the stakeholders.

The observance qualification requirements deal with data collection activities, devices, and facilities. For example, on a consulting project, one of the authors learned that an aircraft manufacturer was developing a detailed qualification plan for a fire suppression system installed in the cockpit of the aircraft. Specific derived requirements for the pressure and concentration of a chosen fire suppression agent existed for the three-dimensional space of the cockpit based upon the distribution of people and critical equipment. These requirements were developed based upon calculations and simulations that had been developed to ensure that the release pressure of the fire suppression system would be great enough to distribute the agent in the correct spatial concentration to suppress the fire but not too great to damage the structural elements of the cockpit. Not all of this analytical work had been done to address a fire suppression agent that had never been used in a cockpit before, so there was a great deal of uncertainty about the validity of the calculations. Observance requirements were developed to identify places in the cockpit to measure the concentration of the fire suppression agent at specific times during tests of the fire suppression system.

The verification plan was to activate the fire suppression system several times and take measurements of pressure and concentration at the spatial locations for which requirements had been defined. Note, for verification, there was no test of the fire suppression system's ability to extinguish a real fire. This verification plan also addressed the examination of the structural elements of the cockpit to verify the requirement that there be no structural damage. The final part of the verification plan defined the criteria for determining whether this verification test was passed or failed. (Note this level of detail would not be in the stakeholders' requirements for the aircraft system but would be in the specification for the fire suppression system, a component of the aircraft. Nonetheless, analogous system-level qualification information would be in the stakeholders' and system requirements for the aircraft system.) The data collection activity here was part of the observance qualification requirement.

Next, validation tests for the fire suppression system were defined based upon three safety scenarios that could be traced to the operational concept for the specification of the fire suppression system, if not the aircraft system. The safety scenarios were defined for three different potential causes of a fire. The observance qualification requirement stated that a fire be started *in* the cockpit based upon each of three causes, and the test would determine whether the fire suppression was activated and effectively suppressed the fire. The validation test requirement defined what was meant by effectively suppressing the fire. A fourth cause of fire is a ballistic hit from a weapon fired at the aircraft (this was a military aircraft). As a result, the test requirement called for several test cockpits to be hit by a weapon, a fire started either spontaneously or through whatever means were necessary (a fire is not guaranteed with a ballistic hit), and the fire suppression system's ability to suppress this fourth type of fire tested. Again, the observance qualification requirement defines that these ballistic tests will be conducted, and the validation requirement defines what successful performance is.

The acceptance test requirement provides the stakeholders' definitions of what acceptable performance is for the system as a whole. Sometimes, this is based upon the validation tests and is synonymous with the validation test plan. At other times, the acceptance test requirements call for additional tests, simulations, or inspections with acceptance criteria that are different than those of the validation criteria. *These qualification requirements, for each phase of the life cycle, are used to design the qualification system to be used during integration for each phase of the life cycle.*

As a final note, the aircraft manufacturer had designed the fire suppression system so that detailed design changes could be made as part of this integration phase activity of testing. Since the fire suppression system agent was new, the manufacturer needed the flexibility to adjust the design of the fire suppression system if the fire suppression was either less or more effective than expected. In fact, two locations had been designed for additional agent distribution tanks in case the design did not meet the requirements. In addition, the tank pressure in the planned tanks could be increased or decreased as needed. In an aircraft, the total system weight is so important that the manufacturer was planning additional verification tests to remove as much concentrated agent from the tanks as possible while meeting the pressure and concentration output requirements.

6.15 REQUIREMENTS MANAGEMENT

"Requirements Management is the identification, derivation, allocation, and control in a consistent, traceable, correlatable, verifiable manner of all the system functions, attributes, interfaces, and verification methods that a system must meet including customer, derived (internal), and specialty engineering needs" [Stevens and Martin, 1995, p. 11]. This definition of requirements management is inclusive of everything discussed in this chapter. For example, requirements management addresses which requirements have been changed, when and by whom, to what documents each requirement traces, and to which components each requirement has been allocated. Requirements management is considered a key element of systems engineering, as shown by INCOSE Pragmatic Principle 3.

A more limited, and perhaps more common, definition is the "care and feeding" of the requirements, sometimes called requirements traceability. More formally, requirements traceability "refers to the ability to describe and follow the life of a requirement, in both a forwards and backwards direction" [Gotel and Finkelstein, 1994, p. 95]. Numerous techniques for tracing requirements and their sources and destinations are semantic networks, assumption-based truth maintenance networks, constraint networks, cross-referencing schemes, hypertext, integration documents, key phrase dependencies, matrices, and templates. Relational and object-oriented databases are used to implement requirements traceability tools.

Pragmatic Principle 3 [DeFoe, 1993] Establish and Manage Requirements

1. *Identify and distinguish between specified (fundamental or essential), allocated, implied, and derived requirements.*
2. *Carry analysis and synthesis to at least one level broader and deeper than seems necessary before settling on requirements and solutions at any given level. (Top-down is a better recording technique than it is an analysis or synthesis technique.)*
3. *Write a rationale for each requirement. The attempt to write a rationale for a "requirement" often uncovers the real requirement.*
4. *Ensure the customer and consumer understand and accept all the requirements.*

5. *Explicitly identify and control all the external interfaces the system will have – signal, data, power, mechanical, parasitic, and the like. Do the same for all the internal interfaces created by the solution.*

6. *Negotiate interfaces with affected engineering staff on both sides of each interface and get written agreement by the two parties before the customer approves the interface documentation.*

7. *Document all requirements interpretations in writing. Don't count on verbal agreements to stand the test of time.*

8. *Plan for the inevitable need to correct and change requirements as insight into the need and the "best" solution grows during development.*

9. *Be careful of new fundamental requirements coming in after the program is underway. They invariably have a larger impact than is obvious.*

10. *Maintain requirements traceability.*

6.16 SUMMARY

Requirements are generally considered the cornerstone of the systems engineering process because requirements define the design problem. Stakeholders' requirements are those requirements initially established by the system's stakeholders with the help of the systems engineering team. The systems engineering design process is a mixture of establishing requirements to define the design problem and partitioning the physical resources of the system into components that perform functions that meet the requirements (the solution to the design problem). This partitioning process is decision-rich in that many important decisions are made by the systems engineering team that will ultimately affect the performance of the system and the satisfaction of the stakeholders.

This chapter defines requirements and the characteristics that these requirements should satisfy. In addition, this chapter provides a method or process for developing these requirements. This process includes the concepts and associated models of an operational concept, external systems diagram, and objectives hierarchy, all of which are extremely valuable aids in the definition of requirements.

The key points made in this chapter concerning the systems engineering design process are that (1) all stakeholders have stakeholders' requirements that, taken together, address every phase of the system's life cycle. Capturing the complete set of stakeholders' requirements ensures a concurrent engineering process. (2) The set of stakeholders' requirements should ensure a decision-rich design process by not over-constraining the design. The following attributes of requirements are meant to ensure the process is not over-constrained: traced, correct, unambiguous, understandable, design independent, attainable, comparable, and consistent. (3) At the same time, the stakeholders' requirements should not under-constrain the design because the stakeholders should be happy with the system that is created. Complete, verifiable, and traceable requirements should guarantee this.

The systems engineering design process defined in this chapter includes the development of an operational concept for each stakeholder group, external systems diagram for each life cycle phase, and an objectives hierarchy for each stakeholder group. These three concepts are then used to develop the stakeholders' requirements, organized by the life cycle phase, see Figure 6.11. Wymore's [1993] partition of requirements was adopted and modified: input/output requirements, technology and system-wide requirements, trade-off requirements, and system qualification requirements. In particular, the trade-off information defining stakeholder values that are needed to support design decisions includes performance trade-offs, cost trade-offs, and cost–performance trade-off information. This initial systems engineering phase is complete when the existence of at least one feasible

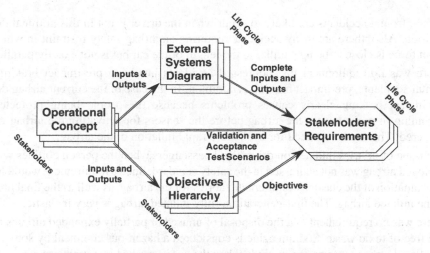

FIGURE 6.11 Summary of stakeholders' requirements development.

solution is verified, the acceptance requirements for the qualification system are defined, and the stakeholders have approved the StkhldrsRD.

CASE STUDY: AIRBAG RESTRAINT SYSTEM

Airbags, a safety device appearing in automobiles in the early 1990s, became the cause of death for a noticeable number of individuals. This severe, undesirable impact can be traced to the requirements for the airbag system. The following requirements issues are paraphrased from those published in 1984 by the National Highway Traffic Safety Administration (NHTSA) as part of Federal Motor Vehicle Safety Standard 208, Occupant Crash Protection (see Buede [1998]):

1. The requirements defined a single safety scenario on which to base the design. This single scenario could only be justified if there was a single worst-case situation. Note, this was not the approach with seat belts for which requirements were defined for the 50th percentile 6-year-old, 5th percentile adult female, and 95th percentile adult male.

2. The single, worst-case scenario for safety protection was the 50th percentile male not wearing a seat belt in a 30-mile-per-hour frontal collision. No specific attention was directed toward children and women and small or large adults. As the results show, this is the root of the problem.

3. While there was a requirement that the airbag not deploy on a very rough or bumpy road or when the car hits a small pole, there was no requirement that the airbag remains undeployed during accidents at sufficiently slow speeds that no lives are in danger. A number of people have lost their lives in accidents in which the car was only moving at 5 or 10 miles per hour, speeds at which there was almost no chance of a fatality.

4. The test condition was defined such that the test dummy is only in an upright position with its hands at the 3 and 9 o'clock positions on the steering wheel, and a frontal accident with the crash force parallel to the length of the car occurs into a fixed barrier at 30 miles/h.

In fact, frontal accidents are likely to occur when the driver is not in this nominal driving position. Also, there are many accidents requiring an airbag safety restraint in which the crash force is close to being parallel to the length of the car but is not exactly parallel.

5. There was no requirement that addressed accidents involving pre-impact braking. For frontal accidents, pre-impact braking is common. In the case of the current airbag design, pre-impact braking clearly causes problems because the people being protected are beginning to move toward the airbag before the sensors for activating the airbag can be triggered. This leads to a need for even more rapid inflation of the airbag.

6. The issue of injuries inflicted on drivers and passengers when the person collides with the deployed airbag was not addressed in the safety standard. Such a requirement would lead to an evaluation of the elasticity of alternate fabrics for the airbag, as well as the final pressure in the inflated airbag. The first-generation, fully inflated airbag is very inelastic.

7. There was no requirement that the disposal of unused or partially expanded airbags is safe and free of toxic waste. Sodium azide is considered a hazardous chemical by some. Also, uninflated airbag systems can explode when the car is crushed in a junkyard.

The requirements for airbags were placed in a federal regulation. It takes 16 months on average to change these regulations. "From 1970 until 1991, federal statutes requiring air bags were debated, imposed, revoked and reinstated as consumer and safety groups battled it out with reluctant automobile manufacturers and mostly Republican administrations. It took a Supreme Court decision in 1983, overturning a Reagan administration revocation of the standard, before the campaign took on real momentum" [Ottaway, 1996, p. 48]. Unfortunately, while so much attention was being paid to the concept of airbags, the requirements for the airbags were overlooked and remained unchanged.

CASE STUDY: *APOLLO 13* DISASTER

This case study is excerpted from Lovell and Kluger [1994], the book associated with the movie titled *Apollo 13*.

Every major component in an Apollo spacecraft, from gyros to radios to computers to cryogenic tanks, was routinely tracked by quality control inspectors from the moment its first blueprints were drawn to the moment it left the pad on launch day; any anomaly in manufacturing or testing was noted and filed away. Generally, the thicker the file any part amassed by the time it was ready to fly, the more headaches it had caused. Oxygen tank two, it turned out, had quite a dossier.

The problems with the tank began in 1965, around the time Jim Lovell and Frank Borman were deep in training for the flight of Gemini 7, and North American Aviation was building the Apollo command-service module that would ultimately replace the two-man ship … One of the most delicate of the delegated tasks was the construction of the spacecraft's cryogenic tanks, a job assigned to Beech Aircraft in Boulder, Colorado.

The Apollo spacecraft's electrical system was designed to operate on 28 volts of current *[derived requirement] – the* amount of juice provided by the service module's three fuel cells. Of all the systems inside the cryogenic tanks that would be driven by this relatively modest power system, none required more rigorous monitoring than the heaters. Ordinarily, cryogenic hydrogen and oxygen were maintained at a constant temperature of minus

340 degrees *[derived requirement]*. This was cold enough to keep the frigid gases in a slushy, non-gaseous state, but warm enough to allow some of the slush to vaporize and flow through the lines that fed both the fuel cells and the atmospheric system of the cockpit. Occasionally, however, the pressure in the tanks dropped too low, preventing the gas from moving into the feed lines and endangering both the fuel cells and the crew. To prevent this, the heaters would occasionally be switched on, boiling off some of the liquid and raising the internal pressure to a safer level.

Beech and North American knew that the tanks the new ship needed would have to be more than just insulated bottles. To handle contents as temperamental as liquid oxygen, the spherical vessels would require all manner of safeguards, including fans, thermometers, pressure sensors, and heaters, all of which would have to be immersed directly in the supercold slush that the tanks were designed to hold, and all of which would have to be powered by electricity.

Of course, immersing a heating element in a pressurized tank of oxygen was, on its face, a risky business, and in order to minimize the danger of fire or explosions, the heaters were supplied with thermostat switches that would cut the power to the coils if the temperature in the tank climbed too far. By most standards, that upper temperature limit was not very high; 80 degrees was about as hot as the engineers ever wanted their supercold tanks to get *[derived requirement]*. But in insulated vessels in which the prevailing temperature was usually 420 degrees lower, that was a considerable warm-up. When the heaters were switched on and functioning normally, the thermostat switches remained closed – or engaged – completing the heating system's electrical circuit and allowing it to continue operating. If the temperature in the tank rose above the 80-degree mark, two tiny contacts on the thermostat would separate, breaking the circuit and shutting the system down.

When North American first awarded the tank contract to Beech Aircraft, the contractor told the subcontractor that the thermostat switches – like most of the switches and systems aboard the ship – should be made compatible with the spacecraft's 28-volt power grid, and Beech complied. This voltage, however, was not the only current the spacecraft would ever be required to accept. During the weeks and months preceding a launch, the ship spent much of its time connected to launch-pad generators at Cape Canaveral, so that preflight equipment tests could be run *[missed operational concept scenario]*. The Cape's generators were dynamos compared to the service module's puny fuel cells, regularly churning out current at a full 65 volts.

North American eventually became concerned that such a relative lightning bolt would cook the delicate heating system in the cryogenic tanks before the ship ever left the pad, and decided to change its specs, alerting Beech that it should scrap the original heater plans and replace them with ones that could handle the higher launch-pad voltage. Beech noted the change and modified the entire heating system – or almost the entire heating system. Inexplicably, the engineers neglected to change the specifications on the thermostat switches, leaving the old 28-volt switches in the new 65-volt heaters. Beech technicians, North American technicians, and NASA technicians all reviewed Beech's work, but nobody discovered the discrepancy.

Although the 28-volt switches in a 65-volt tank would not necessarily be enough to cause damage to a tank – any more than, say, bad wiring in a house would necessarily cause a fire the very first time a light switch was thrown – the mistake was still considerable. What was necessary to turn it into a catastrophe were other, equally mundane oversights. The Cortright Committee soon found them.

The tanks that eventually flew aboard Apollo 13 were … installed in service module 106. Module 106 was scheduled to fly during 1969's Apollo 10 mission, … and the engineers decided to remove the existing tanks from the Apollo 10 service module and replace them with newer ones ...

Removing cryogenic tanks from an Apollo spacecraft was a delicate job ... Rockwell engineers unbolted the tank itself in spacecraft 106 and began to lift it carefully from the ship.

Unknown to the crane operators, one of the four bolts had been left in place. When the winch motor was activated, the shelf rose only two inches before the bolt caught, and the crane slipped, and the shelf dropped back into place ... The tanks on the dropped shelf were examined and found to be unharmed. Shortly afterward, they were removed, upgraded, and reinstalled in service module 109, which was to become part of the spacecraft more commonly known as Apollo 13 ...

One of the most important milestones in the weeks leading up to an Apollo launch was the exercise known as the countdown demonstration test ... To make the dress rehearsal as complete as possible, the cryogenic tanks would be fully pressurized, the astronauts would be fully suited, and the cabin would be filled with circulating air at the same pressure used at liftoff.

During Apollo 13's countdown demonstration test, with Jim Lovell, Ken Mattingly, and Fred Haise strapped into their seats, no significant problem occurred. At the end of the long dress rehearsal, however, the ground crew did report a small anomaly. The cryogenic system, which had to be emptied of its supercold liquids before the spacecraft was shut down, was behaving balkily ... oxygen tank two seemed jammed, venting only about 8 percent of its 320 pounds of supercold slush and then releasing no more.

... When the tank was dropped eighteen months earlier, they now suspected, the tank had suffered more damage than the factory technicians at first realized, knocking one of the drain tubes in the neck of the vessel out of alignment ...

At its present supercold temperature and relatively low pressure, the liquid in the tank wasn't going anywhere. But what would happen, one of technicians wondered, if the heaters were used? Why not just flip the warming coils on now, cook the slush up, and force the entire load of O_2 out of the vent line? ...

But the wrong thermostat switch – the 28-volt switch – was in the tank, and as it turned out, the heaters stayed on for a long, long time ... Given the huge load of O_2 trapped in the tank, the engineers figured it would take up to eight hours before the last few wisps of gas would vent away. Eight hours was more than enough time for the temperature in the tank to climb above the 80-degree mark, but the technicians knew they could rely on the thermostat to take care of any problem. When this thermostat reached the critical temperature, however, and tried to open, the 65 volts surging through it fused it instantly shut.

The technicians on the Cape launch pad had no way of knowing that the tiny component that was supposed to protect the oxygen tank had welded closed ...

Unfortunately, the readout on the instrument panel wasn't able to climb above 80 degrees ... the men who designed the instrument panel saw no reason to peg the gauge any higher, designating 80 as its upper limit. What the engineer on duty that night didn't know – couldn't know – was that with the thermostat fused shut, the temperature inside this particular tank was climbing indeed, up to a kiln like 1000 degrees.

... At the end of eight hours, the last of the troublesome liquid oxygen had cooked away as the engineers had hoped it would – but so too had most of the Teflon insulation that protected the tank's internal wiring. Coursing through the now empty tank was a web of raw, spark prone copper, soon to be immersed in the one liquid likelier than any other to propagate a fire: pure oxygen.

[Lovell and Kluger, 1994, pp. 372–378] The words in italics inside the braces were inserted by the author of this text.

PROBLEMS

6.1 Use IDEF0 to develop an ecosystem model for an information system to advise undergraduate systems engineering students on the development of their plans of study. The information system is the software and hardware system that the undergraduate systems engineering students will use. Assume the systems engineering faculty will maintain the accuracy of the courses and prerequisites. Assume the information system can obtain schedule information over a network from the registrar's office. Assume that the information system produces a written plan of study for each student.

6.2 Use the following operational concept for the operational phase of the Automatic Teller Machine (ATM):

    i. Create one additional scenario for the operational concept.

    ii. Develop an ecosystem model represented as an IDEF0 diagram.

    iii. Create an objectives hierarchy for the ATM system.

    iv. Develop a set of stakeholders' requirements. Use the format of the Stakeholders' Requirements Document and the taxonomy of four types of requirements from this chapter. Make every effort to develop as complete and unambiguous a set of stakeholders' requirements for the operational phase as possible using only the information provided in the following scenarios. Then, add three system-wide requirements and four qualification requirements.

       ATM for Money Mart Corporation. The ATM system is to provide a cost-effective service to bank customers that is convenient, safe, and secure 24-hour access to a common set of banking transactions and reduces the cost of providing these basic transactions. The ATM system shall provide a number of the most common banking transactions (deposit, withdrawal, transfer of funds, and balance query) without the involvement of bank personnel.

       The operational concept is comprised of a group of scenarios that are based upon the stakeholders' requirements and relate to both the bank's customers and employees.

Customer Scenarios

1. Customer makes deposits

    a. Customer provides valid general identification information.

    b. ATM requests unique identification information.

    c. Customer enters unique identification information.

    d. ATM requests activity selection.

    e. Customer selects deposit.

    f. ATM requests account type.

    g. Customer identifies account type (i.e., savings, checking, bank credit card).

    h. ATM requests type of deposit (cash versus check).

    i. Customer identifies the type of deposit – cash/check.

    j. ATM provides a means to physically insert cash/check into ATM.

    k. Customer enters the deposit

        ATM transmits the transaction to the main bank computer, gives customer receipt, returns to main menu.

2. Customer requests cash to be withdrawn from an account

    a. Customer provides valid general identification information.

    b. ATM requests unique identification information.

    c. Customer enters unique identification information.

    d. ATM requests activity selection.

e. Customer selects withdrawal.

f. ATM requests account type.

g. Customer identifies account type (i.e., savings, checking, bank credit card).

h. ATM requests amount of withdrawal.

i. Customer identifies amount of withdrawal (C_{req}).

j. ATM contacts the main bank computer and requests the amount of available funds from the selected account (F_{max}).

k. If $C_{req} > F_{max}$, ATM denies request.

l. If $C_{req} > C_{lim}$, ATM denies request. (C_{lim} is the maximum cash withdrawal allowed)

m. $C_{req} > C_{left}$ ATM apologizes for inability to satisfy request and sends message to bank for more funds. (C_{left} is amount cash ATM has left).

n. Else, ATM transmits the transaction to the main bank computer, gives customer receipt, gives the customer money, and returns to the main menu.

3. Customer requests transfer of funds from one account to another

a. Customer provides valid general identification information.

b. ATM requests unique identification information.

a. Customer enters unique identification information.

c. ATM requests activity selection.

d. Customer selects transfer of funds.

e. ATM requests account type for source of funds transfer.

f. Customer identifies source account type.

g. ATM requests account type for destination of funds transfer.

h. Customer identifies destination account type.

i. ATM queries the main bank computer to determine the availability of funds from the source account (F_{max}).

j. ATM requests the amount of the funds transfer.

k. Customer identifies the amount of funds to be transferred (F_{trns}).

l. If $F_{trns} > F_{max}$, the ATM denies the request.

m. Else the funds are transferred, ATM transmits the transaction to the main bank computer, gives the receipt, and returns to the main menu.

4. Customer requests the status of balance of an account

a. Customer provides valid general identification information.

b. ATM requests unique identification information.

c. Customer enters unique identification information.

d. ATM requests activity selection.

e. Customer selects balance status of an account.

f. ATM requests account type for balance query.

g. Customer identifies account type.

h. ATM queries the main bank computer to obtain the needed information, gives customer receipt, and returns to the main menu.

5. Customer cancels request

a. Customer provides valid general identification information.

b. ATM requests unique identification information.

c. Customer enters unique identification information.

d. ATM requests activity selection.

e. Customer selects withdrawal.

f. ATM requests account type.

g. Customer identifies account type (i.e., savings, checking and bank credit card).

h. During the course of a transaction, the customer indicates the desire to cancel the current transaction.

i. ATM returns to the main menu and gives the customer the choice to begin another transaction.

j. Customer chooses to end the session.

k. ATM resets for the next customer.

6. Customer input device is not working

a. Customer attempts to provide valid general identification information.

b. ATM informs the customer that the input device is not working.

c. If this is the third straight customer for whom the input device is not working, then the ATM sends a message to the bank about this problem.

7. ATM cannot verify the customer identification scheme

a. Customer provides valid general identification information.

b. ATM requests unique identification information.

c. Customer enters unique identification information.

d. ATM checks unique identification, finds the identification incorrect, and requests customer to re-input identification.

e. Customer enters unique identification information.

f. ATM checks unique identification, finds the identification incorrect, and requests customer to re-input identification.

g. Customer enters unique identification information.

h. ATM checks unique identification, finds the identification incorrect, and alerts the customer that any attempts to re-input identification will result in an alarm to the bank.

i. Customer leaves.

j. ATM resets for the next customer.

8. ATM does not have receipts.

a. When only 25 receipts remain, ATM sends message to bank to resupply receipts.

9. Hostile situations

a. Robber attempts to break into ATM.

b. ATM sends message to bank and sounds alarm.

c. ATM shuts down operation.

Bank Employee Scenarios

1. Routine resupply operation

a. Employee enters code into ATM.

b. ATM provides access to valid employee.

c. Employee opens ATM.

d. Employee loads ATM with cash.

e. Employee loads ATM with blank receipts.

f. Employee removes deposits from ATM

g. Employee shuts ATM and initializes for operation.

2. Malfunction operations

a. Employee enters code into ATM.

b. ATM provides access to employee.

c. Employee opens ATM.

d. Employee runs built-in diagnostic tests to determine problem.

e. ATM responds to diagnostic tests.

    f. Employee fixes ATM.

    g. Employee runs built-in diagnostic tests to determine if problem is solved.

    h. ATM responds to diagnostic tests.

    i. Employee shuts ATM and initializes for operation.

6.3 Use the following operational concept for the operational phase of an automobile system called OnStar:

    i. Create one additional scenario for the operational concept.

    ii. Develop an external system model using IDEF0.

    iii. Create an objectives hierarchy for the OnStar system.

    iv. Develop a set of stakeholders' requirements. Use the format of the Stakeholders' Requirements Document and the taxonomy of four types of requirements from this chapter. Make every effort to develop as complete and unambiguous a set of stakeholders' requirements as possible for the operational phase using only the information provided in the following scenarios. Then, add three system-wide requirements and four qualification requirements.

OnStar System for Cadillac. The OnStar system is an information system for Cadillac owners to provide emergency help and a wide range of support. Generally, the operational concept involves a satellite communications link between the car and a control center run by Cadillac.

The operational concept is comprised of a group of scenarios that are based upon the stakeholders' requirements and relates to both the OnStar's users and maintenance personnel.

User Scenarios

1. Driver uses cellular phone to contact control center to find directions

    a. Driver pushes a single button on the OnStar cellular phone.

    b. OnStar calls control center.

    c. Control center person responds and inquires where driver wants to go via the OnStar cellular phone.

    d. Driver responds with location (tourist landmark, restaurant, hotel, ATM, Cadillac dealer, and gas station) via the OnStar cellular phone.

    e. Control center person responds with address and block-by-block directions via the OnStar cellular phone.

    f. Driver uses OnStar to record these directions and plays them back as needed.

2. Driver loses car in parking lot

    a. Driver calls control center using a toll-free number from a pay phone.

    b. Control center person sends signal to OnStar.

    c. OnStar activates flashing lights and honking horn on driver's car.

    d. Driver goes to car and deactivates lights and horn.

3. Driver locks keys in car

    a. Driver calls control center using a toll-free number from a pay phone.

    b. Control center person requests identification information.

    c. Driver provides identification information.

    d. Control center person sends signal to OnStar.

    e. OnStar unlocks your car.

4. Emergency support when an accident occurs

    a. Car is involved in an accident in which the airbags are activated.

    b. OnStar sends a priority signal to the control center, with the exact location.

c. Control center person calls driver on the OnStar cellular phone.

d. If contact is not made, control center person contacts the appropriate 911 number.

e. Control center person provides information on driver's location, car, and license number.

f. Police respond to driver.

5. Vandals/thieves break into driver's car and steal the car

a. Vandals/thieves break into driver's car and drive the car away.

b. The security system of car is activated and sends a signal to OnStar.

c. OnStar sends a signal to the control center.

d. The control center person calls 911 and reports the break-in and provides information on driver's car to the police.

e. OnStar sends signals to the control center, allowing the car to be tracked.

f. The control center person provides this tracking information to the police.

6. Carjackers steal car and kidnap driver and passengers

a. Thieves carjack the car with the driver (and possibly passengers).

b. Driver pushes a red button on the cellular phone.

c. OnStar sends a carjacking signal to the control center with an open phone line so that any conversations can be monitored.

d. OnStar sends signals to the control center, allowing the car to be tracked.

e. The control center provides information about the situation and the location of the car to the police.

7. OnStar is deactivated.

a. OnStar receives its power from the car's battery.

b. The car's battery is dead or disconnected, causing the deactivation of OnStar.

Maintainer Scenarios

1. Maintainer checks emergency carjacking capability

a. Maintainer tests emergency button on the cellular phone to determine that contact with the control center is made. If tests show a problem, adjustments are made, or cellular phone is replaced to correct any deficiencies.

b. Maintainer tests link to control center to make sure that the conversation can be heard and that car's location is transmitted. Adjustments or replacements are made as necessary to correct any deficiencies.

2. Maintainer tests ability of OnStar to unlock car

a. Maintainer checks that unlock signal is received by OnStar.

b. Maintainer checks that OnStar unlocking signal is activated when control center unlock signal is received.

c. Maintainer checks that car locks are unlocked when OnStar unlocking signal is sent.

d. Maintainer makes repairs as needed.

6.4 Use the following operational concept for the *development* phase of an airbag system

i. Create one additional scenario for the operational concept.

ii. Develop an external system model using IDEF0.

iii. Create an objectives hierarchy for the airbag development system.

iv. Develop a set of stakeholders' requirements. Use the format of the Stakeholders' Requirements Document and the taxonomy of four types of requirements from this chapter. Make every effort to develop as complete and unambiguous a set of stakeholders' requirements as possible for the operational phase using only the information provided in the following scenarios. Then, add three system-wide requirements and four qualification requirements.

Vision and Mission Requirement: The systems engineering team for an upgraded airbag safety restraint system shall design an airbag system that saves as many lives as possible while not subjecting any drivers or passengers to unneeded injuries or deaths. Cost of the airbag system will be kept within bounds, and designs will be tailored to various automakers' needs.

Scenarios

1. The systems engineering team (SET) will review all safety regulations published by the NHTSA, send questions and comments to NHTSA on a timely basis, receive responses, and incorporate these regulations into the airbag design
2. The SET will seek out and review all research findings available on airbag systems, formulate questions and comments to the research teams on a timely basis, receive and review responses, and ensure that the airbag design is consistent with the best research available.
3. The SET will send their requirements documents on the airbag system and the manufacturing system for the airbag system to the appropriate corporations for comments and respond to any comments received from these corporations. Comments related to the cost of the systems and the fit of the designs will be of special interest.
4. The SET will send the entire set of required test results on its designs to the NHTSA for review and comment; any questions from the NHTSA will be answered, and further tests will be conducted as needed.
5. The SET will send all safety findings and liability issues and analyses of their designs to corporate headquarters and respond to corporate guidance concerning safety and liability issues.
6. The SET will receive "built to" CIs from the airbag manufacturer, will integrate these items into a test automobile, and will test the integrated airbag against the test requirements. Design changes will be identified and incorporated into the requirements documents as needed based upon the tests. The revised requirements documents will be sent to the automakers and manufacturers for comment.
7. The SET will use additional "built to" CIs to build and forward operational test items to the automakers for integration testing into the automobiles of the automakers. Based upon these operational tests, the automakers will forward additional comments on the airbag design. These comments will be incorporated into the requirements documents.
8. The airbag manufacturers will submit engineering change proposals (ECPs) to the SET as problems are encountered during production. The SET will adopt those ECPs that are warranted, reject those that are not warranted, and comment on the remaining so that an acceptable solution can be found to manufacturing problems.

6.5 Use the following operational concept for the *manufacturing* phase of an airbag system

   i. Create one additional scenario for the operational concept.

   ii. Develop an external system model using IDEF0.

   iii. Create an objectives hierarchy for the airbag development system.

   iv. Develop a set of stakeholders' requirements. Use the format of the Stakeholders' Requirements Document and the taxonomy of four types of requirements from this chapter. Make every effort to develop as complete and unambiguous a set of stakeholders' requirements as possible for the operational phase using only the information provided in the following scenarios. Then, add three system-wide requirements and four qualification requirements.

Vision and Mission Requirement: The Manufacturing Division for an upgraded airbag safety restraint system shall design the airbag manufacturing system to produce the airbag system with as low a long-term cost as possible. Long-term cost includes the discounted cost of producing acceptable airbags as well as providing free parts due to manufacturing flaws. The manufacturing system shall be capable of producing tailored designs for various automakers.

Scenarios

1. The Manufacturing Division will review all safety regulations published by the NHTSA, send questions and comments to the NHTSA on a timely basis, receive responses, and incorporate these regulations into the manufacturing design for airbags.
2. The Manufacturing Division will receive requirements documents on the airbag system from the development team on a periodic basis. The Manufacturing Division will provide comments on these documents as regards any difficulties being forced on the manufacturing of airbags. These comments will be provided on a timely basis.
3. The Manufacturing Division will produce the appropriate number of "built to" CIs based upon the design documentation and schedule requirements of the development team. In order to produce these "built to" CIs, the Manufacturing Division will procure the necessary tools, parts, and supplies.
4. The Manufacturing Division will submit ECPs to the development team as problems are encountered during production. The development team will adopt those ECPs that are warranted, reject those that are not warranted, and comment on the remaining so that an acceptable solution can be found to manufacturing problems. The Manufacturing Division will modify its production process and equipment in accordance with the accepted ECPs.
5. The automakers will send orders for airbags to Corporate Headquarters; Corporate Headquarters will send sales orders to the Manufacturing Division with delivery instructions; the Manufacturing Division will produce the needed airbags and send them to the appropriate automaker; and the Manufacturing Division will send documentation of delivered airbags to Corporate Headquarters.
6. Corporate Headquarters will send periodic projections of airbag production requirements to the Manufacturing Division along with additional corporate guidance regarding cost and quality issues. The Manufacturing Division will send periodic reports on cost and performance data regarding the production of airbags.
7. The Manufacturing Division will send requests for quotations (RFQs) to other corporations for the needed tools and parts (CIs) that comprise the airbag system; the Manufacturing Division will receive and review quotes from various corporations and select those quotes providing the best value to the Manufacturing Division; and the Manufacturing Division will then send orders for the delivery of the tools and parts on a timely basis and receive these tools and parts.
8. The Manufacturing Division will send RFQs to other corporations for the needed consumables and supplies; the Manufacturing Division will receive and review quotes from various corporations and select those quotes providing best value to the Manufacturing Division; and the Manufacturing Division will then send orders for the delivery of the consumables and supplies on a timely basis and receive these consumables and supplies.
9. The Manufacturing Division will send that material (unused consumables and supplies, used tools and parts) that needs to be disposed of to Corporate Headquarters.

Chapter 7

Functional Architecture Development

7.1 INTRODUCTION

Time-tested engineering of systems has shown that the design process for a system has to consider more than the physical side of the system; the tasks, functions, or activities that the system has to perform are critical elements for the design process to be successful on a consistent basis. This is not to say that the designs of functions and physical resources for the system proceed independently; they cannot. However, for success, these two design elements must be equal partners in the design process, providing checks on each other and complementing each other's progress. The functional architecture of a system contains a hierarchical model of the functions performed by the system, the system's components, and the system's configuration items (CIs); the flow of informational and physical items from outside the system through the transformational processes of the system's functions and on to the waiting external systems being serviced by the system; a data model of the system's items; and a tracing of input/output requirements to both the system's functions and items. Note that functional architecture may be called logical architecture by people invoking object-oriented systems engineering.

There are a number of key terms that need to be defined as part of this chapter. Early in the chapter, distinctions are drawn between the modes, states, and functions of a system. There is a considerable difference in meaning in the literature on systems and software related to the terms of mode, state, and function; to be clear in our discussions, these terms have to be defined specifically for use in this book. A system mode is a distinct operational capability of the system; this capability may use either the full or a partial set of the system's functions. An example is the initialization mode versus the full operational mode for your personal computer. A state is a modeling description of the status of the system at a moment in time, as defined by the values on a set of state variables. A function is an activity or task that the system performs to transform some inputs into outputs. Later in the chapter, distinctions are drawn between failure, error, and fault. Failure is a deviation between the system's

The Engineering Design of Systems: Models and Methods, Fourth Edition. Dennis M. Buede and William D. Miller.
© 2024 John Wiley & Sons, Inc. Published 2024 by John Wiley & Sons, Inc.
Companion website: www.wiley.com/go/engineeringdesignofsystems4e

behavior and the system's requirements. An error is a problem with the state of the system that may lead to a failure. A fault is a defect in the system that can cause an error.

After the initial definition of key terms for describing a functional architecture, Section 7.3 defines the method for developing a functional architecture using an Integrated Definition for Function Modeling (IDEF0) model. This model of the development process for a functional architecture is explained, followed by a discussion of using a decomposition process versus a composition process.

Section 7.4 discusses approaches, examples, and issues for defining a system's functions; this discussion is very important because the modeling of a system's functions is not a common skill that is found in engineers. Section 7.4.1 describes several approaches for developing functional decompositions. Section 7.4.2 addresses an important theme of this entire book; namely, there is always more than one system involved in the engineering of a system; examples of functional decompositions for several phases of the life cycle of a system are presented. Third and perhaps most importantly, the concepts of feedback and control in a system's functions are introduced in Section 7.4.3. A common hypothesis of many systems engineers is that most systems fail because of inadequate design of the feedback and control functionality in the system. Finally, Section 7.4.4 provides a discussion of evaluation topics that are useful for critiquing a functional architecture; critical examination of any model is important for engineers, and Section 7.4.4 provides some measures for doing so.

Section 7.5 defines the data collection activities associated with developing the functional model of a system. This section provides some guidance on the types of data to collect, on the need to try alternate modeling ideas, and then on the evaluation of these model alternatives in terms of the need to capture the system's capabilities and communicate these capabilities to both the stakeholders and the discipline engineers.

Then, in Section 7.6, the introduction of fault tolerance functionality in terms of the functional architecture is described. Adding fault tolerance functionality is very important to the success of most systems and is critical to the success of some systems, for example, air traffic control and life support. Error detection, damage confinement, error recovery, and fault isolation and reporting are the types of functions discussed here.

Finally, tracing input/output requirements to functions and items in the functional architecture is described in Section 7.7. This last activity is critical to the process of developing specifications for each component that comprises the system in such a way that the component specifications are directly related and traceable to the System's Requirements Document.

The methods described in this chapter relate to the development of the functional architecture. The method relating to defining the elements of the functional architecture is described in detail and presented as an IDEF0 model with modeling of the elevator system represented with Systems Modeling Language (SysML) activity diagrams. In addition, the chapter provides a data collection process for defining the functional architecture based upon the fundamental approaches behind the structured analysis and design technique that led to IDEF0.

The primary modeling technique relied upon in this chapter is IDEF0, as presented in Chapter 3. In addition, feedback and control models are introduced for evaluating the state of the system and improving the system's performance.

The exit criterion for the development of the functional architecture is the coherent matching of the input/output requirements with the functions and items in the functional architecture. Every input/output requirement should be traced to at least one function and one item in the functional architecture. In addition, every function associated with an external item in the functional architecture should have at least one input/output requirement traced to the function, as should every external item. Recall that all elements of the system's architectures are developed in increasing layers of detail, so the exit criterion for the functional architecture will be applied with each completion of a layer of detail.

7.2 DEFINING TERMINOLOGY FOR A FUNCTIONAL ARCHITECTURE

This section defines the concepts of system modes, states, and functions, followed by simple and complete functionalities. Modes and functionalities have long been thought to be critical to the establishment of an understanding of the logical aspects of a system.

A system *mode* is defined to be a distinct operating capability of the system during which some or all the system's functions may be performed to a full or limited degree. Other authors [Wymore, 1993] define the modes of a system to be functions of the system; that is not the definition presented here. All systems have at least one standard or fully operational mode. Most systems have operating modes during which they are partially operational. For example, an elevator system has a maintenance mode during which one or more of the elevator cars can be stopped for maintenance, while the others continue in operation. Often, systems have start-up and shutdown modes. The laptop computer, on which this author is writing this paragraph, has several modes of operation that correspond to the power that is being supplied; all the laptop's functions are available in each of these modes, but not with the same performance characteristics. Finally, systems often have a number of unwanted failure modes; car manufacturers have installed switches to enable the use of an extra gallon of gasoline to try to avoid the failure mode of no gas.

The *state of the system* is commonly defined to be a static snapshot of the set of measures or variables needed to describe fully the system's capabilities to perform the system's functions. The system is progressing through a constantly changing series of states as time progresses. In other words, the state of a system is the values of a long list of variables, called *state variables*, at a specific point in time. This list of state variables contains all the information needed to determine the system's ability to perform the system's functions at that point in time. The list of state variables does not change over time, but the values that these variables take do change over time. The variables can be continuous or discrete. As an example, the state variables for a laptop computer might include power input rate from the outside, power level of the battery, input rate for each input source (keyboard, modem, and network), output rate for each output device (parallel port, serial port, modem, network, and screen), central processing unit (CPU) usage, and free hard disk space.

A *function*, on the other hand, is a process that takes inputs in and transforms these inputs into outputs. A function is a transformation, including the possible changing of state one or more times. Every function has activation and exit criteria. The activation criterion is associated with the availability of the physical resources, not necessarily with the start of the transformation activity. The function is activated as soon as the resource for carrying out the function is available. When the appropriate triggering input arrives, the function is then ready to receive the input and begin the transformation process. The activation criterion for the function, then is the combination of the availability of the physical resource and the arrival of the triggering input. The exit criterion of a function determines when the function has completed its transformation tasks.

Chapter 3 and the graphical modeling techniques on the companion website cover a number of behavioral modeling techniques that address issues related to the activation and deactivation of functions, both as the result of the natural transformation processes associated with functions as well as the control structure that controls the functional processing and causes the system to change modes. Included in the graphical modeling techniques on the companion website are behavior diagrams, finite-state machines (state-transition diagrams), statecharts, control flow diagrams, and Petri nets. Note that state-transition diagrams and statecharts are related to the definition of mode being used here rather than the definition of state.

Must a function represent a dynamic process? Can a function be used to represent a constant process? All the functions that are shown in this book for the elevator case study represent a dynamic function, that is, inputs enter the function over a given time period, and sometime later, the outputs emerge. Does a pedestal that is holding a vase perform a function? The perspective taken here is that the pedestal does perform a function in this case; if the pedestal fails due to fatigue or an earthquake, then a dynamic process that the system is trying to prevent will occur (the vase will crash to the ground and be ruined).

A *functionality* is a set of functions that is required to produce a particular output. Now we define simple and complete functionalities:

> *Simple Functionality:* an *ordered* sequence of functional processes that operates on a single input to produce a specific output. Note there may be many inputs required to produce the output in question, but this simple functionality is only related to one of the inputs. As a result, the simple functionality may not include all of the necessary functional processes needed to produce the output. Nor does this simple functionality trace the only possible sequence of these functional processes. Note, each simple functionality has a specific order associated with the functions that define the simple functionality; for this reason, we cannot say that a simple functionality is an element of the power set of functional processes because there is no order associated with an element of the power set. Also, we cannot say that this simple functionality is a mathematical function since a given input may be mapped into more than one output.

> *Complete Functionality:* a complete set of coordinated processes that operate on all the necessary inputs for producing a specific output. There is usually no specific order associated with the complete set of functional processes; however, a *partial order* of the functional activities can be established because some functions will usually have to be activated and completed before others. The complete functionality cannot be an element of the power set of functional processes because there is still some order information associated with the functions in the complete functionality. There is no order information in the sets of functions that comprise the power set of functions. There is a well-defined set of inputs, which is one element of the Cartesian product (or *n*-tuple) of inputs and is uniquely associated with the output. This output is also an element of the Cartesian product, or *m*-tuple, of outputs.

A *functional architecture* can be defined at several levels of detail:

1. A functional architecture that defines what the system must do, a decomposition of the system's top-level function. This very limited definition of functional architecture is the most common and is represented as a directed tree. Note that the Object Management Group (OMG) SysML standard uses the term "logical architecture" for what we denote as the "generic physical architecture" in Chapter 8. The OMG SysML standard usage of "physical architecture" is what we denote as the "instantiated physical architecture" in Chapter 8.

2. A functional model that captures the transformation of inputs into outputs using control information. This definition adds the flow of inputs and outputs throughout the functional decomposition; these items that comprise the inputs and outputs are commonly modeled via a data model (see the graphical modeling techniques on the companion website). A SysML model without any allocated components in the activity diagram representation of the functional architecture behavior is used as the modeling technique in this chapter to represent the functional architecture at this level of detail. Other graphical modeling techniques on the companion website for data and process modeling could also be used.

3. A functional model of a functional decomposition plus the flow of inputs and outputs, and the tracing of input/output requirements to specific functions and items.

An example of a functional architecture for the elevator case study can be downloaded from the companion web site.

7.3 FUNCTIONAL ARCHITECTURE DEVELOPMENT

IDEF0 is used here as the graphical process-modeling technique to represent the first elements of the functional architecture defined above. The companion web site content on graphical modeling techniques presents several alternate graphical process-modeling techniques that can be used in place of or in addition to IDEF0. IDEF0 was chosen for process modeling because IDEF0 has well-defined, standardized syntax and semantics that distinguish between the inputs to be transformed into outputs and the control information that guides the transformation process. In addition, IDEF0 has a place to represent the physical architecture, namely the mechanisms. Later, the allocated architecture can be illustrated using the mechanisms within IDEF0. The SysML activity diagram based on the legacy enhanced function flow block diagram (EFFBD) used to model the elevator case study also has well-defined, standardized syntax and semantics distinguishing inputs, outputs, control information, and allocation to the generic and instantiated physical architectures. The activity diagram, like its parent EFFBD, has the additional benefit of being executable and is covered in Chapter 9, Allocated Architecture Development.

It is possible to complete the functional architecture without resorting to any graphical techniques. Text and tables are sufficient to represent all of the information conveyed by any of the graphical techniques. However, Jones and Schkade [1995] provide convincing evidence that most systems and software professionals resort to graphical techniques during the system or software engineering process. The graphical techniques contain much greater information in a format that can be communicated more effectively and efficiently. Note, to really complete a functional architecture, it must be integrated with the physical architecture and simulated dynamically.

7.3.1 Functional Architecture Process Model

Figure 7.1 shows the IDEF0 model for the development of a functional architecture. See the full IDEF0 model for engineering a system on Wiley's Companion Website for this book. The approach shown in this figure begins by creating many function sequences, or simple functionalities, that satisfy the scenarios in the operational concept. These functionalities are created by shining a light into the black box of Chapter 6, thus turning the black box into a "white" box. Now, the functions that are needed to transform system inputs into system outputs become visible.

Then, the engineer synthesizes these many simple functionalities into a functional decomposition; this synthesis can be accomplished via a top-down decomposition or a bottom-up aggregation. Section 7.4 examines these two approaches in more detail. In practice, this second step of defining the functional decomposition combines both aggregation and decomposition. The flow of inputs and outputs from outside the system is added, and the necessary internal items are added, creating a functional model. Before distributing this functional model widely for comment (step 3), the scenarios from the operational concept are used once more to test the draft decomposition and ensure that the functional model is consistent with these scenarios.

The third step addresses the data or items that serve as inputs or outputs to the various functions of the functional architecture. For computer-intensive systems, developing a data model is critical (see the material on graphical modeling techniques on the companion web site) so that the relations among the various items flowing through the system are understood at the level needed for a successful design.

The fourth step is the solicitation of the opinions of other engineers and stakeholders about missing functions or alternate decompositions that are more meaningful than have been produced during the second and third steps. During this third step, the allocated architectural activity, which combines the functional and physical architectures, is proceeding. Feedback from the development of the allocated architecture often causes changes to the functional model, changes that enable the

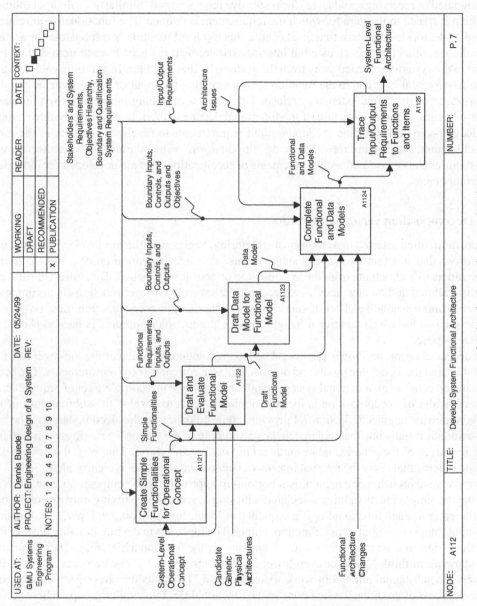

FIGURE 7.1 Process for developing a functional architecture.

181

functional model and the physical architecture to match more closely. (Chapter 9 discusses these issues in more detail.)

The final step in the development of the functional architecture addresses the tracing of input/ output requirements to both the functions in the data model and the items (data elements) flowing through this functional model. Each input (output) requirement is traced to those functions that have been designated as receiving (producing) the respective input (output). Similarly, each input (output) requirement is traced to the item for which the requirement is defined. The functional requirements are traced to the top-level system function because this top-level function is responsible for accomplishing these subfunctions. Each external interface requirement is traced to each item that will be delivered to the system (or carried away from the system) by that interface. In addition, each external interface requirement is traced to the function that is receiving the input or sending the output that has been traced to that same external interface. This process of tracing input/output requirements often raises issues about the structure of the functional decomposition, leading to possible changes in this decomposition. By tracing the input/output requirements to functions and data in the functional architecture, these requirements are being "flowed down" so that the allocated architecture will have all requirements associated with the elements of specifications that are developed for individual system components.

7.3.2   Decomposition versus Composition

Decomposition, often referred to as top-down structuring, begins with the top-level system function and *partitions* that function into several subfunctions. This decomposition process must conserve all of the inputs to and outputs from the system's top or zero-level function. By conserve, we mean use/produce all and add no new ones. Next, each of the several first-level functions is decomposed (partitioned) into a second-level set of subfunctions. Note that not every function must be decomposed; only those for which additional insight into the production of outputs is needed should be further decomposed.

The success of decomposition is predicated on having a sound definition of the top-level function of the system and the associated inputs and outputs, that is, a complete set of requirements. The benefit of having an ecosystem or external systems model is to achieve this complete set of requirements. A major difficulty of decomposition is the partitioning process to develop the subfunctions of the system is somewhat unguided. Section 7.4 provides some guidance for this decomposition. The best decomposition is usually one that will match the partitioning of the system's generic and instantiated physical resources of the generic and instantiated physical architectures. This way, the flow of data and physical items that cross the internal interfaces between components is clearly identified.

The opposite approach, composition, is a bottom-up approach. With composition, one starts by identifying the simple functionalities associated with simple scenarios involving only one of the outputs of the system. Each functionality is a sequence of input, function, output–input, ... , function, output–input, function, output–input, function, output. The functions in the functionality are all functions of the system and are relatively low-level functions in the functional hierarchy. These functions usually show up in third, fourth, or even lower levels of the hierarchy. For complex systems, this initial step is a substantial amount of work. After the many functionalities have been defined, one begins the process of grouping the functions in the functionalities into similar groups. These groups are aggregated into similar groups; this process continues until a hierarchy is formed from bottom to top.

The advantage of the composition approach is that the composition process can be performed in parallel with the development of the generic and instantiated physical architectures so that the functional and physical hierarchies match each other. Second, this approach is so comprehensive that the approach is less likely to omit major functions. The drawback is that the many functionalities must be easily accessible during the composition process so that all of this work can be successfully used; the simple functionalities are often pasted on the walls of a large conference room. The composition method dates to the 1960s and 1970s, when systems engineering was still in its infancy; many systems engineers continue to prefer this approach. There is no empirical evidence that either the composition or decomposition approach is better than the other.

Ultimately, using a combination of decomposition and composition approaches is wisest. This is sometimes referred to as middle-out. Often, one makes use of simple functionalities associated with specific scenarios defined in the operational concept to establish a "sense" of the system. Then, positing a top-level decomposition that is likely to match the top-level segmentation of the physical architecture is common before proceeding to do decomposition that is reinforced by periodic reference to the functionalities to assure completeness.

Decomposition is efficient and often successful when the system is an update or variation of an existing system. Composition is strongly recommended when the system is unprecedented or a radical departure from an existing system.

Before proceeding, it is important to discuss some valuable properties of the functional hierarchy. Besides the obvious design implications that are embodied in this hierarchy, the hierarchy is also important as a communication tool. This communication is important for both other engineers and the stakeholders. For this reason, limiting the number of functions at each node in the functional tree to a number that enhances communication is advisable; large numbers of functions at a given level of a decomposition turn any graphical technique into a "bowl of spaghetti," where the functions are the "meat balls," and the arrows are the "spaghetti."

7.4   DEFINING A SYSTEM'S FUNCTIONS

Assigning functions in the functional architecture in total to one and only one resource in the physical architecture has its basis in the era of systems engineered with discrete components. Today's systems and systems of systems are computer and software-centric and complexly integrated. The computing components within a system may be off in a "cloud" or embedded systems on a chip (SoC) with multiple cores based on instruction set or reduced instruction set architectures (ISA/RISC), graphic processing units (GPUs), and specialized neural networks with software threads touching multiple cores. Software applications are often composed of multiple processes from standard libraries concatenated together. This poses significant challenges in assuring system performance.

Clearly, the functional and physical architectures cannot be developed independently and satisfy the property of one-to-one allocations all the way through the architectural hierarchies. In fact, there are times when the decision to allocate a particular function to one of several resources has substantial performance implications and is the subject of one or more trade studies. The bottom line is that the functional architecture may be revised several times as the allocated architecture is finalized. Therefore, focusing on getting the functional hierarchy right the first time is improper since this is an impossible task.

7.4.1   Approaches for Defining Functions

There are a number of keys one can use to partition a function into subfunctions. At the top of the hierarchy, we would expect to see functions devoted to the system's operating modes, if there are any. For functions that have multiple outputs, we could partition the function into subfunctions that correspond with the production of each output. Similarly, we could key on the inputs and controls to find a partition of the function. More appropriate than either of these is to decompose on the basis of stimulus–response threads that pass through the function being decomposed. Finally, there

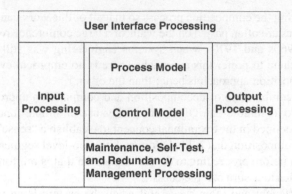

FIGURE 7.2   Architecture template of Hatley and Pirbhai [1988, p. 195].

is often a natural sequence of subfunctions for a particular function. For example, at the bottom of the functional architecture, we would expect to see functions such as receive input, store input, and disseminate input or retrieve output, format output, and send output.

Hatley and Pirbhai [1988] developed an architectural template for representing the physical architecture of the system that supports computer and software-centric systems; Figure 7.2 shows the physical segments of the template. This template suggests the creation of a generic partition of six subfunctions, one for each of the Hatley–Pirbhai components. These six generic functions could be used in any functional architecture:

- *Provide User Interface*: those functions associated with requesting and obtaining inputs from users, providing feedback that the inputs were received and allowed, identifying unintended/unallowed inputs, providing outputs to users, and responding to the queries of those users.
- *Format Inputs*: those functions needed to receive and assure allowed inputs from external interfaces (nonhumans), and other nonhuman system components, to identify unin- tended/unallowed inputs and to process (e.g., analog-to-digital conversion) those inputs to put them into a format needed by the system's processing functions.
- *Transform Inputs into Outputs*: the major functions of the system.
- *Control Processing*: those functions needed to control the processing resources or the order in which these processing functions should be conducted, including mitigations for unintended inputs and undesired outputs, as well as protective actions for unallowed inputs.
- *Format Outputs*: those functions needed to assure and convert the system's outputs into the format needed by the external interfaces or other nonhuman system components and then place those outputs onto the appropriate interface.
- *Provide Structural Support, Enable Maintenance, Conduct Self-test, and Manage Redundancy Processing*: those functions needed to perform internal support activities, respond to external diagnostic tests, monitor the system's functionality, detect errors, and enable the activation of standby resources.

This partition is a very valid approach at the top of the functional architecture; the authors have used this approach several times to initiate decomposition with success. Most systems would have all or nearly all of these functions as an initial partition. Figure 7.3 uses the Hatley–Pirbhai template to show the four top-level functions of the elevator case study, which can be downloaded from the companion website.

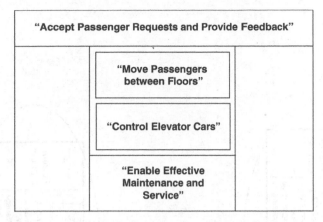

FIGURE 7.3 Elevator functions within the Hatley–Pirbhai template.

As the decomposition of system functions proceeds, we would expect to find smaller subsets of these six generic functions being embedded within each of the higher-level functions. Figure 7.4 renames the Hatley and Pirbhai [1988] partition as functions and illustrates the functional decomposition by showing likely decompositions within the top-level functions; the top-level decomposition of the system function is in the middle of the figure.

McMenamin and Palmer [1984] describe a system's functions as being composed of essential or fundamental activities and custodial activities. All but one of the functions implied by the Hatley and Pirbhai [1988] template are fundamental activities. The function, "enable maintenance, conduct self-test, and manage redundancy processing," performs custodial activities. Additional custodial activities that could be embedded in this function are the provision of structural support, maintenance of information archives, provision of security services, and so forth. In addition, custodial activities maintain the system's memory so the system knows what it needs to know to perform its fundamental activities. This knowledge is called the essential memory of the system; examples include the storage of data items between the time they become available and the time they are used by the fundamental activities. McMenamin and Palmer [1984] recommend separating the custodial activities and the fundamental activities. This separation is not completely possible at the top-level with the taxonomy suggested by the Hatley and Pirbhai [1988] template, nor is this separation often desirable at this high level. However, achieving this separation at lower levels of the functional decomposition is possible and desirable.

Baylin [1990] provides a number of interesting insights into modeling the functional aspects of a system by focusing on the system's objectives. The purpose of any system is to achieve the objectives that have been defined for that system. As a result, the engineer of a system would be foolish not to use the system's objectives as a guide for defining the top-level functions of the system.

Many engineers involved in developing systems have read and suggested Miller's [1978] classic titled *Living Systems* as a guide for defining the functions of a system. Miller examines seven levels of systems that range from a cell through a supranational system and include an organ, an organism, a group, an organization, and a society. One of Miller's claims is that there are 19 subsystems that must be part of any of these living systems. In fact, Miller defines these 19 subsystems in terms of the function that each performs (see Table 7.1), leading the reader of *Living Systems* to believe that it is the 19 functions that are most useful to engineers of human-designed systems. One key to Miller's study or living systems is his assertion that these systems either process matter-energy or information or both. The top two functions in Table 7.1 address the processing of both matter-energy

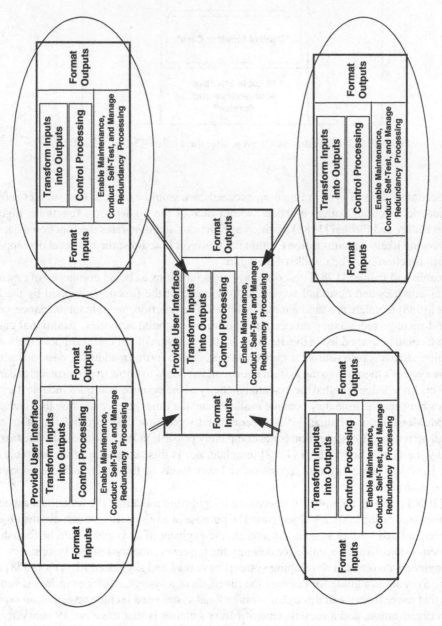

FIGURE 7.4 Exemplary functional decomposition.

and information. The functions on the left half of Table 7.1 process matter-energy, while those on the right process information. There are blanks left in the table so that functions on the left and right that are similar can be opposite each other. This assertion is key to understanding the two columns of subsystems and related functions in Table 7.1.

The common concepts for defining a partition of a function are system modes, function outputs, function inputs and controls, system objectives, stimulus–response threads, and the functional template based upon the Hatley and Pirbhai [1988] architecture template.

7.4.2 Typical Functional Decompositions by Life Cycle Phase

This section suggests functional hierarchies and segments of functional hierarchies for the enabling systems for the system of interest over its life cycle: development, qualification, supply chain, manufacturing, training, and deployment. The previous section dealt with the operational phase of the life cycle.

In particular, development and qualification are systems in their own right. Duffy and Buede [1996] suggest structuring the management portion of the *development phase* into three major activities – formulate the development strategy, execute the development strategy, and evaluate the results of the development activity. Formulating the development strategy has as many elements of a development strategy as needed. Common elements of the development strategy are the procurement, engineering or technical, financing, communication, technology development, and testing strategies. Other elements may include regulatory and risk mitigation strategies. This strategy drives the related enabling systems. The IDEF0 model of the systems engineering design and integration process in Appendix B demonstrates the execution of the engineering elements of the development strategy. The development and qualification systems functional and physical architectures are discussed in Chapter 11, Integration and Qualification.

Dietrich [1991, p. 886] defines *manufacturing* as "using resources to perform operations on materials to produce products." A *manufacturing system* is a "set of resources used to manufacture some product, together with the associated information system and any behavioral requirements imposed by the owners of the resources." The products being produced are the primary outputs of this phase; inputs are defined to be bulk material; internal items are called work in progress (WIP). WIP is material upon which some value-added operations have been performed. Seven types of generic manufacturing functions are defined based upon the types of bulk material, WIP, and primary outputs:

- *Bulk Operation:* manipulate bulk material to produce other bulk material.
- *Kitting Operation:* transform one or more bulk materials into one or more units of WIP.
- *Fabrication Operation:* fabricate a WIP from another unit of WIP and bulk material.
- *Assembly Operation:* assemble two or more units of WIP and bulk material into a subassembly (higher-level WIP).
- *Byproduct Operation:* transform two or more WIPs of different types into two or more WIP types that are not identical to the input WIPs.
- *Distribution Operation:* divide one or more units of a single WIP into two or more units of possibly different types of WIP.
- *Consumption Operation:* consume one or more WIPs yielding bulk, dissipated, or useless material. (Note that shipping finished products and stockpiling subassemblies are considered consumption operations.)

TABLE 7.1   Subsystems and Functions of Living Systems

SUBSYSTEMS WHICH PROCESS BOTH MATTER-ENERGY AND INFORMATION

1. *Reproducer*, the subsystem which is capable of giving rise to other systems similar to the one it is in.
2. *Boundary*, the subsystem at the perimeter of a system that holds together the components, which make up the system, protects them from environmental stresses, and excludes or permits entry to various sorts of matter-energy and information.

SUBSYSTEMS WHICH PROCESS MATTER-ENERGY	SUBSYSTEMS WHICH PROCESS INFORMATION
3. *Ingestor*, the subsystem, which brings matter-energy across the system boundary from the environment.	11. *Input transducer*, the sensory subsystem, which brings markers bearing information into the system, changing them to other matter-energy forms suitable for transmission within it.
	12. *Internal transducer*, the sensory subsystem, which receives, from subsystems or components within the system, markers bearing information about significant alterations in those subsystems or components, changing them to other matter-energy forms of a sort which transmitted within it.
4. *Distributor*, the subsystem, which carries inputs from outside the system or outputs from its subsystems around the system to each component.	13. *Channel and net*, the subsystem composed of a single route in physical space, or multiple interconnected routes, by which markers bearing information are transmitted to all parts of the system.
5. *Converter*, the subsystem, which changes certain inputs to the system into forms more useful for the special processes of that particular system.	14. *Decoder*, the subsystem, which alters the code of information input to it through the input transducer or internal transducer into a "private" code that can be used internally by the system.
6. *Producer*, the subsystem which forms stable associations that endure for significant periods among matter-energy inputs to the system or outputs from its converter, the materials synthesized being for growth, damage repair, or replacement of components of the system, or for providing energy for moving or constituting the system's outputs of products or information markers to its suprasystem.	15. *Associator*, the subsystem, which carries out the first stage of the learning process, forming enduring associations among items of information in the system.
7. *Matter-energy storage*, the subsystem, which retains in the system, for different periods of time, deposits of various sorts of matter-energy.	16. *Memory*, the subsystem, which carries out the second stage of the learning process, storing various sorts of information in the system for different periods of time.
	17. *Decider*, the executive subsystem, which receives information inputs from all other subsystems and transmits to them information outputs that control the entire system.
	18. *Encoder*, the subsystem, which alters the code of information input to it from other information processing subsystems, from a "private" code used internally by the system into a "public" code, which can be interpreted by other systems in its environment.

TABLE 7.1   *(continued)*

8. *Extruder*, the subsystem which transmits matter-energy out of the system in the forms of products or wastes.

9. *Motor*, the subsystem which moves the system or parts of it in relation to part or all of its environment or moves components of its environment in relation to each other.

19. *Output transducer*, the subsystem which puts out markers bearing information from the system, changing markers within the system into other matter-energy forms which can be transmitted over channels in the system's environment.

10. *Supporter*, the subsystem which maintains the proper spatial relationships among components of the system, so that they can interact without weighting each other down or crowding each other.

Source: Adapted from Miller [1978].

7.4.3   Feedback and Control in Functional Design

It is important to emphasize the use of feedback in the design of the system. *Feedback and control* is the comparison of the actual characteristics of an output with the desired characteristics of that output for the purpose of adjusting the process of transforming inputs into that output (see Sidebar 7.1). Open-loop control processes may or may not make this measurement, but in either case, make no adjustments to the process once started. See Figure 7.5. The heating and air-conditioning systems in all but the most expensive cars allow the driver to set the output temperature of the heater and the fan speed; this is an example of an open-loop control system. The driver serves as the feedback process that adjusts the heat and fan speed when a deviation from the desired temperature is noticed. Closed-loop control processes use measurements of the output as feedback for the purpose of adjusting or controlling the transformation process. Heating and air-conditioning systems in most houses have a thermostat for setting the desired temperature; this thermostat adjusts the length of time that the heating or air conditioning is left on in order to reach the desired temperature. This is an example of a closed-loop control system.

SIDEBAR 7.1: HISTORY OF CONTROL SYSTEMS

Mayr [1970] traced the earliest example of a control system to the second century BC; this control system was a water clock that operates on the same principles as current flush toilets and is not dissimilar to numerical integration on a digital computer.

In about 1620 Cornelis Drebbel, a Dutch mechanic and chemist, designed a system to control the temperature in a furnace used to heat eggs in an incubator.

About 1787 Thomas Mead invented a centrifugal governor, which was adapted about a year later by Matthew Boulton and James Watt, who invented a fly ball governor to control the rotation speed of a grinding stone for a wind-driven flour mill. The first study of feedback control and the stability of such systems was described in a paper titled "On Governors" by J.C. Maxwell in 1868.

A *negative feedback process* attempts to close the gap between the current output and the desired output, thus striving for a stable process. A *positive feedback process* attempts to increase the difference between the current output and the desired output, usually creating an unstable situation.

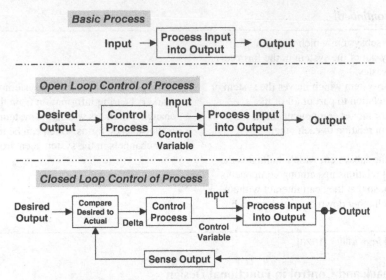

FIGURE 7.5   Open and closed-loop control processes.

In the engineering design process, feedback and control enable the comparison of the current state of the system with the desired state for the purpose of repeating parts of the generation of the current state to obtain a current state that is closer to the desired state. The concept of feedback comes from the engineering of control systems, which has been the training ground for many systems engineers.

Closed-loop control processes contain at least four subprocesses: comparison of current and desired output characteristics, control adjustments to the process based upon the comparison, the transformation process for turning inputs into outputs, and a sensing process for turning the output into measured dimension(s) that can be compared to the desired output. The first element is the comparison process in which the current values of key variables are compared with the desired values of those variables. The comparison process requires definition in advance for what elements of the state of the process are going to be compared. This comparison inevitably introduces a time lag in the process. This element of the feedback process is trivial but, at the same time, is the cornerstone. The second element is the control process for deciding what to do about the difference between the current value of the output and the desired value of the output. The third element of the feedback process is the transformation process that is controlled by the feedback process. This process dictates how a successful feedback process should be created and is often adapted by the feedback process as part of the correction activity. Sensing the output of the process being controlled is the final element of the feedback process. The engineering of feedback control must assure the following: identifiability, observability, and controllability to assure the stability of the process.

While most examples of feedback control systems are in lower-level elements of complex systems, there is no reason why such a concept will not also work at higher levels of abstraction. An example is the "Develop System Allocated Architectures" function of the IDEF0 model of the process for engineering a system in Appendix B and repeated in Figure 7.6. There are three feedback and control loops. The first involves the first and second functions, "Allocate Functions & System-wide Requirements to Physical Subsystems" and "Define & Analyze Functional Activation & Control Structure." Here, the second function performs the measurement, comparison, and control function based upon the output, functional allocations to components, of the first function.

The measurement, comparison, and control (decision making) in the second loop are done in the third function, "Conduct Performance & Risk Analyses," for the output of the second function,

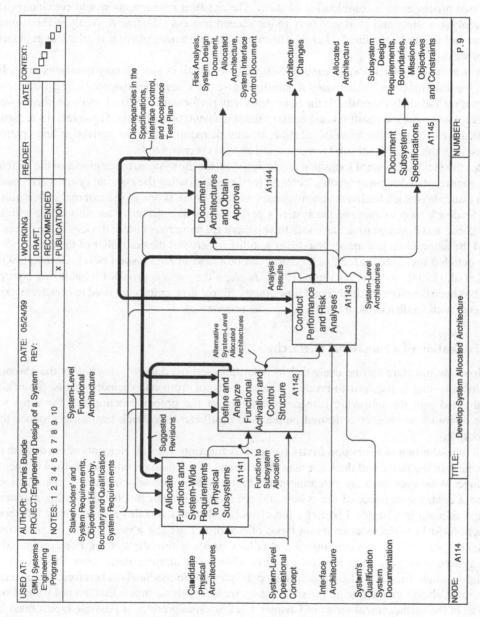

FIGURE 7.6 Illustration of feedback control in the development of the system allocated architecture.

The figure is an IDEF0 diagram with the following elements:

USED AT:

AUTHOR: Dennis Buede   DATE: 05/24/99

PROJECT: Engineering Design of a System   REV:

NOTES: 1 2 3 4 5 6 7 8 9 10

READER   DATE   CONTEXT:

WORKING
DRAFT
RECOMMENDED   x
PUBLICATION

GMU Systems
Engineering
Program

Inputs:
- Stakeholders' and System Requirements, Objectives Hierarchy, Boundary and Qualification System Requirements
- Candidate Physical Architectures
- System-Level Operational Concept
- Interface Architecture
- System's Qualification System Documentation

Controls:
- System-Level Functional Architecture

Boxes:
- Allocate Functions and System-Wide Requirements to Physical Subsystems — A1141
- Define and Analyze Functional Activation and Control Structure — A1142
- Conduct Performance and Risk Analyses — A1143
- Document Architectures and Obtain Approval — A1144
- Document Subsystem Specifications — A1145

Labels:
- Function to Subsystem Allocation
- Suggested Revisions
- Alternative System-Level Allocated Architectures
- Analysis Results
- System-Level Architectures
- Discrepancies in the Specifications, Interface Control, and Acceptance Test Plan
- Risk Analysis, System Design Document, Allocated Architecture, System Interface Control Document
- Architecture Changes
- Allocated Architecture
- Subsystem Design Requirements, Boundaries, Missions, Objectives and Constraints

NODE: A114   TITLE: Develop System Allocated Architecture   NUMBER: P. 9

191

alternative system-level allocated architectures. The analysis process determines whether the system-level allocated architectures contain one that is "good enough" to be the finalized design and then proceeds with documentation. If the decision is that there is not an allocated architecture that is good enough, then analysis results are passed to the first two functional processes as controls for making refinements. The intention here is that the analysis results could be passed to either of these two processes or a combination of them. The smallest refinements would conclude with passing analysis results and guidance only to the second process ("Define & Analyze Functional Activation and Control Structure"). Large refinements would require passing results and guidance to both processes.

There is a final feedback loop during documentation, which is when many questions arise. In this case, the third function is reactivated if questions arise that cannot be answered using the current documentation and analysis results. If the issue deals with performance and risk analysis, the answer can be generated, and the result passed back to the documentation activity. However, if the issue has implications for the allocation of function, tracing of requirements, or activation and control structures, then the initial feedback loop discussed above is reenergized.

Besides the feedback control loops that are designed inside the system, the engineer of the system has to be cognizant of designing feedback control for the system using the external systems. The most common example of such feedback control occurs when a human is one of the external systems and closes a feedback loop to improve the system's performance. The driver of an automobile adjusts the car's speed and direction to achieve safe travel; there are numerous output devices at the driver's station of the automobile to enhance the driver's ability to serve as the controller of the car.

More detailed literature on feedback control can be found in Dickinson [1991], Dorny [1993], Franklin et al. [1994], and Van de Vegte [1994]. A graph-theoretic approach for analyzing control systems has been developed called signal flow graphs. Signal flow graphs are used to transform a set of processes with feedback into a single, composite process.

7.4.4   Evaluation of a Functional Hierarchy

A functional architecture can be evaluated for shortfalls and overlaps. A *shortfall* is the absence of functionality that is required to produce a desired output from one or more inputs. Shortfalls can be divided into the following categories: absence of the proper functionality for some set of inputs, inability to produce a desired output, and insufficient feedback control to produce the desired output.

Recall the definition of a function from Chapter 4. A function maps all elements of the domain to some element in the range and does not map any element of the domain into two distinct elements of the range. Whenever there are potential inputs to the system with which the system's functionality cannot deal, the engineer of the system did not create a system function but rather a system relation. A relation in Chapter 4 includes functions but also includes those entities that fall short of a function. In fact, the most common types of a shortfall are the absence of or inappropriate functional responses to unexpected inputs and failure modes within the system. For example, the elevator system must be able to respond properly when a fire alarm sounds. Less obvious unexpected inputs might be the need for a user to stop the elevator immediately. Therefore, the systems engineer must always enumerate all possible inputs, including those inputs that are not wanted but can arrive. In the mathematical terms of Chapter 4, a Cartesian product of possible inputs must be formed for each function in the functional model of the functional architecture. This is only necessary for the lowest-level functions in the functional decomposition. The Cartesian product of inputs for a function uses each category of input shown in the functional model for a specific function. For each of these categories, there are usually several possible input states, some of which are not desired. For example, if there were three possible input categories to a given bottom-level function and each input category had three possible states, there would be a three-tuple formed by taking the Cartesian product of these three input categories. The three-tuple would have 27 ($3 \times 3 \times 3$) different combinations. The functional definition of this bottom-level function must account for every one of these 27 possible combinations.

The second category of shortfall is the inability to produce a needed output. This type of functionality will be obvious if all of the system's outputs have been defined. This is a major benefit of the ecosystem or external systems model diagram in Chapter 6 and the functional architecture discussed in this chapter. Evaluating for this category of shortfall is not always possible without constructing an overall functional architecture.

The final shortfall addresses the quality of the outputs produced. Often, this quality falls short of that desired by the stakeholders because the engineers have not incorporated sufficient feedback control, either internally to the system or inclusive of the external systems. Missing needed feedback is a common mistake made in the functional architecture. This is true not only for the functional architecture of the system being designed for the operational phase of the life cycle but also for the functional architectures of the developmental and manufacturing systems.

An *overlap* is a redundancy in functionality that is not needed to achieve additional performance, for example, reliability. Functional overlaps, unlike physical overlaps for redundancy, are not needed and therefore can only cause problems.

A common technique for identifying shortfalls and overlaps is to follow each scenario in the operational concept (Chapter 6) through the functional architecture. Each scenario in the operational concept begins with a single input to the system from one of the external systems and continues with a sequence of inputs to and outputs from the system to various external systems. Each scenario was developed by treating the system as a black box. Now is the time to shine a light into that black box (producing a white box) and see what functions the system is performing to transform the inputs into outputs. Start with the first input to the system for a given scenario (see Fig. 7.7); color the line in the context model diagram (metafunction) for that input green (or whatever color you choose). Find an interesting output of the system in the scenario and color that output on the context model diagram green. In Figure 7.7, the input selected was "Request for Elevator Service & Entry Support" by a potential passenger, which is shown as a weighted black line since color is too expensive for a textbook. The output selected was "Elevator Entry/Exit Opportunity" when the elevator arrives at the potential passenger's floor; this output is also shown as a weighted black line.

Now move to the Figure 7.7 first level decomposition diagram and color these same two lines green; see Figure 7.8 for the weighted black lines. Now go to the function on the first level decomposition diagram that received that input (function 1 in Fig. 7.8), find the appropriate output of the function that is needed to get to the output on the context page, and color the line associated with that output green. "Digitized Passenger Request" is shown with a weighted black line in Figure 7.8. Proceed to this next function on the first level decomposition diagram and find the most appropriate output to color. This is like looking through a house for clues to a mystery, searching room by room, finding a clue in each room that leads to the next room, until finally, the room is found with the already identified path outside. In Figure 7.8, "Digitized Passenger Request" led to function 2, "Control Elevator Cars." The appropriate output of this function was "Assignments to Elevator Cars," leading to function 3, "Move Passengers Between Floors," which is where "Elevator Entry/Exit Opportunity" was found.

This process continues for every other page of the functional model. Figures 7.9 through 7.12 show this trace of the input and output from a given scenario throughout the entire functional model of the elevator system in the case study that can be downloaded from the companion website.

In addition, defining failure modes for the system and creating *error detection and recovery functionalities* within the common operating modes as well as the failure modes is critical. These functionalities for error detection and recovery are critical for stakeholder usability. How often has your computer shut down with no warning and little support for saving open files? The more mature an operating system is, the more functionality the operating system commonly has for saving open files as part of the crash, and the more unlikely such crashes are. Details on functionality for addressing error detection and recovery are covered later in the chapter.

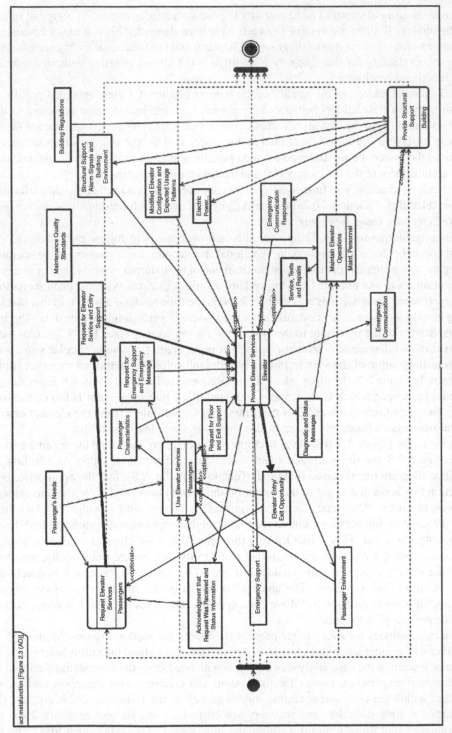

FIGURE 7.7 Scenario trace on the context model diagram.

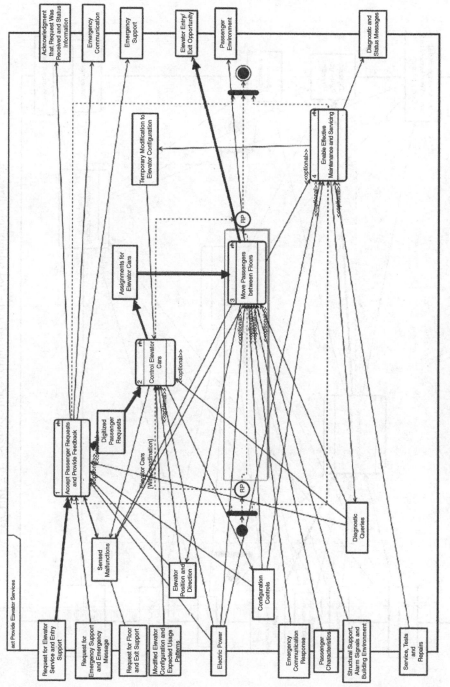

FIGURE 7.8 Scenario trace continued in the first level decomposition diagram.

195

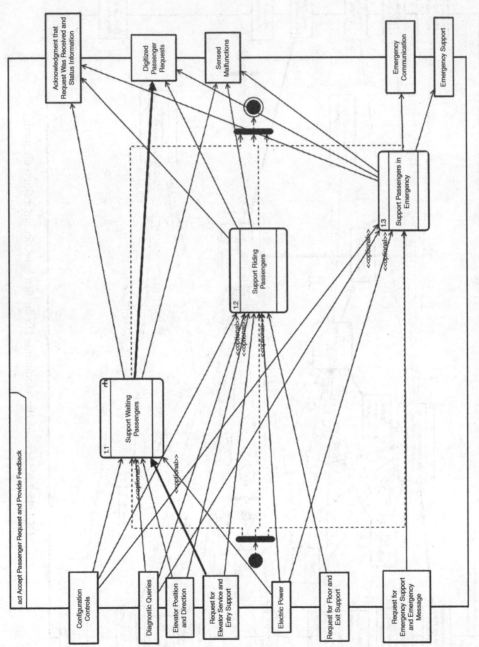

FIGURE 7.9 Scenario trace continued in the first subfunction diagram.

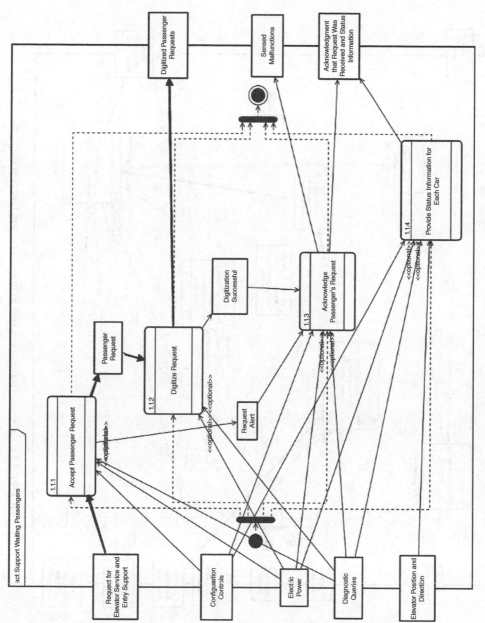

FIGURE 7.10 Scenario trace continued in the first sub-subfunction diagram.

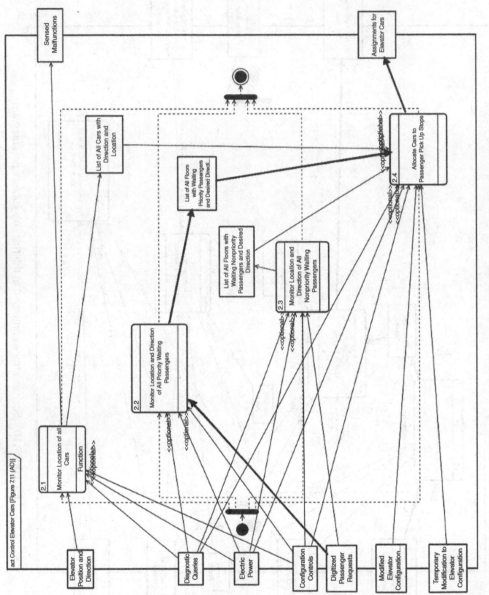

FIGURE 7.11 Scenario trace continued in the second sub function diagram.

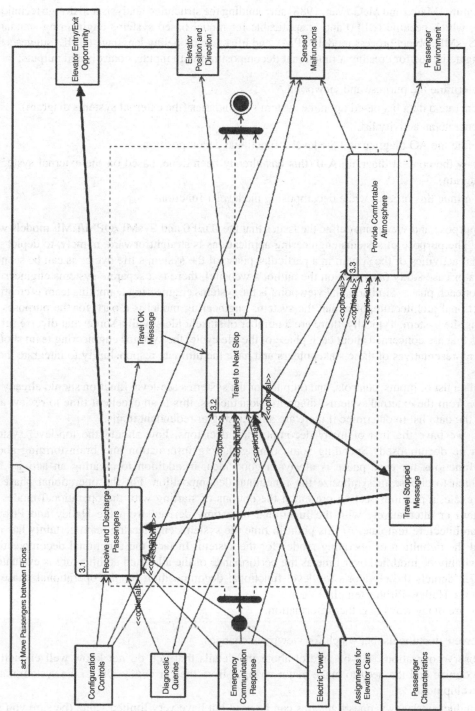

FIGURE 7.12 Scenario trace completed in the third sub function diagram.

7.5 DEVELOPMENT OF THE FUNCTIONAL DECOMPOSITION

The literature [Marca and McGowan, 1988] surrounding the structured analysis and design technique (SADT), which became IDEF0 and is applicable for model-based systems engineering standards such as SysML, object process model (OPM), and life cycle modeling language (LML) suggests the following activities for creating a functional decomposition with inputs, controls and outputs:

- Determine the purpose and viewpoint.
- Generate a data list based upon the system's boundaries (the external systems diagram).
- Generate an activity list.
- Define the AO diagram and the level 1 functional decomposition.
- Draw the context diagram, A-0 (this has already been done, based on the external systems' diagram).
- Continue this process while decomposing the level 1 functions.

The purpose and viewpoint define the issues that the IDEF0 and SysML/OPM/LML models will address. The purpose of systems engineering applications is straightforward, namely, to depict the functional activities of the system in a particular phase of the system's life cycle; as can be seen in the elevator case study (available on the author's website), there is a separate systems engineering model for each phase. Similarly, the viewpoint is the systems engineering team; this team is creating the functional architecture, of which the systems engineering model is a part, for the purposes of designing the system. Typically, there are a number of stakeholders with a somewhat diverse set of opinions that are concerned about each phase of the life cycle; the systems engineering team should include representatives of these stakeholders and have the ultimate responsibility to integrate these opinions.

The data list of inputs, controls, and outputs for the system's top-level function should already be available from the external systems' diagram. Nonetheless, this is an excellent time to review and critique the data list to determine if there are any missing or redundant items.

Next, we have the first of many decomposition decisions; how should the top-level system function he decomposed? Spending some time gathering information and brainstorming about system functions for each phase is always a good idea, in addition to creating an activity list from which to choose or synthesize the functional decomposition. For the operational phase of the life cycle, a previous section presented the options of starting with the operational modes of the system or alternatively with the functional taxonomy derived from the Hatley and Pirbhai [1988] architecture template. At this point in time the systems engineering team certainly has not finalized the definition of operating modes for the system. In fact, the functional decomposition will inevitably be modified over time as the performance of the allocated architecture is evaluated. Figure 7.3 depicts the elevator's top-level functional decomposition for the operational phase in terms of the Hatley–Pirbhai template.

There are many ways to gather information:

- Review documents, but watch for viewpoint changes.
- Observe operations, but be careful about the details that you do not know well enough to recognize and the need to make major changes from the current system to the system under development.
- Conduct interviews; questionnaires can be used but have very limited value (be sure you get the right experts).
- Invent a straw man for the experts to critique.

- Create several alternate decompositions and create a composite straw man based on the best features of each after some critical discussion (this creativity technique is often called the "gallery").

Once a working version of the functional model is created, the functional model should be reviewed by individuals who have substantial knowledge and varying perspectives about the system's functioning in a given life cycle phase. This review process should:

- Try alternate decompositions.
- Disaggregate the functions differently.
- Bundle and unbundle arrows differently.
- Reevaluate functional dominance in terms of feedback and control.
- Catch interface errors.

SIDEBAR 7.2: COMMON MISTAKES IN DEVELOPING A FUNCTIONAL ARCHITECTURE

1. Including the external systems and their functions. The functional architecture only addresses the top-level function of the system in question. The external system diagram establishes the inputs, controls, and outputs for this function. A boundary has been drawn around the system to exclude the external systems and their functions.

2. Choosing the wrong name for a function. The function name should start with an action verb and include an object of that action. The verb should not contain an objective or performance goal such as maximize but should describe an action or activity that is to be performed.

3. Creating a decomposition of a function that is not a partition of that function. For example, a student once decomposed "AO: Provide Elevator Services" into "Al: Transport Users," "A2: Evaluate System Status," and "A3: Perform Security & Maintenance Operations." "Al: Transport Users" was then decomposed as follows: "All: Provide Access to Elevator," "Al2: Transport Users," and "A13: Provide Emergency Operations." Al2 cannot be a child of itself. The sub-functions of a function should all be at the same level of abstraction [Chapman et al., 1992].

4. Including a verb phrase as part of the inputs, controls, or outputs of a function. Verb phrases are reserved for functions.

5. Violating the law of conservation of inputs, controls, and outputs. That is, every input, control, and output of a particular function must appear on the decomposition of that function, and there can be no new ones.

6. Trivializing the richness of interaction between the functions that decompose their parent. Consider many possible simple functionalities that comprise the children of a parent function and then develop the inputs, controls, and outputs that enable these simple functionalities to exist, including the necessary feedback and control.

7. Creating outputs from thin air. The most common mistake is to define a function that monitors the system's status but that does not receive inputs about the functioning or lack of functioning of other parts of the system.

As part of this review process, creating a data model of the inputs, controls, and outputs using an entity–relationship–attribute or higraph model would be wise. These techniques are discussed in the graphical modeling techniques on the companion web site. The data model often introduces critical design issues that have been overlooked in the functional or process model.

How far should the functional decomposition be carried out? Generally speaking, the functional decomposition should proceed to the second, third, or fourth level. At this point, the physical and allocated architectures should be developed and analyzed. The more detailed the operational concept, the more reliably the functional architecture can be developed to the fourth level. Defining the system's functions to line up with the physical components is best so that the inputs, controls, and outputs clearly line up with external and internal interfaces. The level of detail should be appropriate with the viewpoint and purpose, that is, the stakeholders and specified phase of the system's life cycle. Be sure to eliminate details if they are not helping create the allocated architecture. Also, see Sidebar 7.2 for a list of common mistakes made in the development of a functional architecture.

7.6   FINISHING THE FUNCTIONAL ARCHITECTURE

Two key areas of the functional architecture that need to be addressed before the job is finished are (1) defining system errors and the failure modes that result and inserting the functionality to detect the errors and recover them and (2) inserting the appropriate functionalities for some combination of built-in self-test (BIST) and external testability. The functionalities described here are typically not part of the initial drafts of the functional architecture because they depend to a significant degree on the physical architecture; as a result, these functions are often added once the allocated architecture is taking shape.

Fault tolerance is a laudable design goal, meaning that the system can tolerate faults and continue performing. In fact, the design goal of every systems engineering team is to create a system with no faults. However, faults like friction have to be tolerated at best, even after our best efforts to eliminate them. This discussion on fault-tolerant functionality depends greatly on understanding several key terms; see Jalote [1994] and Levi and Agrawala [1994]. Figure 7.13 provides a concept map based on these definitions.

System: an identifiable mechanism that maintains a pattern of behavior at an interface between the system and its environment. [Anderson and Lee, 1981].

Failure: deviation in behavior between the system and its requirements. Since the system does not maintain a copy of its requirements, a failure is not observable by the system.

Error: a subset of the system state, which may lead to a failure. The system can monitor its own state, so errors are observable in principle. Failures are inferred when errors are observed. Since a system is usually not able to monitor its entire state continuously, not all errors are observable. As a result, not all failures are going to be detected (inferred).

Fault: a defect in the system that can cause an error. Faults can be permanent (e.g., a failure of system component that requires replacement) or temporary due to either an internal malfunction or external transient. Temporary faults may not cause a sufficiently noticeable error or may cause a permanent fault in addition to a temporary error.

First, note the difference of the definition of system in the fault tolerance literature and that discussed in Chapters 2 and 6 of this book, which represent the systems engineering community. The fault tolerance community is focused on inferring failures by detecting errors. The notions that are central to this focus are the system's requirements (or specifications), the boundary between the system and the system's environment at which the state of the system is defined, and the interface that connects the system to its environment. The fact that a system has objectives, as defined by

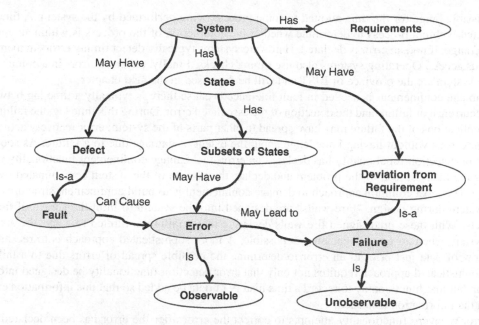

FIGURE 7.13 Concept map for fault tolerance terms.

the stakeholders, and functions (or tasks or activities), as defined by the systems engineers, is not relevant to the fault tolerance community and is therefore not found in their definition of a system.

Achieving fault tolerance in a system means using both the designed functions and physical resources of the system to mask all errors (deviations between actual system outputs and required system outputs) from the system's environment. Fault tolerance can only be achieved for those errors that are observed. The generic system functions associated with fault tolerance are (1) error detection, (2) damage confinement, (3) error recovery, and (4) fault isolation and reporting. The design of physical resources needed for fault tolerance is discussed in the next chapter.

Error detection is defining possible errors and deviations in the subset of the system's state from the desired state, in the design phase before they occur, and establishing a set of functions for checking for the occurrence of each error. Just as with requirements development, defining error checking to be complete, correct, and independent of the design of the system is desirable. Unfortunately, this is not yet possible, so error detection will be imperfect. The most frequent error detection involves errors in data, errors in process timing, and physical errors in the system's components. The most common checks for data errors include type and range errors. Type checks establish that the data is the right type, for example, Boolean versus integer. Range checks ensure that the value of the data is within a specified range. Knowing the correct values of the data is not possible, so type and range checks are approximations of the checking that would be most effective if the truth were known. Semantic and structural checks are also possible on data elements. Semantic checks compare a data element with the state of the rest of the system to determine whether an error has occurred. Structural checks use some form of data redundancy to determine whether the data is internally consistent. A structural check used in coding is to add extra bits to the data bits; these added bits take on values that depend on the values of the data bits. Later, these extra bits and the associated data bits can be checked to ensure an appropriate relationship exists; if not, an error is declared. Similarly, robust data structures in software use redundancy in the data structures to check for data errors. Timing checks are used in real-time or near-real-time systems. Timing checks assume the existence of a permissible range for the time allotted to some process being performed by the system. A timer is activated within a process to determine whether the completion of the process is within an appropriate range; if not, an error is declared. Hardware systems typically detect timing errors in memory and bus access. Operating systems also use timing checks. Finally, physical errors in a component of the system are the province of BIST and will be discussed in the next chapter.

Damage confinement is needed in fault tolerance because there is typically a time lag between the occurrence of failure and the detection of the associated error. During this time lag, the failure or the implications of the failure may have spread to other parts of the system; error recovery activities are dangerous without having knowledge about the extent of damage due to a failure. As soon as the error detection functionality has declared an error, the damage confinement functionality must assess the likely spread of the problem and declare the portion of the system contaminated by the failure. The most common approach to damage confinement is to build confinement structures into the system during design. "Fire walls" are designed into the system to limit the spread of failure impacts. With these predesigned fire walls declaring that a failure is limited to a specific area of the system when an error is declared is possible. A more sophisticated approach is to reexamine the flow of data just prior to an error to determine the possible spread of errors due to a failure; this sophisticated approach requires not only that error detection functionality be designed into the system but that functionality to record a time history of data be added so that this information exists when the information is needed.

Error recovery functionality attempts to correct the error after the error has been declared and the error's extent defined. If the error concerns data in the system, backward recovery is typically employed to reset the data elements to values that were recorded and acceptable at some previous time. These values may not be correct in the sense that they are the values the system should have generated. Rather, these values are acceptable in the sense of type, range, and semantics discussed above in error detection. The purpose of backward recovery is to keep the system from a major failure, not to restore the system to the correct state. As a result, the system's users are typically notified as part of the error recovery process that a failure occurred and are given a chance to attempt to recover the correct data or restart at an appropriate place to generate the correct data. Forward recovery is an attempt to guess what the correct values of the data should have been; this is dangerous but sometimes justified in real-time systems where backward recovery and user notification are not possible. Timing errors are handled by ending a process that is taking too long and asserting a nominal or last computed value for the process output. Physical errors are handled by either graceful termination of the system's activities or switching to redundant (standby) components when they are available. In recovering from physical errors, capturing the last available values of the system's data structure prior to termination or component switching is critical.

Fault isolation and reporting functionality attempts to determine where in the system the fault occurred that caused the failure that generated the error. To isolate faults, the components of the system must be providing information about their current status.

BIST for a specific component incorporates the functionality to test defined functionality and provide feedback about the results. These types of BIST are common during system start-up and routine operation.

The functional architecture must be expanded during the final development of the allocated architecture to include functions for error detection, damage confinement, error recovery, and fault isolation and reporting. In accordance with the fault tolerance community, these functions should be defined for every state variable of the system, which includes the system's outputs. In addition, though, including error trapping for many of the inputs to the system is important. Error trapping includes functions for error detection, damage confinement, and error recovery for user inputs; the system must monitor system inputs to detect unacceptable inputs and alert the user that a given input is unacceptable and to reenter a correct input. For example, the system expects the user to input a number as part of a menu selection or data entry task. However, the user, due to inattention or typing error, enters a letter instead. Most older software would immediately crash, sometimes crashing the entire computer system. However, more recent, well-designed software will monitor the input for such an error and alert the user that this error has been made and request a new input.

7.7 TRACING REQUIREMENTS TO ELEMENTS OF THE FUNCTIONAL ARCHITECTURE

There are two elements of the functional architecture that should have input/output requirements traced to them: the functions and the external items (inputs and outputs). Both tracings can be accomplished in systems engineering tools such as CORE and GENESYS. All elements of the set of input/output requirements should be traced to appropriate functions that have been defined in the functional decomposition. Tracing input requirements and output requirements to functions should be done throughout the functional decomposition, as shown in Figure 7.14; this tracing is guided explicitly by the association of inputs and outputs with functions in the functional architecture. For example, since "calls (requests) for up and down service" is an input of "Support Waiting Passengers," all of the requirements related to this input should be traced to the function "Support Waiting Passengers" and that function's predecessors in the functional decomposition. Similarly, external interface requirements should be traced to the function that is associated with receiving the input or sending or output, respectively. For example, the phone line (external interface) transmits and receives items that are associated with the function "Support Passengers in Emergency"; therefore, the external interface requirement to use a phone line to communicate via the building with maintenance personnel should be traced to this function. Each external interface requirement should also be traced to the predecessors of this function. Finally, all of the functional requirements should be traced to the top-level system function. As discussed in Chapter 6, a preferred convention for the functional requirements is to list the functions in the top-level functional decomposition that define the system function. This tracing of input/output requirements to functions is illustrated in Figure 7.14 for a sample of functions and requirements from the elevator case study, which can be downloaded from the companion website.

The logic for tracing input/output requirements to functions is as follows. The ultimate product of the systems engineering team is a set of specifications for each CI. Intermediate products are specifications for the intermediate components that comprise the system and are built from the CIs. Each of these specifications will contain requirements that are derived from the system-level requirements that are derived from the stakeholders' requirements. In addition, each of these specifications will contain a functional architecture that is relevant to the component or CI of interest. This functional architecture for a component or CI will be a subset of the system's functional architecture and will contain input/output requirements traced to these functions at the system level. These input/output requirements should be contained in the specification. Tracing system input/output requirements to functions is a method for ensuring that the appropriate input/output requirements are contained in each specification that has to be developed during the design process.

In addition, tracing input/output requirements to functions serves as a consistency check. Does each function have requirements traced to it for each Input and output? Is each input/output requirement traced to at least one function?

The input and output requirements are also traced to the external item elements. This tracing is made explicit in the set of input and output requirements for the operational phase of the elevator. The rationale for tracing the input and output requirements to external items is that the external interfaces need to satisfy these requirements. The internal items of the functional architecture will also have the relevant input and output requirements traced to them later in the design phase so that the internal interfaces of the system will have derived requirements that they must meet. This tracing

Functions	Input/output Requirements (A Sample)					
	Input Requirements		Output Requirements		Functional Requirement	External Interface Requirement
	The Elevator System Shall Receive Calls for up and Down Service from all Floors of The Building	The Elevator System Shall Receive Passenger Activated Fire Alarms in each Elevator Car	The Elevator System Shall Provide Adequate Illumination	The Elevator System Shall Open and Close Automatically Upon Arrival at each Selected Floor	The Elevator System Shall Control Elevator Cars Efficiently	The Elevator System Shall Use a Phone Line from the Building for Emergency Calls
0 Provide Elevator Services	X	X	X	X	X	X
1 Accept Passenger Requests + Provide Feedback	X	X				X
1.1 Support Waiting Passengers	X					
1.2 Support Riding Passengers			X			
1.3 Support Passengers in Emergency		X				X
2 Control Elevator Cars					X	
3 Move Passengers between Floors						
3.1 Receive + Discharge Passengers			X	X		
3.2 Travel to Next Stop				X		
3.3 Provide Comfortable Atmosphere			X			
4 Enable Effective Maintenance and Servicing						

FIGURE 7.14 Tracing a sample of input/output requirements to a sample of functions.

can provide a valuable consistency check: Does each item have at least one requirement traced to it? Also, does each input/output requirement trace to some item? If either of these questions is negative for any requirement or item, there has been a breakdown in the requirements development process. Finally, an item will be "transferred by" a link, which "comprises" an interface. The item will have one or more input/output requirements traced to it. In addition, the link will ultimately have derived system-wide requirements traced to it. The interface specifications will be built from the requirements that are traced to the items being transferred by the links comprising the interface as well as the system-wide requirements that ultimately are traced to the interface.

7.8 SUMMARY

The functional architecture of a system, as defined in this chapter, contains a hierarchical model of the functions performed by the system, the system's components, and the system's CIs; the flow of informational and physical items from outside the system through the system's functions and on to the waiting external systems being serviced by the system; and a tracing of input/output requirements to both the system's functions and items.

This chapter introduces quite a few terms that are key to understanding and developing a functional architecture. A system mode is an operational capability of the system that contains either full or partial functionality. A state is a modeling description of the status of the system at a moment in time. A function is an activity or task that the system performs in order to transform an n-tuple of inputs into an m-tuple of outputs. These concepts are key to the development of a functional architecture. The system's modes and functions should be part of the functional architecture, while the system's state should be definable by a set of parameters in any operational mode while performing any set of functions. The parameters that comprise this state may vary based on the operational mode and the functions being performed.

Other key terms addressed in this chapter include failure, error, and fault. Failure is a deviation between the system's behavior and the system's requirements. An error is a problem with the state of the system that may lead to a failure. A fault is a defect in the system that can cause an error. To achieve the desired level of fault tolerance, the system must perform the functions of error detection, damage confinement, error recovery, and fault isolation and reporting.

A method for developing a functional architecture was defined in this chapter. Defining the functional architecture is not easy and is a modeling process that the engineer of a system must learn. The modeling process uses a combination of decomposition and composition. The concepts of feedback and control are critical to defining the system's functions.

The engineering of a system has to rely upon more than the physical design of the system. The functions or activities that the system has to perform are a critical element of the design process, and the design of these functions needs to be given equal importance to the physical design by the engineers. The designs of functions and physical resources for the system are not independent; they must both be done, usually in parallel.

PROBLEMS

7.1 What are the operating modes of your car's stereo system?

7.2 For the ATM of the Money Mart Corporation:

    i. As part of the systems engineering development team, develop a functional architecture using a systems engineering modeling technique. The functional architecture should address all of the functions associated with the ATM. This functional architecture should be at least two levels deep and should be four levels deep in at least one functional area that is most complex. Note that you will be graded on your adherence to proper model semantics and syntax, as well as the substance of your work.

ii. Pick three scenarios from the operational concept and describe how these scenarios can be realized within your functional architecture by tracing functionality paths through the functional architecture. Start with the external input(s) relevant to each scenario and show how each input(s) is(are) transformed by tracing from function to function at various levels of the functional decomposition, until the scenario's output(s) is(are) produced. Highlight with three different colors (one color for each scenario) the thread of functionality associated with each of these three scenarios.

   If your functional architecture is inadequate, make the appropriate changes to your functional architecture.

iii. As part of the systems engineering development team for the ATM, update your requirements document to reflect any insights into requirements that you obtained by creating a functional architecture. That is, if you added, deleted, or modified any input, controls, or outputs for the system, modify your input/output requirements. Also, update your external systems model if any changes are needed.

7.3 For the OnStar system of Cadillac:

i. As part of the systems engineering development team, develop a functional architecture. The functional architecture should address all of the functions associated with OnStar. This functional architecture should be at least two levels deep and should be four levels deep in at least one functional area that is most complex. Note that you will be graded on your adherence to proper modeling semantics and syntax, as well as the substance of your work.

ii. Pick three scenarios from the operational concept and describe how these scenarios can be realized within your functional architecture by tracing functionality paths through the functional architecture. Start with the external input(s) relevant to each scenario and show how each input(s) is(are) transformed by tracing from function to function at various levels of the functional decomposition, until the scenario's output(s) is(are) produced. Highlight with three different colors (one color for each scenario) the thread of functionality associated with each of these three scenarios.

   If your functional architecture is inadequate, make the appropriate changes to your functional architecture.

iii. As part of the systems engineering development team for OnStar, update your requirements document to reflect any insights into requirements that you obtained by creating a functional architecture. That is, if you added, deleted, or modified any input, controls, or outputs for the system, modify your input/output requirements. Also, update your external systems model if any changes are needed.

7.4 For the development system for an airbag system:

i. As part of the systems engineering development team, develop a functional architecture. The functional architecture should address all of the functions associated with the development system for an airbag. This functional architecture should be at least two levels deep and should be four levels deep in at least one functional area that is most complex. Note that you will be graded on your adherence to proper modeling semantics and syntax, as well as the substance of your work.

ii. Pick three scenarios from the operational concept and describe how these scenarios can be realized within your functional architecture by tracing functionality paths through the functional architecture. Start with the external input(s) relevant to each scenario and show how each input(s) is(are) transformed by tracing from function to function at various levels of the functional decomposition, until the scenario's output(s) is(are) produced. Highlight with three different colors (one color for each scenario) the thread of functionality associated with each of these three scenarios.

If your functional architecture is inadequate, make the appropriate changes to your functional architecture.

iii. As part of the systems engineering development team for the development system for an airbag, update your requirements document to reflect any insights into requirements that you obtained by creating a functional architecture. That is, if you added, deleted, or modified any input, controls, or outputs for the system, modify your input/output requirements. Also, update your external systems model if any changes are needed.

7.5 For the manufacturing system for an airbag system:

i. As part of the systems engineering development team, develop a functional architecture. The functional architecture should address all of the functions associated with the manufacturing system for an airbag. This functional architecture should be at least two levels deep and should be four levels deep in at least one functional area that is most complex. Note that you will be graded on your adherence to proper modeling semantics and syntax, as well as the substance of your work.

ii. Pick three scenarios from the operational concept and describe how these scenarios can be realized within your functional architecture by tracing functionality paths through the functional architecture. Start with the external input(s) relevant to each scenario and show how each input(s) is(are) transformed by tracing from function to function at various levels of the functional decomposition, until the scenario's output(s) is(are) produced. Highlight with three different colors (one color for each scenario) the thread of functionality associated with each of these three scenarios.

If your functional architecture is inadequate, make the appropriate changes to your functional architecture.

iii. As part of the systems engineering development team for the manufacturing system for an airbag, update your requirements document to reflect any insights into requirements that you obtained by creating a functional architecture. That is, if you added, deleted, or modified any input, controls, or outputs for the system, modify your input/output requirements. Also, update your external systems model if any changes are needed.

Chapter 8

Physical Architecture Development

8.1 INTRODUCTION

The physical architecture of a system is a hierarchical description of the resources that comprise the system. This hierarchy begins with the system and the system's top-level components and progresses down to the configuration items (CIs) that comprise each intermediate component. The CIs can be hardware or software elements or combinations of hardware and software, people, facilities, procedures, and documents (e.g., user manuals).

Section 8.2 introduces the distinction between a generic and instantiated physical architecture. Note that the Object Management Group (OMG) Systems Modeling Language (SysML) standard uses the term "logical architecture" for what we denote as the "generic physical architecture." The OMG SysML standard usage of "physical architecture" is what we denote as the "instantiated physical architecture."

The generic physical architecture defines the hierarchy in general terms, for example, two processors with associated software, a person, and a building. The instantiated physical architecture lays out the specifics of the processors, software, person, and building in enough detail to permit performance modeling of the system related to the requirements being addressed. The intent of systems engineers should not be to design these components but rather to state representative instantiations for the generic components that are sufficient to model the performance of the system and ensure that the requirements decomposition process makes sense.

Section 8.3 defines a method for developing alternatives for the generic and instantiated physical architectures of the system. The development process proposed here emphasizes multiple alternatives, especially for the instantiated physical architecture, based on the supposition that the design process is quite difficult for even moderate extensions of existing systems. The following quote by Guindon [1990, p. 308] expresses the importance of this approach:

> System design often involves novelty. Even though the designer may be thoroughly familiar with the design process itself, there may not be any precedent in the literature for the system to be designed – it may be a new technology. More frequently, the system may simply involve some novelty in an otherwise

The Engineering Design of Systems: Models and Methods, Fourth Edition. Dennis M. Buede and William D. Miller
© 2024 John Wiley & Sons, Inc. Published 2024 by John Wiley & Sons, Inc.
Companion website: www.wiley.com/go/engineeringdesignofsystems4e

well-understood problem. The novelty may range from a novel combination of requirments for a familiar type of system in a familiar problem domain. As a consequence, there is often no predetermined solution path from the requirements to the finished artifact [Newell, 1969; Nii, 1986; Reitman, 1965; Rittel, 1972; Simon, 1973]. Thus, system design frequently requires the creation of new solutions interleaved with the application of known solutions.

Section 8.4 introduces some creativity techniques to aid in the development of the alternate physical architectures. The morphological box is the primary technique employed and illustrated in this chapter. The morphological box dates back to the 1940s and breaks a system into segments as defined by the generic physical architecture and then provides for the listing of alternate instantiated physical components for each segment. Other techniques that have been proposed and utilized are classified as either brainstorming or brainwriting and are also discussed. See West [2007]. Selecting one or more instantiated components from each component produces an alternative for an instantiated physical architecture for the system.

Engineers commonly resort to describing the system's architecture in a nonmathematics-based graphical format. Block diagrams, the commonly used and non-standardized graphical format, are presented in Section 8.5 to represent the physical coupling of the system's components. A block diagram provides a box or block for each component. The links between the blocks represent the major flows of energy, material, or information between the components represented by the blocks.

Section 8.6 addresses major issues and associated concepts in the development of physical architecture. The concepts of centralized and decentralized, distributed, and client-server architectures are discussed and illustrated. Also, redundancies in hardware, software, information, and time are discussed as ways to achieve fault tolerance via the physical architecture.

The exit criterion for the development of the physical architecture is the provision of a single physical architecture that is satisfactory in terms of detail, quantity, and quality for development of the allocated architecture. This satisfaction with detail, quantity, and quality is typically preceded by the creation of several alternate physical architectures for consideration during the development and refinement of the allocated architecture.

8.2 GENERIC VERSUS INSTANTIATED PHYSICAL ARCHITECTURES

The *physical architecture* provides resources for every function identified in the functional architecture. Since every phase of the life cycle is addressed in the requirements and is being addressed in the functional architectures, there must be a physical architecture for each system associated with the system's life cycle. Recall the sample physical architecture from Chapter 1 (repeated here as Fig. 8.1). Note that this physical architecture includes the vehicle, the support resources for the vehicle during the operational and maintenance phases, and the training resources, which may be training for the operational phase or the training phase. Also, note that even at the third level of the physical architecture, the components are combinations of hardware, software, and other devices.

Military standard MIL-STD-881F [2022] contains a Work Breakdown Structure (WBS) for Defense Material Items. The WBS is often very similar to the physical architecture because the work is organized along the lines of the resources that require development or procurement. For an aircraft system, there are 10 elements that partition the system, as shown in the first column of Table 8.1. These elements span six of the seven life cycle phases (shown in the second column) defined in Chapter 1. The only phase that is absent from this list is retirement, the commonly forgotten phase.

In the same military standard, 17 resource categories, shown in Table 8.2, are defined as a partition of the generic air vehicle. These lists or partitions of the resources for the physical architecture are most useful as memory joggers. For some aircraft, some of these elements are not relevant,

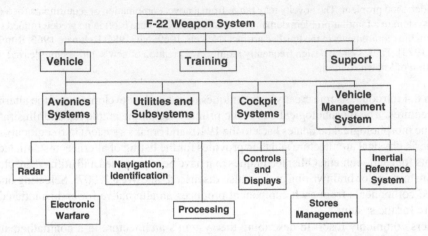

FIGURE 8.1   Sample physical architecture (F-22 Type A Spec). (Adapted from Reed [1993].)

TABLE 8.1   WBS Elements and Related Life Cycle Phases for a Generic Aircraft System (Level 2)

WBS Elements	Life Cycle Phase
Air vehicle	Operational
Payload/mission system	Operational
Ground/host segment	Operational
Aircraft system software release	Operational
Systems engineering	Development
Program management	Development
System test and evaluation	Development
Training	Training
Data	Manufacturing and refinement
Peculiar support equipment	Operational
Common support equipment	Operational
Operational/site activation	Deployment
Industrial facilities	Manufacturing
Initial spares and repair parts	Operational

TABLE 8.2   WBS Elements for a Generic Air Vehicle (Level 3)

Air frame
Propulsion
Vehicle subsystems
Avionics
Armament/weapons delivery
 Auxiliary equipment Furnishings and equipment
Air vehicle software release
Other air vehicle

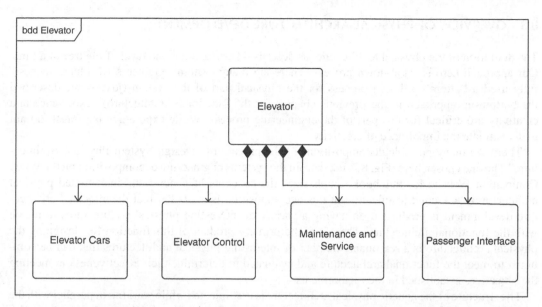

FIGURE 8.2 Generic physical architecture from the elevator case study.

and for example, airlift aircraft do not need armament or antisubmarine warfare capability. More importantly, as technology advances, some of these elements are outdated. With the advent and advance of distributed computing, the central computer element is not relevant or misleading. In addition, at this level of the physical architecture, it is often too early to separate hardware and software.

Common resource categories for an aircraft have been described in Figure 8.1 and Tables 8.1 and 8.2. The resource categories for the elevator's physical architecture from the case study are shown in Figure 8.2. (This elevator material can be downloaded from the companion website.) All these resource categories are examples of a generic physical architecture. A *generic physical architecture* is a description of the partitioned elements of the physical architecture without any specification of the performance characteristics of the physical resources that comprise each element (e.g., central processing unit). An *instantiated physical architecture* is a generic physical architecture to which complete definitions of the performance characteristics of the resources have been added. An instantiated physical architecture for the elevator system would be specific about the call announcement component (e.g., liquid crystal lights), destination control (e.g., push buttons), and the like.

One element that is left out of most physical architectures is the set of procedures that are developed for the users of the system to follow. These *procedures* are explicit operating, maintenance, or support instructions provided in a user's or operator's manual. These manuals usually accompany the system when the system is delivered. These procedures are the focus of attention during the training that is delivered to the users, maintainers, or supporters of the system. Systems engineers should not forget or ignore this element of the system's physical architecture, as was done with the initial airbag system that was described as a case study in Chapter 6. After the serious, and often deadly, effects on children and small adults were noticed, a series of procedures for the placement (or lack thereof) of children and small adults in the front seat were released. Common practice in the development of a system is to accommodate problem issues identified during the qualification of the system (see Chapter 11) by amending and expanding the procedures defining how the system will be used. *Procedures such as these represent the way in which the system's functionality moves from the system under development to the users.*

8.3 OVERVIEW OF PHYSICAL ARCHITECTURE DEVELOPMENT

The definition of the physical architecture, as described here, is done one level of the tree at a time. Our approach here is a top-down process. There are many systems engineers who have successfully used a bottom-up design process for the physical part of the system (just as we described the bottom-up approach in the previous chapter for the functional architecture). Experience and creativity are critical for this part of the engineering process. While experience is a must, do not underestimate the importance of creativity.

There are many possible decompositions of the process of "Design System Physical Architecture." The one chosen here (Fig. 8.3, taken from the systems engineering decomposition on the Wiley Companion Website for this book) emphasizes the concepts of generic and instantiated physical architectures. A second justification of this decomposition is the belief that the allocated architecture development is predicated on having a variety of interesting physical architectures to match with the functional architecture. Therefore, the primary product of this function for designing the physical architecture is a reasonable number of interesting physical architectures that can be combined to meet the functional architecture and evaluated to determine their effectiveness in meeting the objectives established in the requirements.

The structure of the generic physical architecture is first selected while working in parallel with the development of the functional architecture. As discussed in Chapter 7 and elaborated on in Chapter 9, there are great advantages in defining the internal interfaces of the system to have the functional and physical architectures match; that is, enable a one-to-one and onto allocation of functions to components. See Figure 8.4 to review the distinctions between a relation and a function and the additional restrictions for a function that is one-to-one and onto. While there are many advantages to a one-to-one and onto mapping of functions and components, this may not always be possible and should not be forced.

It is generally easier to obtain a one-to-one mapping between components and functions at the top of the design decomposition than at the bottom. It is also more important to do so at the top because each component is going to be handed off to a different design team (usually) as the next phase of the design begins, and it is best that the teams can proceed independently of each other.

It is very difficult to obtain a one-to-one mapping on inherently networked systems. In these systems, an object-oriented bottom-up design usually makes a lot more sense. However, a top-down effort is needed to make sure the bottom-up approach is not leaving anything out, which is a common failure of object-oriented designs. Socio-technical systems tend to be more networked and therefore are going to be harder to achieve the one-to-one goal. The top-down design process can work fine if functions can be aggregated or repartitioned to achieve the one-to-one mapping, but this is still difficult.

First, a generic physical architecture must be developed. The generic physical architecture provides common designators for physical resources in a hierarchical decomposition that partitions the system into greater and greater detail. Although this generic physical architecture has no substance in the sense of specific physical items, this structure is still very important. Some instantiated physical architectures can be eliminated from consideration just based on the division of the system into components. Therefore, serious thought and creativity should be devoted to this initial task.

The second function in the decomposition addresses the creation of a morphological box to assist in generating a set of creative instantiated architectures to analyze during the development of the allocated architecture. A *morphological box* is a matrix in which the columns (or rows) represent the components in the generic physical architecture. The boxes in a given column (or row) then represent alternate choices for fulfilling that generic component. Each option should have well-defined performance (and cost) characteristics. Section 8.4.1 describes the morphological box in more detail and provides several examples.

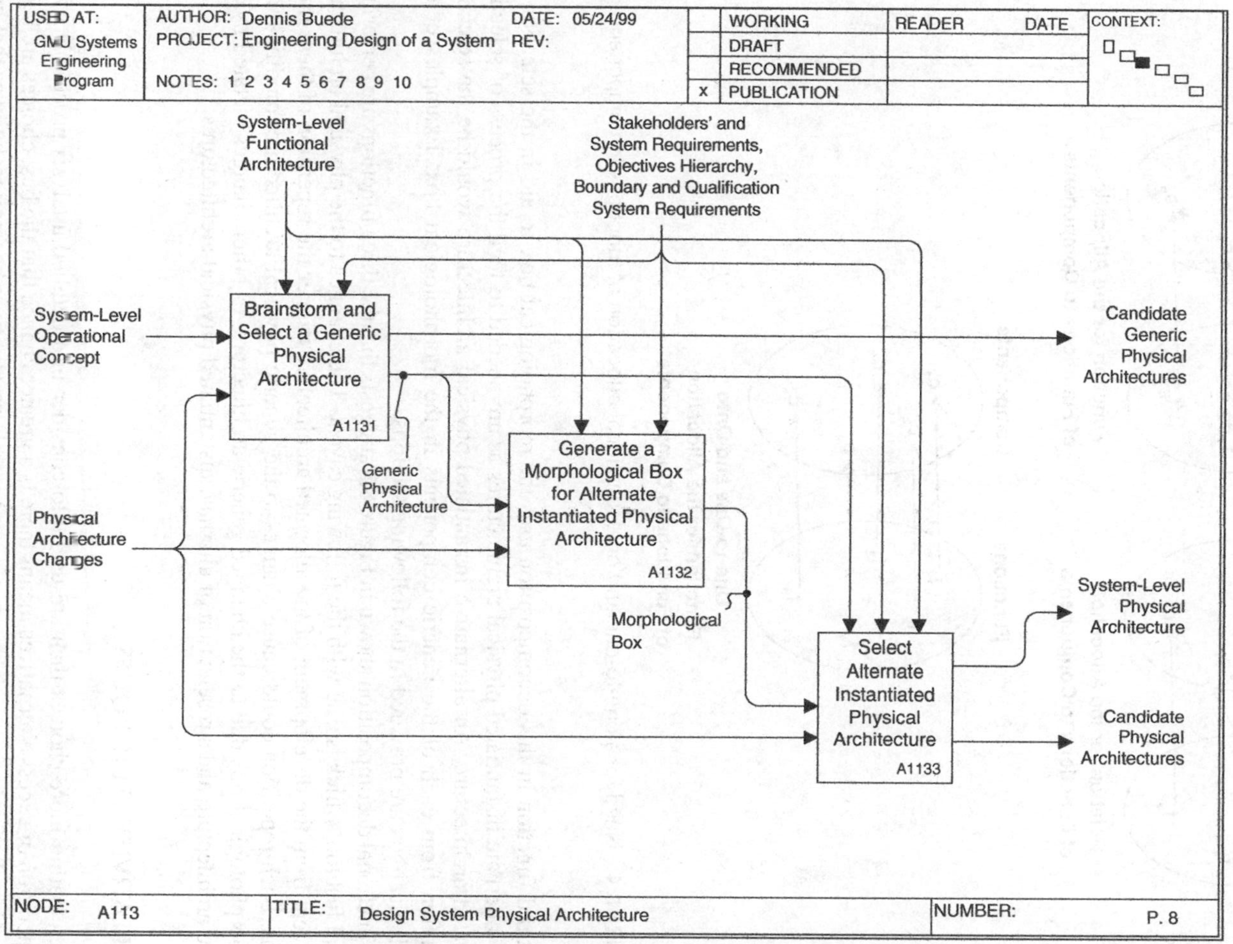
USED AT:
GMU Systems Engineering Program
AUTHOR: Dennis Buede
PROJECT: Engineering Design of a System
NOTES: 1 2 3 4 5 6 7 8 9 10
DATE: 05/24/99
REV:
WORKING
DRAFT
RECOMMENDED
x PUBLICATION
READER
DATE
CONTEXT:
System-Level Functional Architecture
Stakeholders' and System Requirements, Objectives Hierarchy, Boundary and Qualification System Requirements
System-Level Operational Concept
Brainstorm and Select a Generic Physical Architecture
A1131
Candidate Generic Physical Architectures
Generic Physical Architecture
Generate a Morphological Box for Alternate Instantiated Physical Architecture
A1132
Physical Architecture Changes
Morphological Box
Select Alternate Instantiated Physical Architecture
A1133
System-Level Physical Architecture
Candidate Physical Architectures
NODE: A113
TITLE: Design System Physical Architecture
NUMBER:
P. 8

FIGURE 8.3 Development process for the physical architecture.

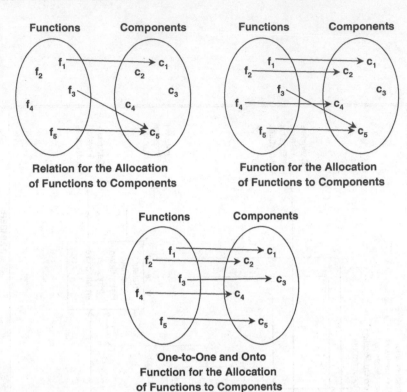

FIGURE 8.4 Need for a one-to-one and onto functional allocation of functions to components.

The third function in this decomposition uses the morphological box to aid in the selection of as many alternate instantiated physical architectures as are needed to feed the process of selecting an allocated architecture. An alternative instantiated physical architecture would be the selection of an option from each of the generic components in the morphological box. Examples of the morphological box are provided in the following sections.

The functional decomposition shown in Figure 8.3 suggests that the three functions are performed in a serial fashion, which is true with the following caveat. The changes to the physical architecture that are sent from the development of the allocated architecture trigger the repetition of these three functions. Each repetition could cause changes to the generic physical architecture, modifications to the morphological box due to the changed generic architecture, or other changes dictated by the allocated architecture and a reselection of alternate instantiated physical architectures.

8.4 CREATIVITY TECHNIQUES

Initially, creating more choices than are useful to consider in a detailed analysis process is wise. This generation of excess alternatives means there is a greater chance that the best choices are being considered in the final analysis. There are many possible creativity-enhancing techniques that have been used by engineers to develop new and interesting solutions to old and new problems. This section begins by focusing on one technique, the morphological box, that has proven useful several times. Then, a larger review of techniques is provided.

8.4.1 Morphological Box

Originally proposed by Zwicky [1969] during World War II and then expanded by Allen [1962], morphological analysis (more commonly known in some disciplines as morphological box) divides a problem into segments and posits several solutions for each segment. In the two-dimensional version, a table is created with columns (or sometimes rows) pertaining to the generic components of the physical architecture. Then, the elements of each column are filled with competing specific instantiations of each component. The instantiations in each column need not fit together; in fact, each column corresponds to a section of a cafeteria (e.g., salads, vegetables, meat, and deserts). A meal would then consist of a selection from each section of the cafeteria. A system's instantiated physical architecture, analogously, is a selection of one box from each column (generic component) of the morphological box. As part of the morphological analysis, each instantiation (one from each column) will be based upon a subset of the system's objectives. For example, one subset of objectives might be low cost; another, high-speed performance; and a third, high usability. Each of these instantiations is, in fact, a theme for the design of the system.

Table 8.3 presents a morphological box (generic components and choices) for a hammer. This morphological box contains five generic components of a hammer: the length of the handle, the material from which the handle is made, the size and surface of the head of the hammer used for striking, the weight or density of the hammer head, and the angle associated with the head of the hammer used for removing nails. Any hammer is one cell from each of the five columns. For example, one hammer design is obtained by taking the top cell of each column: 8-inch handle made of Fiberglass with a rubber grip using a 1-inch diameter flat steel head that weighs 12 oz. and has a steel claw that is nearly perpendicular to the handle. There are $2 \times 5 \times 4 \times 4 \times 2 = 320$ different possible hammers defined in this table, assuming none of the combinations are infeasible. Yet when you go to the hardware store, there may be only a dozen choices. For real systems, there are usually millions of possible combinations. Yet many design teams only consider one or two in any detail, making it very likely that they are missing several creative, high-quality designs. The big advantage of the morphological box is that it forces the design team to recognize that there are many possible solutions to the design problem. The conversation about which design alternative best satisfies the requirements follows naturally.

While the morphological box is a simple concept, there are a number of subtle issues that need to be addressed. First and obviously, there should be at least one column in the morphological box for each generic component in the physical architecture. There are certain situations in which one of the generic components may have two or more columns associated with the generic component; these would be the decomposed generic components of the higher-level component.

TABLE 8.3 Morphological Box for a Hammer

Handle Size	Handle Material	Striking Element	Weight of Hammer Head	Nail Removal Element
8 inches	Fiberglass with rubber grip	1-inch diameter flat steel	12 oz	Steel claw at nearly a straight angle
22 inches	Graphite with rubber grip	1-inch diameter grooved steel	16 oz	Steel claw at a 60° angle with handle
	Steel with rubber grip	1.25-inch diameter flat steel	20 oz	
	Steel I-beam encased in plastic with rubber grip	1.25-inch diameter grooved steel	24 oz	
	Wood			

Second, there is no requirement that each generic component have the same number of options. Clearly, there is value in having at least two choices for any generic component; otherwise, that particular generic component has been fixed. Using some of the brainstorming or brainwriting techniques discussed in Section 8.4.2 is common to develop additional alternatives (boxes) for each generic component (column of the morphological box). There is a great advantage to generating a creative set of choices for any generic component, even if some of the choices are never selected in the final set of alternate instantiated physical architectures.

In addition, there are situations in which it is wise to permit more than one choice from a generic component to be selected for a single instantiated physical architecture. This possibility of selecting several choices in a single generic component for a single instantiated physical architecture usually does not make sense for a central component in the architecture. However, there are often generic components associated with the "bells and whistles" of the system. An example would be the list of peripherals that can be added to a computer or an automobile. There is some efficiency in grouping all of these under one generic component for the system rather than having a generic component for each of the possible peripherals.

Figure 8.5 provides another example of a morphological box; this example describes alternate designs for an automobile navigation support system. A number of automakers provide such navigation support systems as peripherals (or extras). In addition, several peripheral companies provide such navigation support systems that can be added to any automobile. In general, these navigation support systems provide the driver and passengers with information about where they are on the highway and how to get where they want to go. However, there are extras that can be provided, as shown in the last column, "Other System Interfaces." These extras include the ability to have the car doors unlocked when the owner has locked him/herself out, notify the police or emergency service if the airbag deploys, and activate the lights and horn externally if the driver has lost the car in a

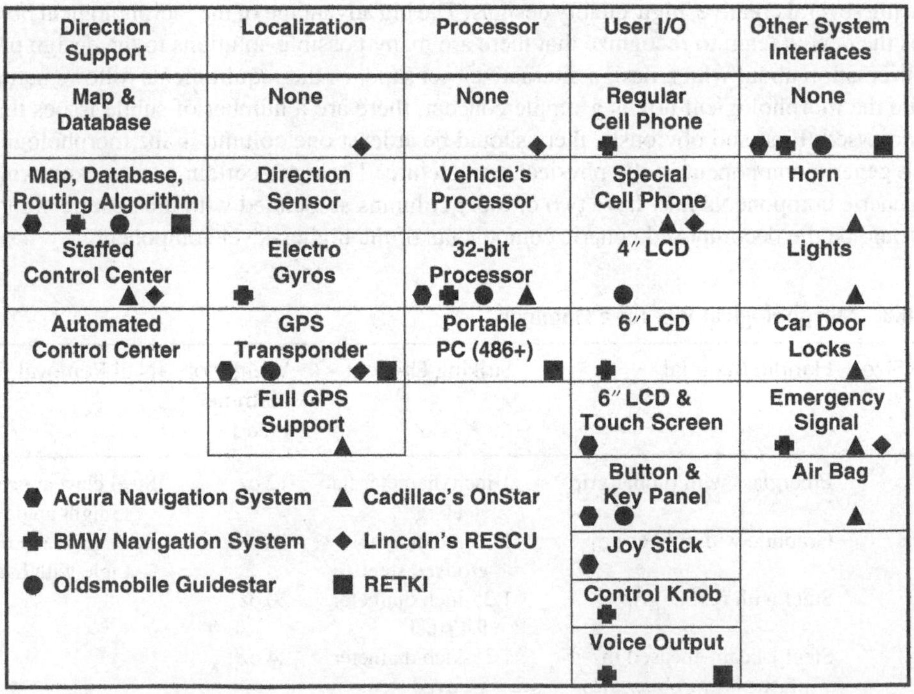

Direction Support	Localization	Processor	User I/O	Other System Interfaces
Map & Database ◆	None ◆	None ◆	Regular Cell Phone ✚	None ◆ ✚ ● ■
Map, Database, Routing Algorithm ◆ ✚ ● ■	Direction Sensor	Vehicle's Processor	Special Cell Phone ▲ ◆	Horn ▲
Staffed Control Center ▲ ◆	Electro Gyros ✚	32-bit Processor ◆ ✚ ● ▲	4″ LCD ●	Lights ▲
Automated Control Center	GPS Transponder ● ● ◆ ■	Portable PC (486+) ■	6″ LCD ✚	Car Door Locks
	Full GPS Support ▲		6″ LCD & Touch Screen ●	Emergency Signal ✚ ▲ ◆
			Button & Key Panel ✚ ●	Air Bag ▲
			Joy Stick ●	
			Control Knob ✚	
			Voice Output ✚ ■	

◆ Acura Navigation System   ▲ Cadillac's OnStar
✚ BMW Navigation System   ◆ Lincoln's RESCU
● Oldsmobile Guidestar   ■ RETKI

FIGURE 8.5 Morphological box for automobile navigation support system.

parking lot. Selecting more than one option in the second to last column is also possible; this column represents the generic component associated with the user interface for the navigation support system. The selection of multiple boxes is also common for user interface generic components.

There is one major caution that must be provided in the development of a morphological box. The system concept has to be narrowed down to some degree before it is possible to define a single morphological box. For example, if the system is a substantial computer system, a morphological box cannot be defined before an architecture for the computer system has been selected. For example, suppose the alternate computer system architectures were a client-server, a mainframe, or a distributed processing architecture connected via several local area networks (LANs). The generic components that are applicable to a client-server architecture may not be consistent with those generic components for a mainframe system or a distributed network. Therefore, the design process should narrow the computer system architecture down to a client-server or mainframe before developing a morphological box.

Once a reasonable number of possible choices for each component of the physical architecture have been identified, identifying infeasible combinations is wise. Friend and Hickling [1987] have defined a graphical representation to highlight pair wise infeasible choices across two generic components. Each generic component is shown as a circular node in a graph. The specific choices for a generic component are shown as pie-shaped wedges in the relevant generic component's node. An infeasible combination of choices from two distinct generic components is shown as a line between those options. Pairwise infeasible combinations may also be identified and documented in table format.

Pairwise examples of infeasible combinations are shown in Figure 8.6 for the morphological box of the hammer, as shown in Table 8.3. In this hypothetical example, the line segment from angled nail removal feature to 22-inch handle denotes an infeasible combination; an angled nail removal claw cannot be placed on a 22-inch handle because too much stress would be focused at the intersection of the handle and hammer's head. The second line segment shown between the 22-inch handle and a wood handle eliminates the ability of the user to apply too much force for the wood handle to absorb.

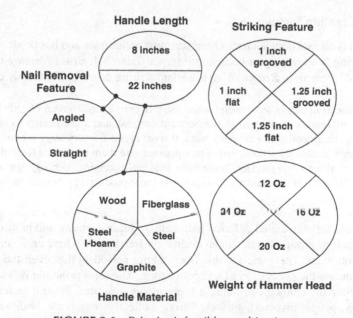

FIGURE 8.6 Pairwise infeasible combinations.

TABLE 8.4 VanGundy's Typology of Brainwriting and Brainstorming

Brainwriting and Brainstorming Categories	Examples
Brainwriting I – an individual works alone to create a list of ideas	Analogy, attribute listing, people involved
Brainwriting II – a group of individuals separated in space generates ideas separately and the ideas are collected but not shared	Collective notebook
Brainwriting III – a group of individuals separated in space generates ideas separately, the ideas are shared and additional ideas are generated	Delphi method
Brainwriting IV – a group of individuals working in the same room generates ideas separately and the ideas are collected but not shared and no discussion takes place	Nominal Group Technique
Brainwriting V – a group of individuals working in the same room generates ideas separately; all of the ideas are shared but none are discussed; additional ideas are generated	Brainwriting pool
Brainstorming I – a group of individuals generates ideas via verbal discussion, no defined procedure is used	Unstructured group discussion
Brainstorming II – a group of individuals generates ideas via verbal discussion within the bounds of predefined procedures	Classical brainstorming
Brainwriting/Brainstorming I – a group of individuals generates ideas via predefined written and verbal procedures	Brainwriting game

These two-line segments reduce the total number of choices from 320 to 224; the 8-inch handle still retains 160 possible combinations, but the 22-inch handle only has 64 possible combinations – any of the four striking surfaces with any of the four weights with the one nail removal generic component with four of the five possible handle material generic components.

8.4.2 Option Creation Techniques

VanGundy [1988] is an excellent source of brainstorming techniques and has produced a typology of techniques involving brainwriting or brainstorming, see Table 8.4. *Brainstorming* is the generation of ideas via verbal interaction. *Brainwriting* is a silent writing process. VanGundy claims:

> Brainstorming, for example, is most useful when there is only a small group of individuals, time is plentiful, status differences among group members are minimal, and a need exists to verbally discuss ideas with others. Brainwriting, on the other hand, is most useful for very large groups, when there is little time available, status differences need to be equalized, and there is no need for verbal interaction. In addition, brainwriting often produces more ideas than brainstorming, although the uniqueness and quality of these ideas might or might not be superior to those produced by brainstorming. [VanGundy, 1988, p. 75]

A common characteristic, called deferred judgment, of brainstorming and brainwriting exercises is that the individual or group operates in an evaluation-free period where criticism and discussion in general are prohibited. The logic for this freethinking period is that even the most preposterous idea may stimulate the generation of a superior idea. A second principle is that the more ideas generated, the better the chance of finding a high-quality solution. Several techniques discussed below are analogy, people involved, attribute listing, collective notebook, brainwriting game, and brainwriting pool.

Analogies are often used in systems engineering because building upon our experiences with previous systems has a great deal of creative power. An example of an analogy would be to use the 17 elements of the generic aircraft in Table 8.2 to develop the physical architecture of an automobile, an air traffic control system, or an elevator system. Using the physical architecture from a system recently developed as an analogy for a new generation product is another example of analogic reasoning. The use of analogies for generating ideas is by far the most common, efficient, and highly recommended; however, left unchecked, analogic reasoning can produce the most disastrous results.

Examining the system's physical architecture considering the stakeholders (*people involved*) affected by the use and maintenance of the system can be useful in defining the physical architecture for the operational phase. Remember, though, that the entire life cycle of the system must be addressed, so there will be physical architectures for the manufacturing, deployment, and training phases as well.

Attribute listing dates to the 1930s and is based on the concept that physical architectures can all be traced to modifications of previous architectures. Once the requirements and objectives of the system have been developed and a generic physical architecture has been created, the individual defines a feasible (or nearly feasible) instantiation of the generic physical architecture. Then, without detailed evaluation, the systems engineer systematically modifies the characteristics of the instantiated physical architecture with key objectives of the system in mind. For example, VanGundy provides the following example for a hammer:

> To develop a better hammer, for example, the following parts could be listed: (1) straight, wooden, varnished handle; (2) metal head with round striking surface on one end and a claw on the other; and (3) metal wedge in the top of the handle to secure the head to the handle. Of these parts, the basic attributes of the handle shape/composition and the metal wedge could be selected for possible modification. The handle could be constructed of fiberglass, wrapped with a shock-absorbing material, and shaped to better fit the human hand; the metal wedge could be modified by replacing it with a synthetic, pressure-treated bonding. [VanGundy, 1988, p. 88]

Morphological analysis (sometimes called matrix analysis) results in a morphological box, which is a systematic extension of attribute listing. This topic was discussed in detail with the examples above.

Haefele [1962] of the Proctor and Gamble Company developed the Collective Notebook. Each participant in this group-oriented technique keeps a notebook of ideas over a relatively long time period to solve a specified problem; Haefele suggested one month. Each participant is to add one idea each day. At the end of the idea collection period, each participant reviews her own ideas and selects the best one; ideas needing more research or other good ideas that may relate to other problems are annotated. A coordinator, who collects this summary information and the notebooks, creates a detailed synopsis of the ideas generated that can then be reviewed by the participants.

The *brainwriting game* uses competition among the participants to create the most improbable solution in hopes that this competition will generate the best solution. First, the design problem is presented to the group. Each participant buys a specified number of blank, numbered cards. The participant places her initials on her cards and then writes an idea that she hopes will win the prize for the most improbable solution. All of the cards are then displayed to the entire group. Participants then individually write more practical solutions based upon concepts taken from the cards detailing improbable solutions. After the practical solutions are collected, the group votes on the winner of the most improbable solution. Finally, subgroups are formed that then work on similar, practical solutions. Finally, the group selects its best idea(s).

The *brainwriting pool* involves a group of five to eight people. The group leader presents the design problem to the group, and each individual begins writing solutions on a piece of paper. As soon as each individual gets four solutions documented, he places his paper in the middle of the table and selects a paper from someone else. He then reviews the ideas on that paper and adds new ideas triggered by reading the list. After placing another few ideas on that paper, he exchanges it for another paper in the middle of the table. This continues for 20–30 min. The group then reviews the ideas.

In addition to the techniques summarized by VanGundy [1988], Altshuller [Arciszewski, 1988; Terninko et al., 1996] began the development of a theory of inventive problem solving (TRIZ) for product development in Russia in 1946. TRIZ is the result of the analysis of approximately 1.5 million patents from across the world. The problem-solving methods employed in TRIZ include Altshuller's inventive principles, a table for engineering contradiction elimination, standard techniques to eliminate conflicts, standard solutions to inventive problems, and an algorithm for inventive problem-solving. This material is still largely proprietary and is marketed by a number of consultants and seminar leaders.

An important creativity concept with which to finish draws upon the notions of value-focused thinking [Keeney, 1992], introduced in Chapter 6. This approach is like the attribute listing method discussed above. The individual selects one or more important key performance requirements and defines an instantiated physical architecture or choices within a single generic component. Then, another single performance requirement or set of performance requirements is selected and used to generate an instantiated architecture or set of choices for a single generic component. After continuing this process for a productive period of time, the results are critiqued and adapted to feasible solutions.

8.5 GRAPHIC REPRESENTATIONS OF THE PHYSICAL ARCHITECTURE

There are many graphical representations of a physical architecture with little standardization. The most common graphical format is called a *block diagram*. Figure 8.7 illustrates a block diagram for the control system of an aircraft. Each box inside the dotted line defining the control system represents a physical component of the control system. The lines between the boxes indicate the flow

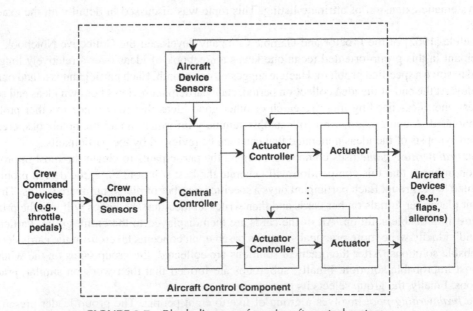

FIGURE 8.7 Block diagram of an aircraft control system.

of electromechanical energy between the boxes. The boxes outside the dotted line represent other components of the aircraft system. This block diagram shows a decentralized controller structure in which there is a central controller and an actuator controller for each device actuator. Note the feedback loops inside the control component, as well as the feedback loop involving most of the elements of the control component and the actuator devices that are part of the aircraft but outside the aircraft control system. Note that the aircraft control component inside the dotted line performs the function "control processing" from the Hatley–Pirbhai template in Chapter 7. This allocation of functions to components is covered in Chapter 9.

There was no accepted convention for block diagrams prior to SysML, which was introduced in Chapter 3. SysML contains two types of block diagrams: block definition diagrams and internal block diagrams. The block definition diagram (see Fig. 3.14) shows the hierarchical decomposition shown in Figure 8.1. The internal block diagram (see Fig. 3.16) presents the information shown in the generic block diagram of Figure 8.7.

8.6 ISSUES IN PHYSICAL ARCHITECTURE DEVELOPMENT

The major issues in designing the physical architecture are (1) functional performance, (2) availability and other "ilities" as achieved through such characteristics as fault tolerance, (3) growth potential and adaptability, and (4) cost. Achieving sufficient functional performance via the development of the physical architecture has been addressed initially in previous sections of this chapter and will be finished in the next chapter during the development of the allocated architecture. Similarly, most of the system-wide (or suitability) factors described in Chapter 6 are often achieved by additional physical resources and associated functionality. Ultimately, many of these additional capabilities as well as cost are issues of tradeoffs. These tradeoffs need to be examined during the evaluation of alternate allocated architectures. Achieving substantial fault tolerance is nearly always important for a system. Finally, there are several issues that impact the ability to grow or adapt a system to needed changes by the stakeholders. The elusive issue of design flexibility is often discussed but difficult to achieve in general. Flexibility is related to such topics as modularity, complexity, and loose versus tight coupling.

Section 8.6.1 addresses the architectural concepts of centralization versus decentralization and distribution of functions and components. Examples from automated systems are used to illustrate these concepts. Section 8.6.2 discusses ideas for design flexibility. Design advantages associated with product platforms are described in Section 8.6.3. Section 8.6.4 focuses on the design issues of a physical architecture associated with increasing fault tolerance and availability through redundancy of physical assets, software assets, information, and time.

CASE STUDY: FBI FINGERPRINT IDENTIFICATION SYSTEM

Since the advent of modern information processing technology, the Federal Bureau of Investigation (FBI) has sought ways to improve and perfect its fingerprint collection, identification, and archival systems. By 1993, the Bureau's Integrated Automated Fingerprint Identification System (IAFIS) consisted of three major interactive segments: the Identification Tasking and Networking (ITN/FBI) segment, the Interstate Identification Index (III/FBI) segment, and the Automated Fingerprint Identification System (AFIS/FBI) Segment. In 1993, proposals were solicited from industry to address the ITN/FBI segment.

Among the many challenges associated with developing a competitive technical solution was the subset of requirements related to processing the fingerprint images. Fingerprint images arrive at the FBI through several means. The most common is the widely recognized set of impressions made on a paper form known as a ten-print card. Since the majority of cards comply with a standard set of dimensions, it is a straightforward matter to determine the expected size of the binary image file created when the cards are processed by a digital scanner; both the front and the back sides are scanned.

The following discussion is concerned with the decompression of the scanned card image, followed by its presentation to an expert fingerprint analyst for classification and identification. The FBI's request for proposal (RFP) included a detailed specification for the segment and all sub-elements including the ten-print processing subelement (TPS). According to the RFP the TPS would consist of workstations organized into workgroups. Each workgroup would thus be analogous to one of the many FBI teams engaged in fingerprint analysis. Typically, a team consists of a supervisor and perhaps a dozen expert fingerprint analysts. The supervisor's role is to manage the classification and identification of the numerous fingerprint card submissions that the FBI handles on a daily basis. The specification also quantified specific processing requirements for the daily influx of 10-print cards, which at the time of the RFP were given to be an average of 30,000 per day. For example, all incoming cards were required to be scanned and converted to binary data so that they could be distributed electronically to the fingerprint analysts for subsequent processing. To minimize any impact to the communications infrastructure, the specification required that the images be compressed at a ratio of 10:1 prior to transmission over the local area network.

Data concerning the processing response time demands on the fingerprint analysts were also included within the RFP. Chief among the critical task processing times are (1) the average time for the analyst to perform a fingerprint image comparison (FIC), given as 60 s, and (2) the time allowed for the display of the human-machine interface screen, including fingerprint images, given as 1 s from the time of the request. Thus, the average processing time that a fingerprint analyst requires to complete the task associated with an individual ten-print card was taken to be 60 s. This meant that the component performing the decompression function needed to be fast enough to sustain an input queue of ready and available images for each fingerprint analyst.

A second complicating fact was the decompression algorithm. At the time the RFP was released, the most popular algorithm available was based upon a high-quality wavelet scalar quantization (WSQ) approach. The popularity was based on common knowledge among the bidders that the National Institute of Science and Technology (NIST) was about to revise the algorithm specification in preparation for a formal certification. Public access to the algorithm specification enabled the competing design teams of the ITN/FBI segment to benchmark the implementation of the WSQ algorithm in order to quantify its processing requirements. In general, the implementations were found to be floating point arithmetic intensive. As a result, it was recognized that such execution behavior is well suited to the latest family of high-performance machines known as reduced instruction set computers (RISC). The specific implementation could be either a software routine or a custom-fabricated large-scale integration (LSI) chip impeded into a math coprocessor card. See Figure 8.8 for a flowchart illustrating the six decision options with an associated block diagram for each option.

Based upon the data provided in the RFP, performance data collected from benchmarks of competing decompression algorithms, and performance data collected from the manufacturers of the computer hardware proposed to host the algorithms, a trade study was conducted to determine how to best implement the function. The study described here analyzed six alternate allocations for decompressing the fingerprint images:

a. Implement software on the workstation within each work group by increasing the TPS workstation processing capacity to enable all decompressions to be performed locally on the individual analysts' workstation.

b. Implement software on the work group's server by increasing the TPS server processing capacity to enable all or some decompression to be performed locally on the TPS server for a given work group.

c. Implement in software by distributing the decompression among under-utilized workstations and server processors enterprise-wide without having to increase the total number of processors or their inherent processing capacity.

d. Implement in software by distributing the decompression among under-utilized workstations and server processors on each local network, without having to increase the total number of processors or their inherent processing capacity.

e. Implement in hardware on the workstation by adding a WSQ coprocessor card in all TPS workstations to perform the decompressions locally.

f. Implement in hardware on the server by adding a WSQ coprocessor hardware card in all TPS servers to perform all or some of the decompressions.

The bidder, based on a thoughtful process, developed the set of six alternatives in Figure 8.8. Table 8.5 shows a morphological box that contains these six options, as well as many other possibilities.

The first row shows the generic components that were part of this segment, as shown in Figure 8.8. The second through fourth rows show possible instantiations of the generic components. The six alternatives defined for the trade study shown on the previous page are designated with the letters a, b, c, d, e, and f at the bottom of each box in the matrix.

The result of producing this morphological box suggested some new alternatives that would have been competitive with the six analyzed in the trade study; these are shown as g and h in Table 8.5.

Provided by Tim Parker

8.6.1 Major Concepts for Physical Architectures

Nearly every physical architecture is either centralized or decentralized. A *centralized architecture* uses a central location for the execution of the transformation and control functions of the system. A *decentralized architecture* has multiple, specific locations at which the same or similar transformational or control functions are performed. The block diagram for an aircraft control system in Figure 8.7 shows a decentralized architecture; note that there is a central controller, but the controllers for each of the aircraft's actuated devices have been decentralized. In the decentralized architecture shown in Figure 8.7, the central controller manages the decentralized device controllers. A centralized architecture would not have individual device controllers; rather, the centralized controller would perform all of the functions.

The advantages of centralized control are the ease of design and testing. Decentralized control requires far more understanding and testing of the various internal use cases associated with

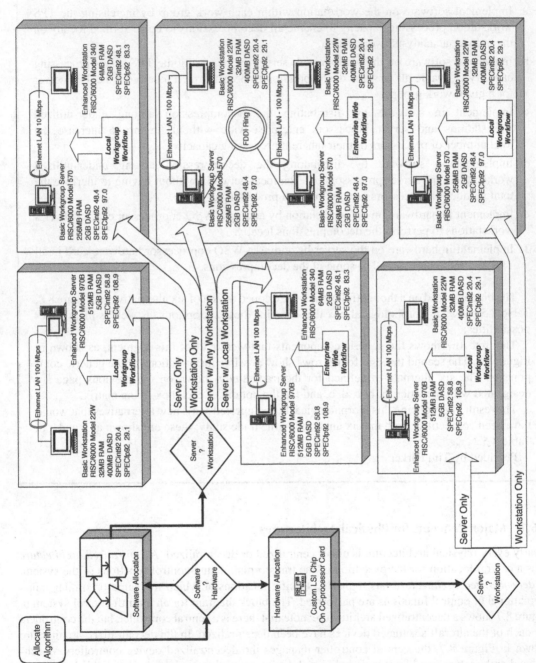

FIGURE 8.8 Flowchart of alternate functional design allocation options with associated block diagrams.

TABLE 8.5   Morphological Box for the Card Image Decompression Component

Workstation	Server	Software	LSI Chip	Workflow Management	Communications
Basic Workstation	Basic Server	No WSQ Algorithm	None	Local Workgroup Workflow	Ethernet LAN (10BaseT) – 10 Mbps
RISC/6000 Model 22W	RISC/6000 Model 570	(e, f)	(a, b, c, d)	(a, b, d, e, f)	(a, e)
32MB RAM	256MB RAM	(g, h)		(g)	
400MB DASD	2GB DASD				
SPECint92 20.4	SPECint92 48.4				
SPECfp92 29.1	SPECfp92 97.0				
(b, c, e, f) (g, h)	(a, c, e) (g, h)				
Enhanced Workstation	Enhanced Server	WSQ Algorithm	WSQ on LSI Chip	Enterprise-Wide Workflow	Ethernet LAN (100BaseT) – 100 Mbps
RISC/6000 Model 340	RISC/6000 Model 970B	(a, b, c, d)	(d, e)	(c)	(b, d, f)
64MB RAM	512MB RAM		(g, h)	(h)	(g)
2GB DASD	5GB DASD				
SPECint92 48.1	SPECint92 58.8				
SPECfp92 83.3	SPECfp92 108.9				
(a, d)	(b, d, f)				FDDI WAN – 100 Mbps (c, h)

the ways in which decentralized control can occur. Centralized control has drawbacks, however, a single point of failure and a processing bottleneck when the control system is most critical. A single point of failure is particularly problematic when the system can be attacked by other systems. A hybrid approach for control that decentralizes some control issues that are not of critical concern at the system level but maintains centralized control for critical system-level activities has some merits.

A *distributed architecture* is one in which there are two or more autonomous processors connected by a communications interface and running a distributed operating system [Coulouris et al., 1994; Shuey et al., 1997]. The distributed operating system enables the processors to coordinate their actions and share the system's resources. The processors can perform the same functions, depending upon the needs of the system. Processing control issues for a distributed system are handling the redistribution of processing functions after partial failures; managing moves, changes, and additions to the processing activities; and synchronizing processing activities to meet performance and efficiency objectives. An important distinguishing feature of a distributed system architecture is that the users are unaware of the distribution of processing.

A distributed system can be either homogeneous or heterogeneous. The earliest distributed systems were *homogeneous*, that is, comprised of identical processors, running identical operating systems and application software, and connected via a single communications network. Users on a homogeneous distributed system view the system as their processor but obtain the benefits of being able to share data with each other over wide geographic regions. Eventually, some processors become much busier than others and the issue of load sharing arises; load sharing distributes computational tasks from one processor to another. Note, *load sharing* is the reallocation of functions to different resources in the physical architecture and is therefore an issue in the allocated architecture. Load sharing requires users to access and share multiple processors, which may result in increased response times in some situations. Finding the best approach to load sharing is quite complex.

Heterogeneous distributed systems have two or more types of processors comprising the processor network, plus operating and application software and one or more communications networks connecting the processors. The Internet is the most common example of a heterogeneous distributed system. Specially designed, heterogeneous distributed systems are, or will, enable medical support in hospitals by both specialists and generalists; financial transactions; fingerprint analysis by both experts and automated assistants; review of tax records by both experts and automated assistants; and analysis of data collected by satellites by a wide variety of researchers. Each architecture shown in Figure 8.8 for the FBI fingerprint identification system case study is a heterogeneous network involving two types of processors, clients and servers.

Flynn [1972] created a taxonomy of single and multiple processor systems. Centralized architectures in the Flynn taxonomy are "single instruction, single data" (SISD) and "single instruction, multiple data" (SIMD). Multiple data means multiple data streams are being used for the processing activity. Distributed architectures include "multiple instruction/single data" (MISD) and "multiple instruction/multiple data" (MIMD). Multiple instruction means multiple processors. MISD is typically used for the fault-tolerant approach called passive hardware redundancy in Section 8.6.3. MIMD is the generic distributed system architecture.

The major reasons that a distributed processing architecture is attractive in designing systems are transparency, openness, scalability, resource allocation, concurrency, and fault tolerance. *Transparency* means that the users view the distributed system as a complete system, *without* any knowledge of how the hardware and software components are performing. An *open architecture* is one for which the hardware and software interfaces are sufficiently well-defined that additional resources can be added to the system with little or no adjustment. *Scalability* means that multiple-sized versions of the system are available. *Resource sharing* exists when more than one hardware and software module can be used to execute the same task with no human intervention. A *concurrent architecture* is one in which multiple tasks are being executed simultaneously. A single processor can perform concurrent operations by interleaving the operations of multiple tasks; however, multiple distributed processors can clearly perform concurrent operations without any direct knowledge of what the other processors are doing. Finally, *fault tolerance* is achieved if the distributed system can adjust its operations when one of the hardware or software elements fails. Details for achieving fault tolerance are discussed in Section 8.6.4.

A client-server architecture is a software architecture that is superimposed on a distributed system to facilitate the processing and management of the system. The *client-server architecture* distinguishes between client processes (requestors) and server processes (task completors). Each distributed processor is performing its assigned task; when one processor needs support from another processor, the processor needing support becomes a client and issues a request across the network. The processor that accepts the request becomes the server, responds that it will complete the request, and uses both hardware and software resources to complete the task and send the result to the client. Note this server may have just issued a client request of its own and may be waiting for a response from some other processor. Servers may be set up for database, file, print, fax, mail, communication, and imaging operations. This client-server architecture will be discussed in more detail in Chapter 10.

The portability of computer operating systems and applications software that are relatively independent of the underlying computer hardware, in conjunction with significant advances in computer and communications performance, have enabled the viability of *service-oriented architectures*

(SOA). Independent, loosely coupled services performing small, elemental functions are engineered to interact to form solutions. These elemental services are reused across applications, providing a level of agility to quickly adapt solutions as needs change. SOA enables service-oriented *cloud computing*.

The Internet has made the concept of a distributed system common place. Distributed systems introduce some complexity into systems and software engineering because new design alternatives exist for specializing parts of the distributed system for specific functions.

An old networked system concept that has been employed with great success by the Internet is that of an overlay network. An overlay network is a virtual network that is designed on top of a distributed system. The nodes in the overlay network are the real computer systems in the distributed system, but the links are virtual since any node can be connected to another node via an IP address that is created as a virtual link using real network interfaces. Initially the Internet was an overlay network built upon the telephone transmission lines network. The Darknet is an overlay network that works on the Internet that uses IP addresses that are not public so that users are anonymous.

8.6.2 Design Flexibility

Many engineers talk and write about design flexibility, modularity, loose coupling, complexity, and other such topics, but it is usually quite difficult to find nuggets that prove useful in the real world. This section will explore some of these ideas.

In Chapter 6, we talked about how much change occurs during the design process and how this change makes success elusive. In addition, most systems are designed to last many years or even decades. The mark of a long-lived system is one that has been upgraded successfully many times. These many upgrades are only possible if the system's architecture has provided an adaptable platform for such upgrades. The Sidewinder missile of the U.S. Navy and Microsoft's Windows NT operating system are two examples of architectures that have supported dramatic changes over many upgrades, such that the original design is no longer present, but the "architecture" remains. So, in addition to working hard to keep track of the changes that are occurring in the requirements, we can also design our systems to be more "changeable" in the future.

Fricke and Schulz [2005] address this problem by defining four aspects of changeability: flexibility, agility, robustness, and adaptability.

- *Robustness* characterizes a system's ability to be insensitive toward changing environments. Robust systems deliver their intended functionality under varying operating conditions without being changed (see Taguchi [1993] and Clausing [1994]). That is, no changes from an external source need to be implemented into such systems to cope with changing environments.

- *Flexibility* represents the property of a system to be changed easily. Changes from an external source have to be implemented to cope with changing environments, but these changes can be made easily due to design choices. Note these changes may not be made rapidly.

- *Agility* represents characterizes a system's ability to be changed rapidly. Changes from an external source have to be implemented to cope with changing environments, but these changes can be made rapidly, which requires flexibility.

- *Adaptability* characterizes a system's ability to adapt itself to changing environments. Adaptable systems deliver their intended functionality under varying operating conditions by changing themselves. No changes from an external source have to be implemented into such systems to cope with changing environments.

Some examples of each of these should help make the points emphasized by Fricke and Schulz. An all-terrain automobile such as a jeep might be an example of a robust vehicle; it can travel reasonably well on many different surfaces. If this all-terrain vehicle can also have a cloth top that can removed and stored, this adds to its robustness. A flexible system is one that can interface easily with many other types of systems, each of which might be changing. For example, laptop computers with many USB ports in the 2007 time frame can interact with nearly all printers, projectors, and control devices. The peripherals or other systems that can plug into the USB ports still have to be changed as the environment changes, but the core computer does not need to change for these reasons. Flexibility is important for future upgrades. An agile system is designed to be changed rapidly. Here a racecar comes to mind. Racecars have to be modified dramatically to run well on different racetracks from one week to the next. A great deal of money is spent on the design to facilitate these rapid changes. Adaptable man-made systems are being designed but with some limitations. Microsoft has designed its operating and office products to learn and adapt to different users to facilitate the performance of these different users. While this has been the goal at Microsoft, many feel (including this author) that their efforts are far from successful.

Fricke and Schulz [2005] describe three basic design principles that support all four types of design for changeability and six extending design principles, each of which supports a subset of the types of design for changeability. The three basic principles are ideality/simplicity, independence, and modularity/encapsulation. The six extending principles are integrability, autonomy, scalability, non-hierarchical integration, decentralization, and redundancy. Aspects of decentralization were discussed above. This next section addresses redundancy for fault tolerance, a form of adaptability.

8.6.3 Design Advantages of Product Platforms

Meyer and Lehnerd [1997] introduce the concept of product platforms, provide many examples in existence at the time, and discuss business strategy, platform design and renewal, and performance measurement. They start with the amazing story of Black and Decker in the early 1970s. Faced with increased competition and the associated lowering of profit margin, as well as new safety regulations for double insulation from the U.S. Congress, Black and Decker undertook an extensive redesign of the products and their manufacturing facilities to achieve a much tighter product platform for 122 power tools across eight power tool product groups (e.g., drills, hedge trimmers, power hammers). The result of this innovation of a product platform by Black and Decker was market dominance for over a decade. Black and Decker were able to reduce their prices for the power tools dramatically while maintaining their desired profit margin. Many competitors went out of business. Cycle times for new derivative products were dramatically decreased, leading to an average of a new product every week for several years after completion. But, over time, Black and Decker provided waivers for deviations from the product platform standards, leading to the return to a disorganized array of power tool products.

Other well-known examples of product platforms are the Sony Walkman and HP Deskjet printers. Today product platforms exist in the auto industry for the chassis, engine and power train (both hardware and software), the commercial airline industry (e.g., DC-3 and Boeing 777), the camera industry, the computer industry (e.g., personal computers), the printer industry, the integrated circuit industry, the software industry (e.g., Microsoft Office), the engine control software (e.g., Bosch) industry, the wireless sensor industry, the pharmaceutical industry (e.g., Tylenol), and the entertainment industry (e.g., Disney World). There is even work in product platforms for augmented cognition systems.

There are several clear benefits to having a sound product platform associated with 5–500 products; the Black and Decker example above demonstrates many of these benefits. Product platform designs often lead to reduced operating costs, improved maintainability, improved ease of use, much more rapid introduction of new products and variations around existing products, reduced manufacturing costs, reduced complexity in manufacturing small numbers of product variations in a timely manner (the job shop problem), and the rapid reconfiguration of the manufacturing system to deliver new products.

However, the Black and Decker success story collapsed because there were also some drawbacks. There was pressure to achieve performance levels in some of the power tools that could not be attained with the platform constraints. There were pressures to have a distinctive look for some of the products, but this could not be accomplished within the constraints of the platform. Significant investments must be made to incorporate new technologies across the platform; this may be done if some of these products do not benefit from the introduction of the new technologies. Implementing design changes may cost more because they affect the entire product platform. It is not always best to upgrade all products in the platform at the same time, but this must be done to keep the platform synchronized.

So clearly, the "bill payers" for new or upgraded product platforms must address the tradeoffs, which span the life cycle. Human decision-makers never feel comfortable feeling they have a good handle on all these tradeoffs, especially when they span a wide range of time. So, part of the systems engineering task is to clarify these tradeoffs to the "bill payers" and communicate the impacts of these tradeoffs on cost, schedule, and performance across the life cycle. This is called Multiple Objective Decision Analysis (MODA), which is described in Chapter 14.

8.6.4 Use of Redundancy to Achieve Fault Tolerance

Fault tolerance was discussed in Chapter 7 from the perspective of functions that need to be performed to detect errors, confine the damage, recover from the damage, isolate the damage, and report the problem. Design issues associated with the physical architecture are just as important in achieving fault tolerance. A primary source of high availability and fault tolerance is redundancy. Often, hardware redundancy receives most of the attention. However, Johnson [1989] identified four elements of *redundancy:* hardware, software, information, and time. *Hardware redundancy* uses extra hardware to enable the detection of errors as well as to provide additional operational hardware components after errors have occurred. This hardware redundancy can be implemented in passive, active, and hybrid forms.

Passive hardware redundancy masks or hides the occurrence of errors rather than detecting them; recovery is achieved by having extra hardware available when needed. The rest of the system and its operators are commonly not even aware that an error has occurred. This approach only works if there are sufficient hardware replicas to continue to mask errors. The most common passive implementation is called triple modular redundancy (TMR) and relies on a majority voting scheme to mask an error in one of the three hardware units. Figure 8.9 (top left) shows TMR; unfortunately, the single "voter" element is a single point of failure in this system. Therefore, TMR is often implemented as triplicated TMR (Fig. 8.9 bottom right). Triplicated TMR implements three voters and produces three versions of the output, which are usually sent to another module that has been implemented as triplicated TMR. Naturally, there is nothing magical about three; *N*-modular redundancy (NMR) is the generalization of TMR. TMR can mask a single error; 5MR can mask two errors, and so on.

Voting is a common conflict resolution technique used inside a computer as well as with groups of people. However, implementing voting inside a system has some unexpected difficulties. Issues in voting implementation are establishing the time at which the computation was done, the precision of numbers achievable in a digital computer, and the need to produce a single answer eventually. Timing of the computations is critical because the hardware and software components producing inputs to the *voter* may be performing repetitive computations on a data stream and be out of synchronization. For repetitive operations, there must be some synchronization mechanism involved to ensure that the vote is being taken on computations from the same samples of data stream of inputs.

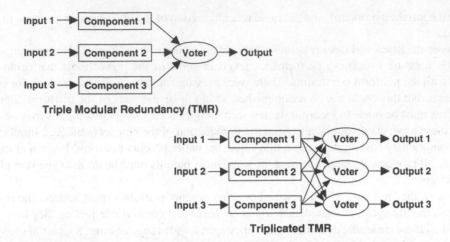

FIGURE 8.9 TMR and triplicated TMR. (Adapted from Johnson [1989].)

The precision issue addresses the concern that there is some imprecision in numerical operations involving digital equipment. Quantization of a number on a digital computer can produce several different valid results. As a result, the voter may see three different outputs from the three components, but the outputs are the result of normal processing operations. In many cases, the majority voting scheme is replaced with either a selection of the median value or truncation of the numerical values to some predefined level of significant digits.

The last issue, the production of a single answer, requires that a single point of failure be introduced. When the final result (e.g., bank account balance or control signal to the rudder) has to be delivered by the system in question, this final answer is determined on a single processor.

Finally, voting for passive redundancy can be achieved via hardware or software. A hardware implementation is faster but usually requires more cost, space, power, and weight. A software implementation (see Fig. 8.10) provides greater flexibility for change but can also require additional cost, space, power, and weight in the form of processors if voting is a major part of the system's redundancy, which is often the case.

Active hardware redundancy attempts to detect errors, confine damage, recover from the errors, and isolate and report the fault, as described in Chapter 7. The basic building block for active hardware redundancy is called *duplication with comparison;* see Figure 8.11 for a hardware

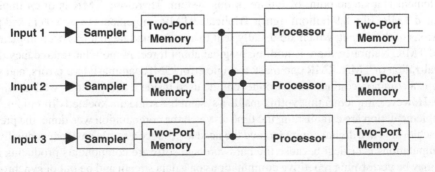

FIGURE 8.10 Software implementation of voting for triplicated TMR. (Adapted from Johnson [1989].)

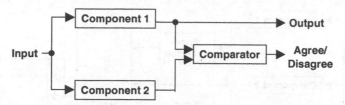

FIGURE 8.11   Hardware duplication with comparison. (Adapted from Johnson [1989].)

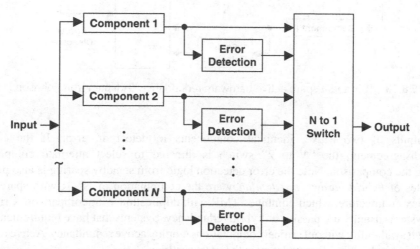

FIGURE 8.12   Standby sparing with N-1 replicas. (Adapted from Johnson [1989].)

implementation. Two identical units are used to compute the same output for the same set of inputs; these outputs are compared in a "comparator." If the outputs disagree by a predefined amount, an error is declared. (Note the issues of synchronization and precision also apply here.) Once an error is declared, functionality to confine the damage, recover from the errors, and isolate the reports is activated.

Hot and cold standby sparing are different than duplication with comparison and are the most common approaches to active redundancy; see Figure 8.12. In *hot standby sparing* multiple replicas of a component are performing identical functions; only one of them is providing outputs, but all are ready to take over with no delay. Error detection in standby sparing is not done by comparing outputs from redundant components but by examining the output for known errors or monitoring the component for inactivity. A watchdog timer is an example of this latter approach; a *watchdog timer* declares a fault if it is not continuously reset by the component with which it is associated.

Cold standby sparing maintains the component replicas in a nonoperational mode until needed. This is useful for applications where short disruptions are acceptable or long life is key, for example, spacecraft operations. For real-time applications, hot standby sparing is critical to success but increases power consumption and decreases the life of the system. Standby sparing is most used by providing multiple, excess processors, any of which can be used to perform necessary system functions. When one processor fails, a controller no longer assigns tasks to that processor, with the slack being absorbed by the remaining processors.

The final example of active hardware redundancy, *pair-and-a-spare*, combines the features of duplication with comparison and standby sparing. Figure 8.13 shows a comparison (far right)

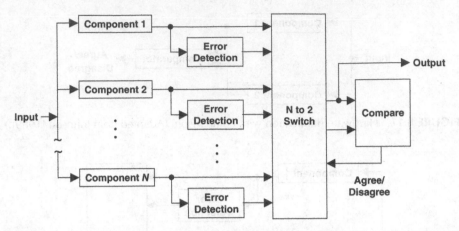

FIGURE 8.13   Pair-and-a-spare active hardware redundancy. (Adapted from Johnson [1989].)

of the outputs of two active, identical components to detect an error. If the comparison yields a disagreement, the "*N* to 2" switch is directed to select alternate components for conducting the comparison. Note the error detection logic from standby sparing is also present.

Examples of *hybrid hardware redundancy* are the combination of NMR with spares, and the triple-duplex architecture, which combines TMR with duplication with comparison. Critical computation systems usually use passive or hybrid redundancy. Systems that have requirements for long life and high availability without critical computations employ active redundancy. Active redundancy is usually less costly; hybrid redundancy is the most costly.

Software redundancy is a second means for detecting and recovering from errors. N-version software redundancy is a seldom-used approach to provide multiple operational software components in the event of a software failure. Each version is programmed by separate groups of programmers, assuming that while each group may make mistakes, no two will make the same mistake. More common forms of software redundancy are consistency and capability checks; both can be used for error detection in standby sparing. *Consistency checks* compare the output of a component with known characteristics of that output, for example, minimum and maximum values. *Capability checks* are software designed to run periodic hardware tasks with known answers.

Information redundancy is achieved by adding extra bits of information to enable error detections using special codes [Johnson, 1989]. Information redundancy is useful to catch *system*-induced errors rather than component faults; however, system-induced errors can be indicative of component faults if the errors occur with sufficient frequency. Information redundancy is a very rich area, having many alternate approaches. Information redundancy is one form of error detection that can be used for standby sparing; see Figure 8.12.

Time redundancy can be used to replace hardware and software in non-real-time systems to achieve error detection. When extra processing time is available, computations can be *performed* multiple times with a single hardware and software combination and compared. If discrepancies exist, an error has been detected. This approach is also used for error detection in standby systems and is quite useful in distinguishing between transient and permanent errors. Time redundancy assumes that additional time exists for functional performance to enable the needed error detection and recovery. On the plus side, time redundancy can save significantly on hardware and software, reducing cost, weight, power, and other key suitability issues.

8.7 SUMMARY

The focus of this chapter has been the resources that comprise the system, called the physical architecture. The system is first segmented into its top-level components, the segmentation progresses down to the CIs, or hardware and software elements, facilities, people, procedures, and user manuals.

The physical architecture can be either generic or instantiated; the generic physical architecture is an abstract separation of the system's resources into components before any key performance decisions are made. The instantiated physical architecture specifies the performance characteristics of each element of the generic physical architecture to the degree needed for the performance modeling of the system.

Creativity techniques are important to aid the generation of alternate, instantiated physical architectures. The morphological box was described in detail and illustrated as an effective technique for gathering creative ideas and increasing the chances of combining these creative ideas into a sound, instantiated physical architecture. The morphological box is defined by the generic physical architecture and then provides slots for alternate ideas for instantiated physical components of each segment.

Representing the physical architecture using a block diagram was presented in this chapter. Block diagrams are completely non-standardized representations of the system's components, showing the major flows of electromechanical energy, material, and information between the components.

Finally, key concepts such as centralized and decentralized and distributed and client-server architectures were presented. The decentralization of transformation and control functions and the distribution of functional and physical elements of the architecture have become the norm in most systems' architectures. These concepts were defined and illustrated.

Redundancy in hardware, software, information, and time was presented since achieving fault tolerance is often a critical design issue that the engineer of the system must address. Hardware redundancy is the most discussed and implemented approach to achieving fault tolerance with the physical architecture. Software redundancy is almost always too expensive to develop. Information redundancy, adding extra bits to data elements for the purpose of checking the meaningfulness of data elements later, is used extensively on communications interfaces that become part of the physical architecture. Utilizing unused data processing time to repeat computations, time redundancy, is not a common approach.

CASE STUDY: COMMERCIAL AIRCRAFT CRASH AT SIOUX CITY, IOWA

On July 19, 1989, United 232 (a DC-10 aircraft) crashed into a cornfield next to the Sioux City airport in Iowa while trying to make an emergency landing after losing one of three engines. In all, 110 passengers and one flight attendant were killed during this emergency landing; 185 people survived the accident, some without a scratch.

Engine failure is the most commonly trained maneuver in simulators. The DC-10 has three engines: one on each wing and one on top of the fuselage in the vertical tail (or horizontal stabilizer). United 232 lost the engine on top of the fuselage due to the loss of a fan disk; the fan disk separated from the engine and crashed through the tail. Pilots fought through the engine loss by porpoising (rotating the thrust levels) the two remaining engines to land in Sioux City. However, the descent rate of the landing was too great; the aircraft caught fire upon landing, tumbled, and broke apart in corn and soybean fields.

The fan disk, about 300 pounds of titanium, on the number two engine, was missing; it had shattered into pieces and crashed through a chamber designed to contain such a break-up. There are three independent hydraulic systems on the DC-10 aircraft, and a unique engine powers each hydraulic system. The hydraulic system on an aircraft provides the forcing function for the aircraft's stabilization systems: the ailerons on the wings that permit the aircraft to bank right and left, the rudder that allows the aircraft to turn right and left, the elevators on the tail that cause the aircraft's nose to rotate up or down, and the flaps and slots on the wings that permit the aircraft to change the amount of lift generated by the wings. Losing engine number two should have only caused the loss of one of the three hydraulic systems. However, the three independent hydraulic systems converge in the tail at the exactly the location that the fan disk ripped out, the single point of failure for all three hydraulic systems.

Experts believe there was a preexisting fracture on the fan disk. Ultrasonic sensors are used to detect fractures during production. However, these sensors do not provide good results when the fracture is near the surface. The National Transportation Safety Board (NTSB) investigators concluded that the fracture had been there since the fan disk was built. The fracture would have grown with use; the maintenance crew was blamed for not finding the fracture during routine maintenance activities. Nonetheless, this does not dismiss the design flaw of a single point of failure for what were considered to be three redundant hydraulic systems [Birnbaum, 1989; Magnuson, 1989].

PROBLEMS

8.1 Create a generic physical architecture for the ATM problem in Chapters 6 and 7. Create a morphological box for your generic physical architecture of the ATM. Identify three instantiated physical architectures based upon the morphological box.

8.2 Create a generic physical architecture for the OnStar system in Chapters 6 and 7. Create a morphological box for your generic physical architecture of OnStar. Identify three instantiated physical architectures based upon the morphological box.

8.3 Create a generic physical architecture for a personal computer. Create a morphological box for your generic physical architecture of a personal computer. Identify three instantiated physical architectures based upon the morphological box.

8.4 Create a generic physical architecture for a stereo system. Create a morphological box for your generic physical architecture of a stereo system. Identify three instantiated physical architectures based upon the morphological box.

8.5 Create a generic physical architecture for the development system of an airbag system. Create a morphological box for your generic physical architecture of the development system. Identify three instantiated physical architectures based upon the morphological box.

8.6 Create a generic physical architecture for the manufacturing system of an airbag system. Create a morphological box for your generic physical architecture of the manufacturing system. Identify three instantiated physical architectures based upon the morphological box.

8.7 Using the information in Figure 8.7, create a block definition diagram and an internal block diagram for the "Aircraft Control Component," which is inside the dotted lines of the figure. Be sure to use the semantics and syntax of SysML. Note, you will have to ignore any arcs coming from or going to components outside the dotted line.

8.8 You are on the elevator design team and have just convinced the team that the block decomposition at the subsystem level (Fig. 3.14) is incorrect. You have convinced the team to add a communications bus so that the communications between the subsystems can be more efficiently routed through the communication bus. Modify the block definition diagram and internal block diagrams shown in Figures 3.14 and 3.16, respectively, for the elevator subsystems to show this design change. Consider the communications bus to be a new component or subsystem.

Chapter **9**

Allocated Architecture Development

9.1 INTRODUCTION

The development process for the allocated architecture is the activity during which the entire design comes together. The allocated architecture integrates the requirements decomposition with the functional and physical architectures. The process of developing the allocated architecture provides the raw materials for the definition of the system's external and internal interfaces and is the only activity in the design process that contains the material needed to model the system's performance and enable tradeoff decisions. The reader should not infer from this discussion that the requirements development is started and finished, followed by the functional architecture, followed by the physical architecture, followed by the allocated architecture. Rather, the design process is like peeling an onion; each of these activities in the design process should be completed at a high level of abstraction (low level of detail), culminating in an allocated architecture at this high level of abstraction for a set of subsystems that comprise the system. Then, the entire process is repeated at a lower level of abstraction (greater detail) for the next tier of components (peel of the onion), consistent with the Vee model discussed in Chapter 1. This repetition at lower and lower levels of abstraction (greater and greater detail) is continued as long as useful to the design process. As details determine problems with the design, decisions are reviewed, and changes are implemented at higher levels of abstraction as needed.

This chapter describes the activities involved in developing an allocated architecture in detail: allocate functions to subsystems, trace non-input/output requirements and derive requirements, define and analyze functional activation and control structure, conduct performance and risk analysis, document architectures and obtain approval, and document subsystem specifications.

The methods introduced in this chapter match the functions that comprise the development of the allocated architecture. Various methods are discussed for allocating functions of the system in question to subsystems and components of the system. The derivation of input/output, system-wide and technology, tradeoff, and qualification requirements is discussed as a key method for providing the material to complete the component specification. Three methods for flowing down system-wide and technology requirements that have been traced to the system are described. Models for defining and

The Engineering Design of Systems: Models and Methods, Fourth Edition. Dennis M. Buede and William D. Miller
© 2024 John Wiley & Sons, Inc. Published 2024 by John Wiley & Sons, Inc.
Companion website: www.wiley.com/go/engineeringdesignofsystems4e

analyzing functional activation and control structures are discussed in the graphical modeling techniques on the companion website and are therefore not presented in this chapter. However, critical system-wide issues associated with functional activation and control are discussed here. A normative model for conducting trade studies and risk analyses is presented in Chapter 14. Examples of common trade studies and risk analyses are discussed and illustrated in this chapter. No new models are introduced in this chapter.

The exit criterion for finishing the allocated architecture is the acceptance of the design by the stakeholders. The acceptance of the design by the stakeholders should involve a detailed understanding that the requirements development process has met the major characteristics of the requirements, as defined in Chapter 6: a thorough understanding of how the allocated architectures of the systems in each lifecycle phase will meet the requirements as defined, belief that the design trades have accurately reflected the tradeoff requirements, and agreement that the test or qualification systems in each phase of the life cycle are adequate for qualification requirements as defined.

9.2 OVERVIEW

The *allocated architecture* provides a complete description of the system design, including the functional architecture allocated to the physical architecture, derived input/output, technology and system-wide, tradeoff, and qualification requirements for each component, an interface architecture that has been integrated as one of the components, and complete documentation of the design and major design decisions.

There are five major activities associated with the development of the allocated architecture:

- Allocate functions and system-wide requirements to physical subsystems
- Allocate functions to components,
- Trace system-wide requirements to system and derive component-wide requirements,
- Define and analyze functional activation and control structure,
- Conduct performance and risk analysis,
- Document architectures and obtain approval,
- Document subsystem specifications.

Figure 9.1 shows these five functions in an IDEF0 (Integrated Definition for Function Modeling) diagram for developing the allocated architecture; see Appendix B on the Companion website for the full model. Note that Sections 9.3 and 9.4 address the two subfunctions under the first function (these were combined to make the diagram easier to read). As can be seen by the flow of information among these activities, substantial interaction and feedback is required among the first four to make sure the design works; this feedback and control was discussed in Chapter 7. However, viewing the development of the allocated architecture in isolation would be inappropriate. The developments of the three architectures (functional, physical, and allocated), which we have been discussing, all have to proceed in parallel because insight or changes in one have repercussions in the others. Figure 9.2 puts the allocated architecture development in context with the other architectures and requirements development.

As discussed in the introduction, the design process proceeds through the steps shown in Figure 9.2 several times at decreasing levels of abstraction. The more complex the system's functionality and tightly coupled the system's components are, the more important is the repetition of the design process at decreasing levels of abstraction (increasing detail). Initially, the design process establishes functional and physical decompositions, which are united to form the allocated

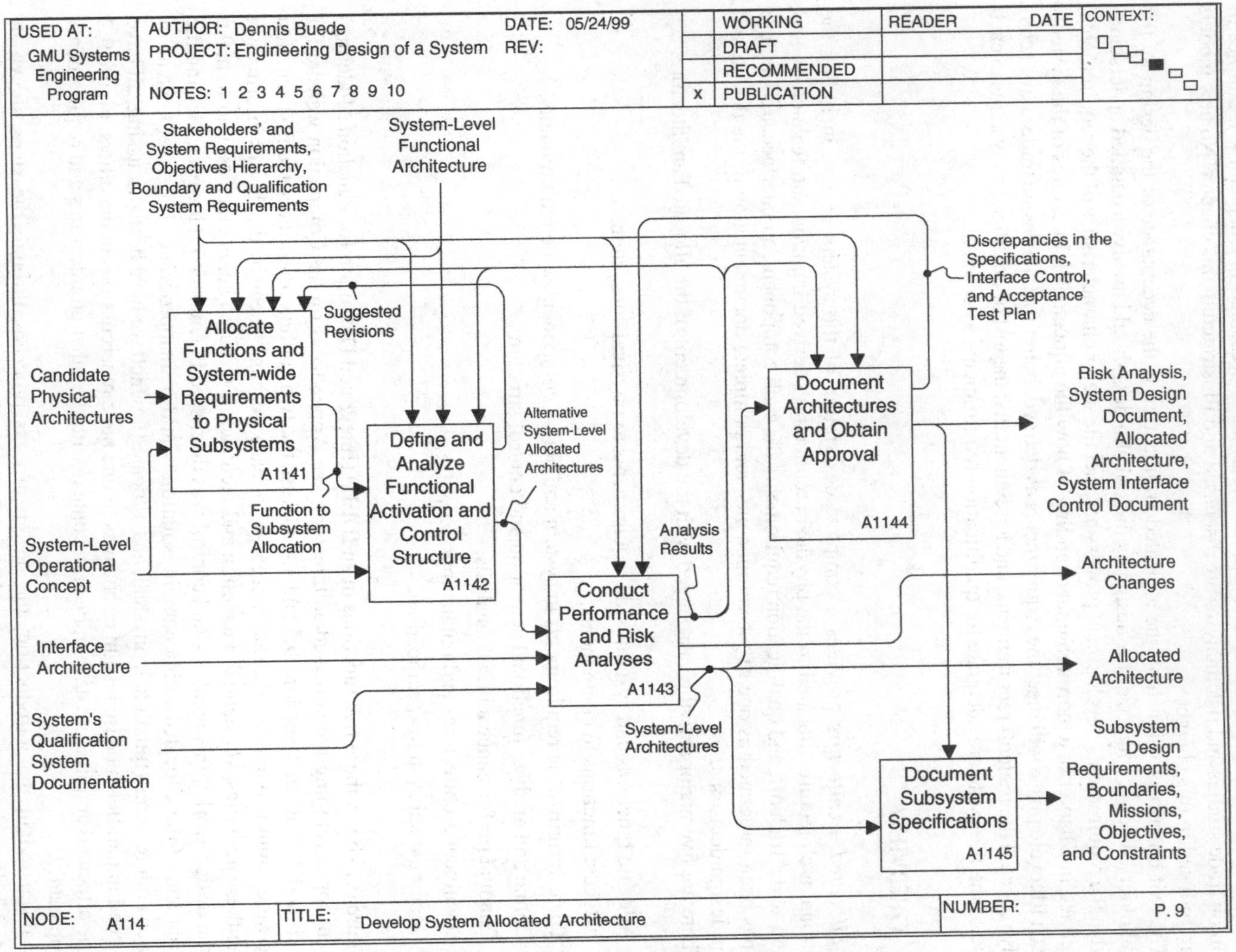
USED AT:
GMU Systems Engineering Program
AUTHOR: Dennis Buede
PROJECT: Engineering Design of a System
DATE: 05/24/99
REV:
NOTES: 1 2 3 4 5 6 7 8 9 10
WORKING
DRAFT
RECOMMENDED
x PUBLICATION
READER
DATE
CONTEXT:
Stakeholders' and System Requirements, Objectives Hierarchy, Boundary and Qualification System Requirements
System-Level Functional Architecture
Suggested Revisions
Discrepancies in the Specifications, Interface Control, and Acceptance Test Plan
Candidate Physical Architectures
Allocate Functions and System-wide Requirements to Physical Subsystems
A1141
Function to Subsystem Allocation
Define and Analyze Functional Activation and Control Structure
A1142
Alternative System-Level Allocated Architectures
System-Level Operational Concept
Interface Architecture
System's Qualification System Documentation
Conduct Performance and Risk Analyses
A1143
Analysis Results
System-Level Architectures
Document Architectures and Obtain Approval
A1144
Document Subsystem Specifications
A1145
Risk Analysis, System Design Document, Allocated Architecture, System Interface Control Document
Architecture Changes
Allocated Architecture
Subsystem Design Requirements, Boundaries, Missions, Objectives, and Constraints
NODE: A114
TITLE: Develop System Allocated Architecture
NUMBER:
P. 9

FIGURE 9.1 IDEF0 representation of developing the allocated architecture.

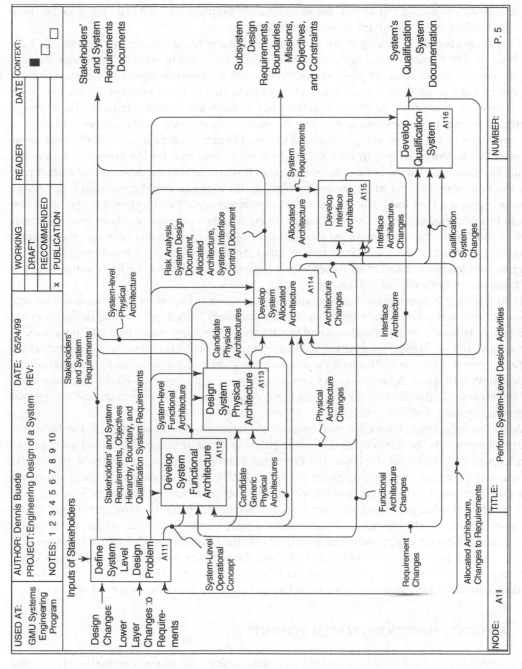

FIGURE 9.2 System-level design activities.

architecture. The allocated architecture divides the design problem into chunks, primarily along the lines of the physical architecture, namely the system's components. Naturally, these design decisions should not be made prematurely; there should be adequate confidence that little or no modifications will be needed. Yet, as the design process evolves through additional repetitions of the activities shown in Figure 9.2, the more detailed simulation models and trade studies may provide justification for modifying earlier design decisions.

The primary benefit of making major design decisions early using models and trade studies built at a high level of abstraction is that these initial decisions are aimed at dividing the design problem into manageable chunks that can proceed concurrently with a reasonable chance of success. Dividing the system's design problem into completely independent chunks is not possible. To accommodate this interaction, there must be design interfaces just as there are system interfaces. These design interfaces are part of the development system for unprecedented systems that are being completed concurrently with the design of the operational system. For precedence systems and platform architectures, the development system may already have been developed for the preceding system but should be revalidated for the to-be-developed system. It is critical that the development system provide the time to review and adjust the design chunks; this time can only be provided if the design process begins at a high level of abstraction. Some engineers argue that this initial peel of the onion should be completed within weeks (6–12) after having written a proposal and been awarded a contract. If the design segmentation is not finalized until each component has been decomposed into several levels of detail, there will be no time to adjust this design decision if the division of the system into components is found to be flawed. There is even less chance that the flaws will be found if too many details are analyzed too quickly.

Distinguishing between good decisions and good outcomes is important. If we were in complete control of our environment, then decisions and the outcomes associated with the decisions could be equated. However, as discussed in detail in Chapter 14, decisions must be made in the face of uncertainty with incomplete information and inadequate control of the outcomes. Therefore, saying that a decision was good or bad because the outcomes associated with that decision were good or bad, respectively, is illogical. A decision can be considered good if the people with the best knowledge and largest stake in the decision are involved in the decision, and these people discuss the relevant alternatives, values, and facts with clarity.

As an example, Ford Motor Company designed and introduced the Edsel in 1957. The Edsel had a large, elongated "0" built into the middle of the grill at the front of the car that caused many people to react negatively on an artistic basis. The Edsel was a complete failure, at least partially because the automobile industry was in a recession in 1957 and 1958. Were the design decisions associated with the Edsel bad? It is not possible to tell without knowing more about what design decisions were made and how the design process was carried out seven years after Edsel's introduction. Ford Motor Company introduced the Mustang, which has been a fantastically successful car and has achieved classic status. Were the design decisions associated with the Mustang good? Again, it is not possible to tell without knowing more about them. With time, it is much easier to tell whether the outcomes associated with a decision are good or bad, but it becomes more and more difficult to tell whether the decisions that were made were good or bad, especially if those decisions are not documented.

9.3   ALLOCATE FUNCTIONS TO COMPONENTS

After the definition of the functional and physical architectures, the systems engineering team must assign functions from the functional hierarchy to the subsystems and components in the physical architecture. When this is done, the first step in defining the allocated architecture is completed. This allocation of functions to components is often the most crucial design decision made by the engineers of the system. Engineers prefer to allocate processing tasks to software if there will be a future need to update the processing algorithms. However, if speed of processing is critical, hardware can perform the computations much faster. Computer manufacturers experiment with moving some processing tasks from hardware to software but often find that the speed of processing suffers too much and revert to designing hardware for the processing tasks. Similar issues arise when considering the decision to allocate a function to people within the system or a combination of hardware and software. This allocation decision is discussed in more detail later.

Figure 9.3 expands upon Figure 8.4 for the allocation of the system's functions to subsystems and components. Clearly, allowing the allocation decision to be represented as a mathematical relation, and not a function, as shown in the top left of Figure 9.3, is inadequate; there will be some functions that are not allocated to any component and some functions that are being processed by two or more components. Forcing the allocation of functions to components to be represented as a mathematical function, as shown in the top right of Figure 9.3, solves these problems. However, there may be some components with no functions to perform; these components should either be dropped from the system or the engineers should revisit their functional architecture to ensure that the functional architecture is complete. There is also the possibility that some functions will be performed by the same component; there is nothing wrong with this because the functions can be aggregated into a single function. If as expected all of the components are needed, the allocation of functions to components will be onto, as shown in the bottom left of Figure 9.3. A functional allocation is one-to-one when the number of functions and components is the same, as shown in the bottom right of Figure 9.3.

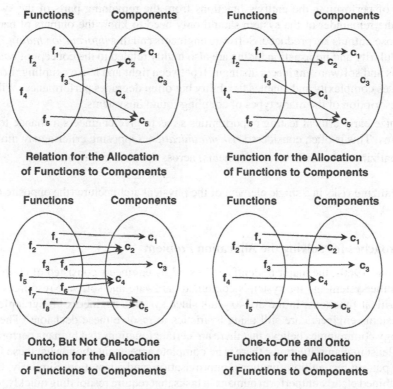

FIGURE 9.3 Mathematical relations and functions for the allocation of engineering functions to components.

Note that the mapping of functions to components was picked consciously rather than the mapping of components to functions. Allowing two components to be mapped to the same function is consistent with the definition of a mathematical function but should be avoided by the engineers of a system. When two components are performing the same function, it will not be possible to segment the responsibilities of the components until the functional and physical architectures are examined in greater detail; this defeats the purpose of iterating through the engineering process as suggested by the Vee model and most engineers of systems.

9.3.1 Define the Allocation Problem

For any single physical architecture and the associated functional architecture, there are many possible allocated architectures that could be defined. The basis on which this allocation is done could be formulated as a multi-objective optimization problem:

1. Maximize the fundamental objective (must be based upon analysis using the fundamental objectives hierarchy). Note that besides common operational performance parameters, there are often other elements of the fundamental objectives concerning performance in other phases of life cycle, for example, maintenance, deployment, and refinement, about which to be concerned.
2. Minimize the number and complexity of interfaces. This is often called *modularization,* which is nearly synonymous with maximizing the ability to encapsulate the functions inside the physical entities of the system. By encapsulation, we mean the ability to hide the implementation details of performing the entity's functions from the remaining parts of the system. Essentially, the remainder of the system should only need to know the outputs of each entity, not how those outputs are produced. Software engineers call this *information hiding.* The concepts of modularity and information hiding are also highly related to the concept of *coupling.* Many systems and software engineers distinguish between tight and loose coupling. Loose coupling decreases complexity and enables flexibility but often degrades performance. Wikipedia has a nice description of the many types of coupling found in systems.
3. Maximize early critical testing opportunities so as to give engineers a chance to find and fix problems. This is often considered *risk minimization.* Opposing criteria may minimize risks:
   a. Equalizing risks (difficult requirements) across the physical architecture
      or
   b. Localizing risks in a single element of the physical architecture (the opposite of equalizing risks)

9.3.2 Approaches for Solving the Allocation Problem

In the 1950s and 1960s, the major tradeoffs addressed by engineers consisted of choosing between the human in the system and the system's combined hardware and software resources for performing certain critical functions. In the 30–40 years since systems engineers first grappled with these decisions, systems engineers are still using heuristics to resolve these decisions. The engineering and psychology communities believe that there are certain functions that humans perform better than machines, at least in many situations; there is no complete agreement about what these functions are, for example, pattern recognition functions, improvisation, and adaptation. Similarly, hardware and software combined clearly outperform humans in tasks that require responding quickly to control signals, performing repetitive tasks, and performing many different activities at once. Paul Fitts [1951] was the first to try to systematize these allocation issues by producing what has come to be known as a "Fitts' list" and later known as "Men are better at – machines are better at" or "MABA – MABA." Fitts' first list is shown in Table 9.1.

Sheridan and Verplanck [1978] developed a taxonomy of 10 possible distribution strategies for allocating the functional responsibility of control between the human and the computational resources of the system. These allocation strategies range from having the human be the planner, scheduler, optimizer, and the like to take the human out of the system's functions completely; see Table 9.2. For example, the first distribution in the table puts the entire cognitive load on the human, which reflects automation in the 1960s and 1970s, such as machine tools. Entries 5 and 6 reflect the computer developing suggestions for actions but letting the human have approval or intervention capability; this reflects much of the automation in military systems today. Entries 7–9 reflect the status quo in autopilots for aircraft and trains.

TABLE 9.1   Original Fitts List from 1951

Humans Appear to Surpass Present-day Machines with Respect to the Following:	Present-day Machines Appear to Surpass Humans with Respect to the Following:
1. Ability to detect small amounts of visual or acoustic energy.	1. Ability to respond quickly to control signals, and to apply great force smoothly and precisely.
2. Ability to perceive patterns of light or sound.	2. Ability to perform repetitive, routine tasks.
3. Ability to improvise and use flexible procedures.	3. Ability to store information briefly and then to erase it completely.
4. Ability to store very large amounts of information for long periods and to recall relevant facts at the appropriate time.	4. Ability to reason deductively, including computational ability.
5. Ability to reason inductively.	5. Ability to handle highly complex operations, i.e., to do many different things at once.
6. Ability to exercise judgment.	

TABLE 9.2   A Taxonomy of the Distribution of Responsibility between Human and Computer

1. Human does all planning, scheduling, optimizing, etc., and turns task over to computer merely for deterministic execution.
2. Computer provides options, but the human chooses between them, plans the operations, and then turns task over to computer for execution.
3. Computer helps to determine options, and suggests one for use, which human may or may not accept before turning task over to computer for execution.
4. Computer selects option and plans action, which human may or may not approve, and the computer can reuse options suggested by human.
5. Computer selects action and carries it out if human approves.
6. Computer selects options, plans, and actions and displays them in time for human to intervene, and then carries them out in default if there is no human input.
7. Computer does entire task and informs human of what it has done.
8. Computer does entire task and informs human only if requested.
9. Computer does entire task and informs human if it believes the latter needs to know.
10. Computer performs entire task autonomously, ignoring the human supervisor, who must completely trust the computer in all aspects of decision-making.

Now that computer-based systems and embedded computer systems are much more sophisticated and prevalent, the most critical functional allocation decision facing systems engineers often relates to the allocation of a function between hardware and software. Allocating a function to hardware has the benefit of reduced development cost and faster processing and response time. The advantages of allocating to the software are the flexibility to modify the function in the future as design problems are found, or new algorithms prove superior in terms of timing, quality, or quantity measures.

Price [1985] developed the principles (Table 9.3) for functional allocation that are primarily related to allocating functions between humans and machines but which, when generalized, relate to all functional allocation decisions. Principles 2 and 4 emphasize the creative nature of design that was emphasized in Chapter 8 on physical architecture; this creativity applies equally to functional architecture and the allocated architecture. Principle 3 supports the use of decision analysis (see Chapter 14) for systematizing the decision process.

Capturing requirements for the refinement phase of the system's life cycle is the point of principle 5. The Vee model of the systems engineering process is compatible with principle 7. The process model for the allocated architecture, shown in Figure 9.1, supports principle 9.

The essence of Price's principles is that the allocation of functions to elements of the physical architecture involves conflicting objectives. Making this selection even more difficult is the fact that the systems engineering team has to evaluate objectives in more than one-time span, for example, short-term performance versus future performance after possible upgrades have been completed. For these types of allocation decisions, the decision analysis approach covered in Chapter 14 is recommended. The core of this approach is the use of an appropriate part of the objectives hierarchy that contains all of the key performance requirements and their stakeholder tradeoffs. Figure 9.4 illustrates such an objectives hierarchy for a hypothetical decision.

TABLE 9.3   Price's Functional Allocation Principles

1. *Allocation is part of design*: allocation is one part of a larger process.
2. *Allocation is invention*: there is no formula for allocation, imagination is crucial to the success of the process.
3. *Allocation can be systematized*: the inclusion of imagination and invention does not preclude formalizing allocation as a rational decision process, combining invention, and systematization yields a superior result.
4. *Make use of analogous technologies*: building upon allocation decisions and their resulting successes and failures expands our allocation expertise.
5. *Consider future technology*: allocation decisions cannot be based on what exists now but must address expected advances in technology.
6. *Consider human optimization (realistic system implementation)*: allocation cannot be based upon idealistic expectations of how the system will be realized but should be based upon the likely capabilities of the system in its environment.
7. *Use cycles of hypothesis and test*: like any other part of system design, we are not smart enough to do it right the first time, so build in stages of and time for iteration.
8. *Provide interaction*: there are three design decisions that cannot be completely separated. The engineering decision of what the physical resources of the system are, the functional allocation of which functions will be performed by each system resource, and the detailed design decision that implements the allocation. There must be interaction among these decisions during the design process.
9. *Provide iteration and decomposition*: do not make the allocation final too quickly.
10. *Develop tools of cognitive analysis* (human–machine allocation only).
11. *Assure interdisciplinary communication*: involve experts from all relevant fields in the allocation process.

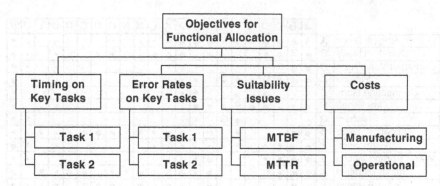

FIGURE 9.4   Sample objectives hierarchy for functional allocation.

Another perspective on this allocation problem involves the use of design structure matrices. See Browning [2001] for more information. The design structure matrix (DSM) is meant to capture interactions of all sorts between functions so that intelligent combinations of functions into components can be derived based on the richness of the interactions. This is a bottom-up approach to the allocation problem, while we have previously been talking about this task as if it could only be approached from a top-down perspective. As discussed in the functional architecture chapter, there are many systems engineers who prefer the bottom-up approach.

As an example of a DSM application, consider the creation of a development system architecture for the small block V-8 engine at General Motors [Eppinger, 1997]. This engine effort called for 90% of the parts and 80% of the manufacturing equipment to be redesigned. As a result, 22 product development teams (PDTs) were created, as shown in Figure 9.5. In an effort to determine the best way to organize the concurrent efforts of these PDTs, the interactions among the teams were documented and categorized as monthly, weekly, or daily. The matrix in Figure 9.5 is an example of a DSM. The three-sized dots represent these three levels of interaction. Note the DSM is not symmetric because the rows represent where the input to a team is coming from, while the columns represent which teams are receiving a given team's outputs. So, the second column of the first row indicates which kind of interaction is needed for an input to PDT A from PDT B. This is the opposite representation of an N^2 diagram.

The main analytic concept behind DSMs is that the information in the matrix provides a clue as to how to rearrange the rows and columns so that clusters form along the diagonal of the reorganized matrix. These algorithms date back to the 1970s. Figure 9.6 shows such a rearranged matrix with four clusters along the diagonal for four aggregations of the PDTs that should prove very useful. Note the last PDT is the assembly PDT; it interacts with so many PDTs that it does not belong to any aggregate team.

So far, the functional allocation decision process has been addressed as if the decisions had to be made during the design process and could only be modified during system upgrades. However, the computational resources that are now available for insertion into systems permit the design to include the real-time reallocation of functions to predefined resources. Typically, this reallocation is between human and computer (hardware and software), or between one hardware resource and another, each running the same set of software. Examples of this dynamic reallocation include distributed processing architectures, parallel processing architectures, flexible manufacturing systems, and sophisticated command and control systems. This material is beyond the scope of this book; the interested reader is referred to Chu and Tan [1987]. Gobinath and Gupta [1990], Levis et al. [1994], and Perdu and Levis [1993]. Jackson [2007] makes a strong case for an adaptive allocation of functions to components to develop more adaptive and resilient systems.

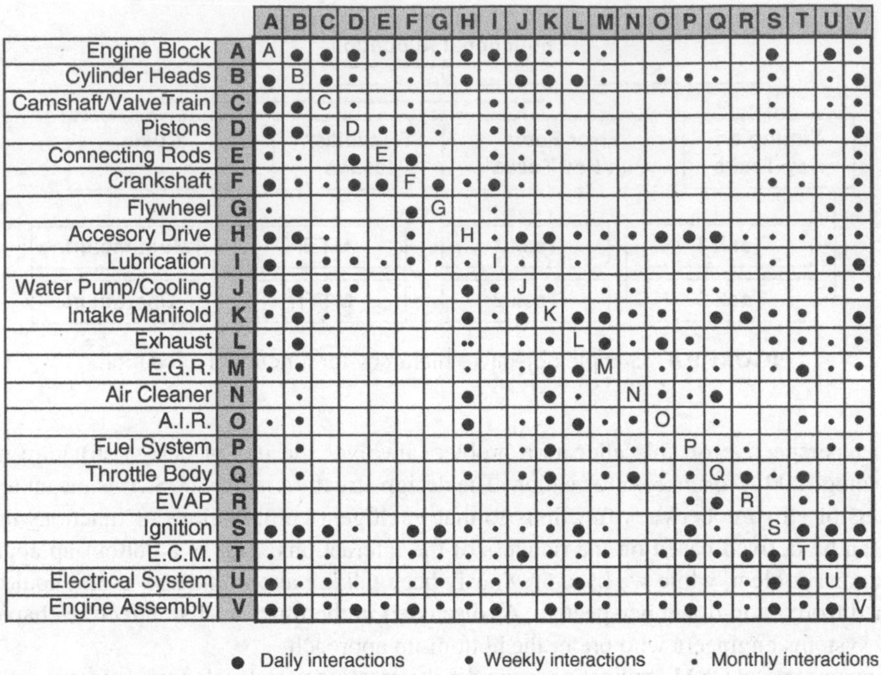

FIGURE 9.5 Interactions among PDTs for the small V-8 Engine Project at General Motors. (Adapted from Eppinger [1997].)

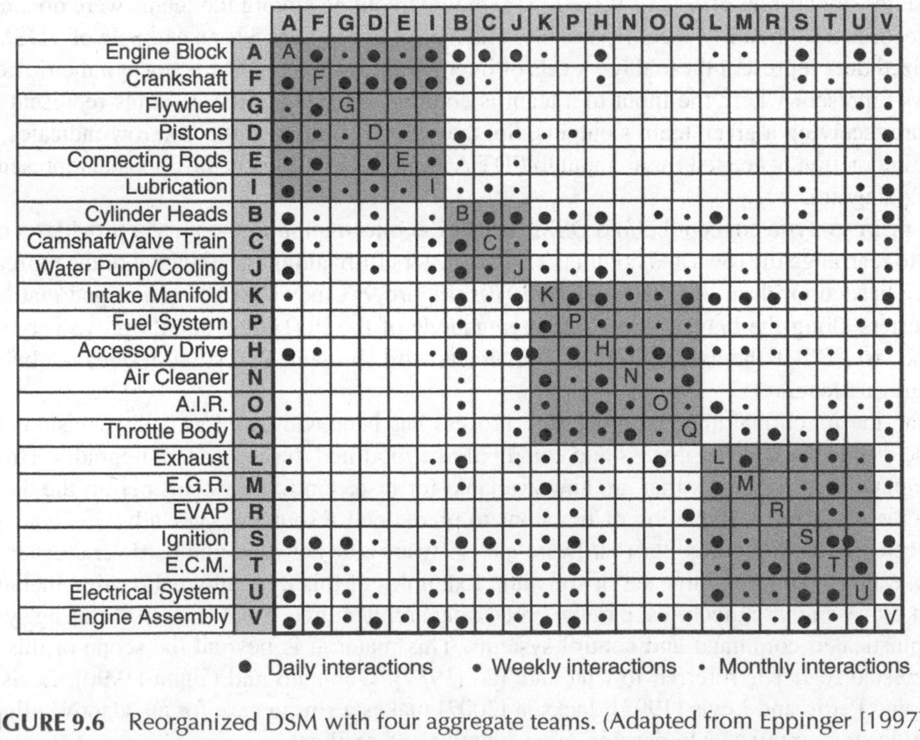

FIGURE 9.6 Reorganized DSM with four aggregate teams. (Adapted from Eppinger [1997].)

9.3.3   Finishing the Allocation Problem

Part of the critical documentation that is part of systems engineering is capturing the allocation of functions to the system and the system's components. Every bottom-level function in the functional decomposition should be allocated to one component of the physical architecture, or physical decomposition, as discussed in Figure 9.3. This physical decomposition begins with the system as the root of the tree. The top-level system function, or root of the functional decomposition, is allocated to the system. The functions at the first level of functional decomposition are then allocated to one component on the first level of the physical decomposition. This allocation of the first level of functions may be the level of detail achieved in the first iteration through the engineering of the system (or first peel of the onion). In Systems Modeling Language (SysML), this allocation of functions to components is shown by adding the components as mechanisms to the function blocks, thus creating a representation of the allocated architecture. See Figure 9.7 for an example of this depiction using SysML. Both SysML and IDEF0 models for the elevator are available on the companion website; see the section called allocated architecture. See the graphical modeling techniques on the companion website to see the allocation of functions to the system and the system's components utilizing the entity-relationship diagram. Each iteration through the engineering of the system process adds another layer of bottom-level functions and components to the functional and physical architectures, respectively. Each bottom-level function will then be allocated to one component.

To obtain an executable model of the allocated architecture, later discussions will make it clear that the only allocation of functions to components that matters is the allocation of functions at the bottom of the functional architecture to components at the bottom of the physical architecture. However, it is highly recommended that an executable model be created of the allocated architecture at several stages in the engineering of the system. Therefore, it is highly valuable to have a running record of the allocation of functions to components, so that this executable model is available at any level of abstraction needed.

As discussed in Chapters 6 and 7, there are tremendous benefits obtained by having the functional decomposition match the physical decomposition on a one-to-one basis. That is, for each function in the first level of the functional decomposition, there is one and only one component to which to allocate the function. In addition, every component must be allocated to one and only one function. This one-to-one mapping of functions to components must continue to the second and all subsequent levels of both the functional and physical architectures. (Note this definition of a one-to-one allocation of functions to components is consistent with the definition of a one-to-one function in Chapter 4.) Such a convenient mapping of functions and components can only occur if the functional and physical architectures are developed in concert with each other. The benefit of this one-to-one mapping is the ease with which input and output items can be allocated to external and internal interfaces. The true value of this matching is covered in the next chapter.

9.4   TRACE NON-INPUT/OUTPUT REQUIREMENTS AND DERIVE REQUIREMENTS

In Chapter 7, on the functional architecture, the discussion of tracing requirements addresses the input/output requirements. These input/output requirements are traced to specific functions in the functional architecture. When the functions are allocated to the components as described above, these input/output requirements are associated with components. There remain several issues, though, to complete the derivation of requirements for each component in the allocated architecture: deriving additional input/output requirements for each function based upon internal items that the architecture needs, tracing system-wide and technology requirements to the system, and deriving appropriate component-wide and technology requirements for each of the components, tracing tradeoff requirements to the system and deriving tradeoff requirements that are appropriate for each component, and tracing test requirements to the system, followed by the derivation of verification requirements for each component.

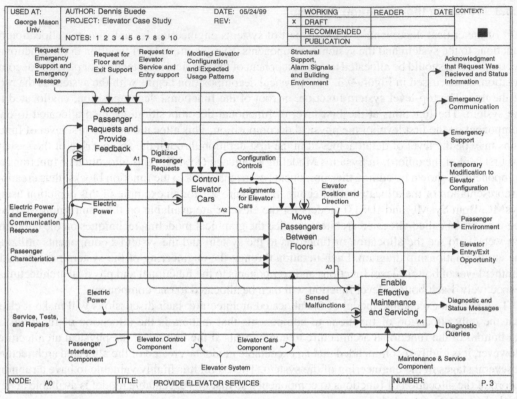

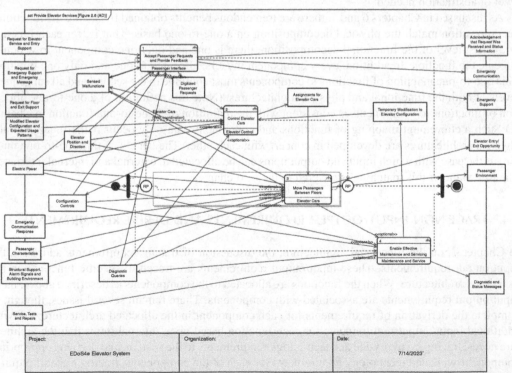

FIGURE 9.7 Allocating functions to components using SysML.

9.4.1   Derive Internal Input/Output Requirements

Deriving input/output requirements based on internal items that the system must create and use is not a difficult process if a graphical model (e.g., IDEF0, data flow diagram, or N^2 chart) of the functional and allocated architectures exists. Once the functions have been allocated to the components, derived input/output requirements can be created based upon internal items (inputs and outputs) appearing in the functional architecture. Figure 9.7 shows the allocated architecture for the elevator case study that can be downloaded. There are five internal items that are created by one function and consumed by another function at this first level of the allocated architecture: digitized passenger requests, assignments for elevator cars, elevator position and direction, sensed malfunctions, and temporary modification to elevator configuration. A derived input and output requirement would have to be created for each of these items. Each of these derived input and output requirements would be traced to both the item and the functions responsible for consuming and creating the item, respectively. For example, Figure 9.7 shows that "Digitized Passenger Requests" is an internal item produced by the first top-level subfunction and sent to the second top-level subfunction. For this one internal item, two derived requirements would be created:

> The elevator system shall produce digitized passenger requests.
> The elevator system shall consume digitized passenger requests.

Each of these derived requirements would be traced to the item "Digitized Passenger Requests"; the first derived requirement would be traced to the function "Accept Passenger Requests & Provide Feedback," while the second derived requirement would be traced to the function "Control Elevator Cars." Additional performance requirements for "Digitized Passenger Requests" would be created if appropriate.

9.4.2   Trace System-Wide Requirements and Derive Subsystem-Wide Requirements

Tracing the system-wide and technology requirements to the system is a very easy process. Almost all these requirements will be traced to the system, although it is possible that some of these requirements should be traced to specific components that comprise the system. The most common example of this is a technology requirement such as "the system shall employ 'abc' technology." A technology requirement that can be traced to a subset of the components of the system should be traced to just those components to which it applies.

However, the difficult portion of this task is the derivation of new requirements for the components based upon the system-wide requirements traced to the system. For example, there may be a cost requirement that says, "the system shall cost $1000 or less to use per month during its operation." How do we allocate, or "flowdown," this requirement among the components of the system?

Grady [1993] identifies three techniques that are used for flowdown: apportionment, equivalence, and synthesis. Apportionment spreads a system-level requirement among the system's components of the system, maintaining the same units. Apportionment is appropriate for cost requirements; the system-level cost requirement is divided or apportioned out to the system's components, not necessarily in equal increments. Keeping a margin, 5–10%, in reserve as a risk mitigation strategy is not uncommon. For example, if the operating cost for the system is to be $1000 or less as suggested above for the elevator, the four components of the elevator shown in Figure 9.7 may be apportioned operating cost requirements of $40, $60, $800, and $50, respectively, with $50 held as risk mitigation.

Other examples for which apportionment is used are reliability, availability, and durability. In fact, the suitability (or quality or "ilities") requirements are commonly apportioned from the system to the components. Note that it is not required that the apportioned values sum to the system-level requirement, as is the case of cost when the margin is included. If the system's components work in series, the component values for reliability will be larger than the system reliability. For the elevator case study, the minimum threshold for reliability is 0.9, with a design goal of 0.99. The four components identified in Figure 9.7 all have to be operational for the elevator to be operational, so they are working in series. The apportioned reliability thresholds for these components may then be 0.96, 0.995, 0.96, and 0.99; the product of these four numbers is 0.91, which provides a margin of a bit less than 0.01 for risk mitigation. Similarly, there would be design goals apportioned to the four components of 0.996, 0.9995, 0.996, and 0.999, respectively. An example of a derived reliability requirement is:

> The elevator component, Passenger Interface, shall have a reliability of 0.96 or greater. The design goal is 0.996.

Equivalence is a simple flowdown technique that causes the component requirement to be the same as the system requirement. An example of a requirement to which equivalence is appropriate is "the system shall be olive green in color." Requirements for which equivalence is appropriate for flowdown are almost always constraints.

The more complicated technique for flowdown is synthesis. *Synthesis* addresses those situations in which the system-level requirement is comprised of complex contributions from the components, causing the component requirements that are flowed down from the system to be based upon some analytic model. The system-level requirement will have significantly different units than the derived component requirement. In this case, an analytic or simulation model must be developed and analyzed to determine how to take the system-wide requirement and derive component requirements. In fact, this approach is most often used to derive requirements associated with outputs or inputs of the system, such as accuracy, range, or thrust. For the elevator case study, there is an output requirement relating to the average time between the passenger making a request and being delivered to the requested floor. This system-level requirement would be flowed down via synthesis to all four components shown in Figure 9.7.

9.4.3   Trace Tradeoff Requirements and Derive Subsystem Tradeoff Requirements

Deriving tradeoff requirements that are appropriate for each subsystem follows tracing the system's tradeoff requirements to the system. This derivation is based upon the system-wide tradeoff requirements. This step is the third element of requirements derivation that is part of finishing the allocated architecture.

The tradeoff requirements developed for the system all address tradeoffs for cost, for schedule and performance, and for cost with schedule and performance; tracing all of these requirements to the system is therefore appropriate. Each of these tradeoff requirements is related to an individual input, output, or system-wide requirement. Based upon the derivation of requirements for each of these input/output or system-wide requirements, it is straightforward to develop an objectives hierarchy for each component, as shown in Figure 9.8. Generally, every element of the system's objectives hierarchy that is related to a system-wide requirement will also become part of the objectives hierarchy for each component; cost, schedule, and suitability requirements are generally flowed down to every component, as discussed above. Similarly, it is inappropriate to create a component-wide requirement when there is no system-wide requirement from which the component-wide requirement can be derived.

Before moving on to input/output requirements, the derivation of ranges for each system-wide requirement, the associated value curve over the derived range, and the weight to be assigned to that range must also be addressed. First, the two extremes of the value range must flow down from the

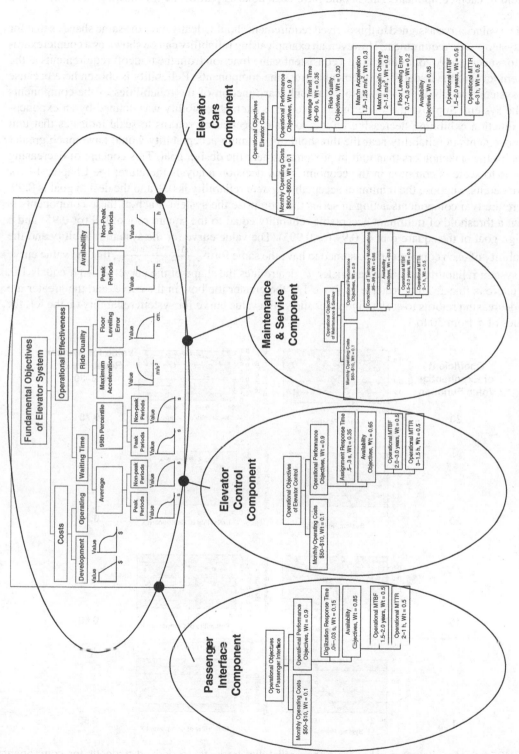

FIGURE 9.8 Derived objectives hierarchies for the elevator case study.

system to each component. This should have been done as part of the flowdown process described above.

The value curve assigned to this derived requirement should ideally have the same shape as that for the system-wide requirement. However, an example using reliability can be shown as a counterexample for successfully communicating a consistent value function from the tradeoff requirements at the system level to the tradeoff requirements across the components. Reliability is chosen here because the system's reliability is known to be a nonlinear function of the reliabilities of the components of the system. Suppose the value function for the system's reliability was defined by an exponential function exhibiting decreasing returns to scale. Decreasing returns to scale indicates that unit improvements in reliability near the threshold of minimum acceptability would have much greater value to the stakeholders than unit improvements near the design goal. This concept of decreasing returns to scale is common in the economics and decision analysis literature; see Chapter 14 for more details. Suppose the minimum acceptable system reliability is 0.9, and the design goal is 0.99. There are two components acting in series that comprise the system. Each of these components is given a threshold of minimum acceptable reliability equal to the square root of 0.9 (or 0.95) and a design goal of the square root of 0.99 (or 0.995). The value curve for the system reliability and the reliability of each component is assumed to have the same form, $\frac{(1-e^{-\alpha(r-r_{min})})}{(1-e^{-\alpha(r_{max}-r_{min})})}$, that the value curve for system reliability had. The parameter, α, determines the shape of the curve. When α equals 1.0, the curve is linear. The greater α is above 1.0, the greater the bow in the curve, and the greater are the decreasing returns to scale. Figure 9.9 shows the value curve for system reliability on the left for values of α from 30 to 1.

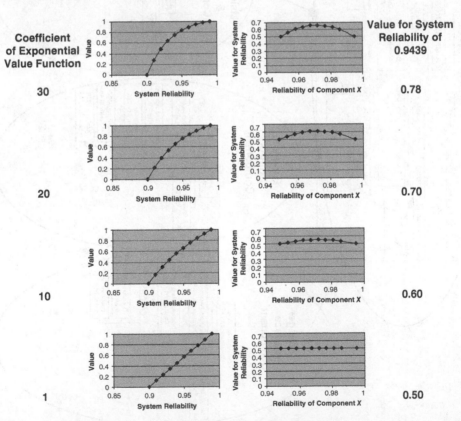

FIGURE 9.9 Sensitivity of value for system reliability tradeoffs to derived tradeoffs for component reliabilities.

The right-hand graphs in Figure 9.9 show the value for system reliability as a function of the reliability for the first component, X, when the system reliability is held constant at 0.9439. In each of the graphs on the right, the value is computed by the weighted average of the values for the reliabilities of the two components:

$$\text{Value} = 0.5 \, v \, (\text{reliability of component X}) + 0.5 \, v \, (\text{reliability of component Y})$$

The weights for the two components are assumed to be equal since the distance from the threshold to the goal is the same for both. As can be seen in Figure 9.9, the value for the reliabilities of the two components is not constant over the range of values for the reliability of the first component, even though the reliability of the system is being held constant. The numbers to the far right of Figure 9.9 show the value for the system's reliability when the system's reliability is held constant at 0.9439. The values in the right-hand graphs are also not equal to these numbers except for the case of the linear value curves. This suggests that only linear value curves should be used for tradeoff requirements.

The final issue in deriving tradeoff requirements for each component concerns those tradeoff requirements that address the quality, quantity, or timeliness of the system's inputs or outputs. Each of these input and output requirements will already have been traced to a function that was allocated to a component. Therefore, each tradeoff requirement for an input or output can already be associated with one component, assuming the allocation mapping of the input/output requirement to functions was one-to-one. A complicating issue, however, is that there may be good reasons to create a tradeoff requirement for an input or output requirement that was derived on the basis of the need for an internal item produced and consumed by the functional architecture. An example of this in Figure 9.7 is the "Digitized Passenger Requests" for the first sub-function. This internal item is related to the elevator objective of "Waiting Time," as shown in Figure 9.8. Such a tradeoff requirement must be traceable to a performance aspect of a stakeholders' input/output requirement; nonetheless, it is the only case when the objectives hierarchy will have an element that is not identical to an element of the system's objectives hierarchy.

9.4.4 Trace Qualification Requirements and Derive Subsystem Qualification Requirements

The final element of completing the requirements development for each individual component is tracing the qualification requirements to the system and then deriving qualification requirements for each component. Recall that the four categories of qualification requirements are observance, verification plan, validation plan, and acceptance plan. These last two categories only apply to the system. Therefore, after all qualification requirements have been traced to the system, derived requirements for the components are developed only from the first two categories (observance and verification). This derivation process is quite straightforward; observance requirements relate to specific input/output and system-wide and technology requirements. Therefore, deriving observance requirements follows the derivation process of input/output and system-wide and technology requirements. Deriving a verification plan for each component should be relatively straightforward, given the verification for the system.

9.5 DEFINE AND ANALYZE FUNCTIONAL ACTIVATION AND CONTROL STRUCTURE

When discussing IDEF0 and SysML (Chapter 3) and functional decomposition (Chapter 7), the need for activation and termination criteria is mentioned. That is, there are criteria that need to be established for each function; these criteria determine what set of inputs (and associated values) will activate the function and what set of outputs (and associated values) are sufficient to terminate the function. The bottom-level functions in the functional architecture must have their activation and termination criteria completely specified. The intermediate- and top-level functions are aggregates of the bottom-level functions and, as such, are for modeling purposes only; intermediate- and top-level functions do not have or need activation and termination criteria. However, recall the previous discussions of peeling the onion and the fact that the bottom-level functions of an early peel of the onion become intermediate-level functions during later peels of the onion.

In addition to the activation and termination of a function, the conditions under which one function precedes or follows another function's processing must be clearly defined. Examples of approaches to defining such precedence conditions are found in the graphical modeling techniques on the companion website under behavioral modeling. Most of these behavioral modeling methods allow the dynamics of the system to be explored by providing an executable model of the system's functions. These executable models are either discrete-time or discrete-event simulations when implemented on a computer. The reader is referred to the graphical modeling techniques on the companion website for more detailed discussions on this subject.

Before discussing the dynamic issues associated with the performance of a system, the balancing or aligning [Yourdon, 1989; Schmekel and Wingard, 1993] of multiple models of a system should be addressed. At this point, the functional architecture contains a data model and a process model of the system in question. The generation of activation and termination conditions for each function plus the control structure associated with the concurrent or asynchronous behavior of functions with respect to each other is contained in the behavioral model of the system. Yet each of these models contains overlapping data elements: Inputs and outputs are in all three models, and functions are in the process and behavior models. These models better be consistent and coherent representations of each other, or their results will be worthless to the engineers of the system; in essence, the engineers will have modeled several different systems while thinking they were addressing only one system. Schmekel and Wingard [1993] present the most complete treatment of this topic known to the authors.

There are several benefits of executable models. First, the design can be explored to find major design flaws that are manifested as deadlocks, livelocks, starvation, surge or race conditions, or oscillatory conditions. The second major benefit is to permit the systems engineering team to assess the degree to which the design meets various timing, throughput, and other performance requirements.

Deadlock, livelock, starvation, surge (race), and oscillation are dynamic characteristics that are not desired in dynamic, time-varying systems. *Deadlock* is an undesired state of the system in which activity ceases, and throughput is nonexistent. Deadlock can occur for two reasons: contention over resources and waiting for communication [Levi and Agrawala, 1994]. Contention over resources occurs when each of several components requires the same resource for a task, but none of the components is willing to free the resources it has accumulated; as a result, activity stops while the components wait for additional resources to complete their assigned tasks. Waiting for communication occurs when various components are attempting to synchronize their actions or verify their status; in either case, each component enters a state called "wait for communication," but the communication never arrives because the components are in a strongly connected wait state.

Deadlock associated with resources is often described using the "dining philosophers" problem. There are five philosophers sitting around a circular table, preparing to eat spaghetti. There are five forks, one between each of the adjacent philosophers. Before eating the spaghetti, each philosopher requires two forks to move the spaghetti from the bowl in the middle of the table onto her/his plate. If each philosopher grabs (and locks) the fork on the left, no philosopher will be able to eat; this is a deadlock. The solution requires the creation of a conditional locking mechanism on the forks by the philosophers that ensures that each philosopher obtains both forks for a limited time to move the spaghetti to her/his plate. After completing this initial task, each philosopher then releases both forks for a period of time. Once each philosopher has spaghetti on her/his plate, only one resource is required by each and all five philosophers can eat simultaneously.

Graph theory is often used to depict the resource-sharing problem with what is called a "wait-for-resource" graph. Define each component as a node. Define the relation R to be "awaits a resource possessed by." Figure 9.10 shows a system with four components in which there is a potential deadlock involving the first three components. Mathematically, it can be shown that any system having a wait-for-resource graph with a cycle can become deadlocked if several other conditions apply [Levi and Agrawala, 1994]. If there are many components and the wait-for-resource graph is complex, the existence of a cycle may not be obvious by inspection. Typical solutions to eliminating or reducing the chance of deadlock due to resource contention are to oversize buffers and resource pools, reduce the concurrency of operations, add delays, institute a manual or automated deadlock detection and recovery process, and allow preemption of locked resources. Ferrarini and Maroni [1997] define three generic categories of options: avoidance, prevention, and recovery.

A "wait-for-communication" graph can be used to examine the possibility of deadlock due to communication. In this case, a cycle (with other conditions) is not sufficient to guarantee deadlock; a strongly connected, cyclic graph is necessary. Deadlocks have been studied in communication systems [Duato et al., 1997] for a long time, and procedures have been embedded into most communication protocols to break communication deadlocks when they occur.

Livelock is a dynamic condition with the same result as deadlock but for a different reason. In deadlock, the system (or part of the system) halts activity because various activities are holding or utilizing resources needed by other activities. In *livelock* the resources are being routed in cycles (oscillating) while waiting for the proper allocation of resources to enable the completion of necessary activities; unfortunately, this proper allocation of resources is never achieved, and the system cycles continuously, never reaching the desired outputs. In communication networks, livelock can only occur when information packets are permitted to traverse paths that are not minimal.

Starvation occurs when a function needs a particular resource for execution, but the resource is always allocated to other functions due to a poorly designed resource assignment algorithm. This condition is one that can be found with little trouble as long as a reasonable effort is made to model the dynamics of the system. However, it can easily be overlooked if no effort is devoted to examining the system's dynamic properties.

The dynamic condition called *surge* or *race* occurs in relatively uncontrolled systems when components are competing with each other to perform a task. A common example is found in older elevator systems during nonpeak times; a potential passenger pushes the up button and *observes* that all of the stationary elevator cars are converging on her floor. She gets into one of the elevator cars. The next passenger now pushes the down button, and the remaining elevator cars surge to that passenger. The surge condition is a waste of resources while it is occurring and can leave the system in an undesirable state for future tasks; all of the elevator cars but one will end up waiting on the same floor for future passengers.

These negative dynamic conditions can be designed into a system inadvertently without the engineers' knowledge *unless* the designers undertake a detailed study of their design. Discrete-event simulations involving Petri nets, queuing theory, behavior diagrams, or extended function flow block

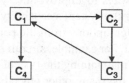

FIGURE 9.10 Wait-for-resource graph depicting deadlock.

diagrams are needed to investigate the design of the system via mathematics and simulation and to understand the degree and extent of such negative behaviors. Naturally, if negative behaviors exist, design changes can be examined to eliminate or minimize them.

9.6 CONDUCT PERFORMANCE AND RISK ANALYSES

A wide range of quantitative analyses is commonly performed during the system development process that fits within the categories of performance, tradeoff, and risk analyses. The parametric diagrams of SysML can be used to design and document these analyses. In fact, these analyses can be considered a system in their own right.

Risk analyses are often completed at the beginning of the development process to examine the major design options under consideration. For example, at the earliest stage of development, the systems engineering team should consider a range of divergent concepts. A *risk analysis* examines the ability of the divergent concepts to perform up to the needed level of performance across a wide range of operational scenarios. At this time, there remains substantial uncertainty about the stakeholders' needs, the state of technology under consideration, and the details of the allocated architecture. The relative costs and schedule implications of the various concepts also have to be taken into account. This is where the stakeholders have to debate how much money and time they are willing to pay for increased performance in selected operational scenarios. Addressing uncertainty and multiple objectives in these early risk analyses is critical; see Chapter 14.

Performance analyses are for the purpose of discovering the range of performance that can be expected from a specific design or a set of designs that are quite similar. The performance parameter in question can be associated with an output of the system or with a system-wide metric; in either case, there is almost always a related objective in the objectives hierarchy and an associated performance requirement. These performance analyses usually take the shape of engineering models and simulation models. The simulation models may be deterministic or stochastic, depending on the issue involved and the experience level of the design team with the technology.

Common system-wide performance analyses address operational feasibility issues such as reliability, availability, maintainability, usability, supportability, durability, and affordability. Similarly, performance analyses are conducted to address concurrent engineering issues related to the impact of the operational system design on the manufacturing, deployment, training, and disposal systems. Blanchard and Fabrycky [1998] provide detailed discussions of many of these topics: design for reliability, maintainability, usability, supportability, producibility and disposability, and affordability. References for detailed analysis of cost, reliability, maintainability, and availability include Blanchard and Fabrycky [1998], Frankel [1988], Pages and Gondran [1986], Pohl [2007], Pohl and Nachtmann [2007], and Sage [1992].

Some organizations have dictated that the system be designed to cost; that is, there is a cost constraint, and the engineering design team has to guarantee that the system will meet this cost constraint. Design-to-cost works best by designing a reduced-capability system with various optional features that can be added if the cost estimates are low.

A *trade study* focuses on finding ways to improve the system's performance on some highly important objectives while maintaining the system's capability in other objectives. Trade studies are focused on comparing a range of design options from the perspective of the objectives associated with the system's performance and cost. For example, aircraft manufacturers always do trade studies focused on the aircraft's weight while maintaining the system's cost, safety, and so forth. Similarly, safety, reliability, and cost are among the many other objectives that are commonly the focus of a trade study.

9.7   DOCUMENT ARCHITECTURES AND OBTAIN APPROVAL

Documenting the system design completely is important. Not only should the key elements of the requirements process (operational concept, external systems diagram, objectives hierarchy, and requirements), and the three architectures (functional, physical, and allocated) be documented, but also the audit trail for how the results were obtained and why they are what they are. In every system development activity, there are many occasions during the life of the system when engineers will want to find out why a particular part of the design is the way it is. This curiosity usually arises because the engineers want to change the design and need to understand the original rationale for the current configuration; there may have been some issues that the current engineers have not thought of that would keep them from making the change they are contemplating. Unfortunately, it is rare to talk to an engineer who went looking for design rationale on any type of system and was successful. The design decisions that are made intuitively and on the spur of the moment (often without even realizing that a key decision is being made) are seldom documented. The design decisions that are made consciously with an explicit analytical approach, such as decision analysis (see Chapter 14), will be very well documented as long as the analysis material is archived properly.

Obtaining approval of the system's design, or allocated architecture, typically requires long meetings with many members of the engineering team and representatives of the stakeholders. A number of key design decisions are revisited, arguing for the value of the systematic development and archiving of the rationale for these decisions. Once the system's allocated architecture is approved, it is quite simple to develop a specification for each subsystem with the information that is available.

9.8   DOCUMENT SUBSYSTEM SPECIFICATIONS

At this point, the system design is complete, and each major subsystem of the system can be documented in terms of its own operational concept, external component diagram, objectives hierarchy, and requirements document. The requirements document for each component, commonly called a specification (or spec for short) includes input/output, technology and subsystem-wide, tradeoff, and qualification requirements.

Shortly after the subsystem design activities are initiated, a preliminary design review should be held with the stakeholders to obtain their input and approval for proceeding further with the subsystem design.

9.9   SUMMARY

The allocated architecture combines the physical and functional architectures so as to meet the stakeholders' requirements and related derived requirements. This combination of the physical and functional architectures requires the allocation of functions to physical resources; at this point, the system's design can be simulated and analyzed in terms of the stakeholders' requirements and operational concepts of the stakeholders. As the physical and functional architectures are integrated, the interfaces of the system (both external and internal) can also be defined and designed.

The processes that comprise the development of the allocated architecture are the allocation of functions to components, the tracing of system-wide requirements to the system, the derivation of requirements, the definition and analysis of functional activation and control structures, the conduct of performance and risk analyses, documenting the allocated architecture, and documenting the specifications.

The allocation of functions to physical resources was addressed in terms of the appropriate objectives for this major decision. From a historical perspective, the most difficult allocation decision is machine versus human. The allocation between hardware and software is also discussed. Ultimately, this allocation process requires tradeoffs between fast and accurate performance of tasks versus the ability to upgrade and change the processes for performing the tasks. As such, decision analysis (see Chapter 14) should he used to evaluate alternate allocation options in terms of the objectives of the stakeholders.

To complete the component specifications, additional requirements (input/output, system-wide and technology, tradeoff, and qualification) must be derived from those that are already available. Examples of these derivations are provided. Three methods for flowing down requirements that were initially traced to the system are also described.

Critical system-wide issues associated with functional activation and control are discussed here. These issues include deadlock, livelock, starvation, and surge (or racing) of the system.

Decision analysis is discussed as a normative model for conducting risk analyses, performance analyses, and trade studies. An illustration of a risk analysis was provided.

The design process has been likened to peeling an onion throughout this book. The development of the allocated architecture should proceed as though an onion were being peeled. The first allocated architecture developed should be for the subsystems of the system at a high level of abstraction (low level of detail). Then, the entire process is repeated at a lower level of abstraction (greater detail) for the components of the subsystems, consistent with the Vee model discussed in Chapter 1. This repetition at lower and lower levels of abstraction yields allocated architectures at higher and higher levels of detail. The advantage of this approach is that as each new peeling begins, the engineers for each component can work their design processes in relative seclusion from the engineers for other components. Each group of engineers has interfaces between their components and other components and the external systems that have been defined at an appropriate level of detail, yielding a coherent set of requirements with which to work. The work of these several teams of engineers will need to be integrated and coordinated at the newest level of detail before the allocated architecture can be completed for this more detailed level of abstraction.

CASE STUDY: WIDE AREA AUGMENTATION SYSTEM OF THE FEDERAL AVIATION ADMINISTRATION (FAA)*
* PROVIDED BY TIM PARKER

The objective of the U.S. FAA Wide-Area Augmentation System (WAAS) is to provide a navigation aid, for use by commercial and general aviation that is derived from the global positioning system (GPS) standard positioning service (SPS). (GPS employs a constellation of 24 satellites, each of which continually broadcasts its position at the time of broadcast.) The GPS satellites provide the radio frequency equivalent of a navigator's optical star fix. However, the accuracy and integrity of the SPS broadcast are not the ultra-high quality that the Federal Aviation Administration (FAA) requires to ensure the safety of civilian aircraft passengers and operators. Therefore WAAS determines the position of the GPS satellites more precisely than the SPS, and broadcasts "corrections" in real-time.

To validate the competing designs, the FAA required each bidder to develop a special analysis tool known as a service volume model. The goal is that specific aspects of the performance of a given system design could be easily synthesized and simulated using computers. The results of the simulations are then useful for understanding the effects of flowing down certain performance allocations as requirements on lower-tier system components such as the placement and number of ground monitoring antennas used for observing the GPS satellites. Because the simulation is capable of representing the dynamic nature of the spacecraft orbits, the tool can analyze the effects of outages resulting from individual or combinations of component failures (i.e., satellites and antenna monitor sites). In the case of this particular procurement, the FAA included a task in the statement of work that described the use of the simulation tool for determining the exact number and location of the monitoring antennas.

The top-level requirements, that the WAAS simulation helps to explore, are the selection and geographic location of the ground monitor sites used to observe the GPS satellites, the number and location of geostationary satellites used to broadcast the corrections, and the coverage area or service volume where the WAAS service is available for use. Additionally, the simulation accounts for certain a priori aspects of the models used to represent the effect on system performance from such phenomena as pseudo-range measurement error due to receiver noise, signal propagation delay due to the ionosphere, satellite clock estimation error, and satellite clock dither prediction error, that is, selective availability [Braasch, 1990; Kee et al., 1991]. The flowdown to the components is quite involved since there is a dynamic relationship among the number and geographic location of the monitor sites, the GPS satellites, and the a priori characteristics of the systems algorithm. Suffice it to say that an acceptable result based upon specific a priori assumptions could flow to several components, the allowable receiver noise at the ground monitor site, the location and number of ground monitor sites (i.e., ground monitor site geometry), the number and location of the geostationary satellites for broadcasting the corrections, and the resulting coverage area or expected service volume, which is a function of both the geosatellite antenna pattern (i.e., footprint) on the surface of Earth as well as the geometry of the ground monitor sites.

To support the precision approach phase of aircraft flight operations. WAAS must deliver data to the user in the form of corrections for each GPS satellite's position and clock. This data, when applied to determine the position of a given user, should yield an answer that is accurate to better than 7.6 meters (in both the vertical and horizontal dimensions) 99.9% of the time throughout the coverage area.

A simple way to recognize how this relates to the problem of determining the number and placement of the monitor sites is to first understand that the problem that WAAS solves is essentially the navigation satellite user's problem inverted. By this we mean that normally the user of the GPS is concerned with tracking at least four satellites whose spatial relationship to each other and to the user, represented by a unit less value known as geometric dilution of precision (GDOP), satisfies the expression GDOP < 7. Visualize this relationship as an inverted pyramid with the user at the apex and each of the four vertices of the base representing a GPS satellite. Simultaneously solving the equations for the range measurement between the user and each of the observed GPS satellites yields the user's position.

Now recall that the problem that WAAS must solve is to correct the broadcast position and clock of each observed GPS satellite based upon the precisely known location of a set of ground monitor stations. Imagine the ground monitor sites as independent observers of the GPS satellites sharing a universal clock. For a given satellite's position, the ground monitor stations become the vertices of the base of a polyhedron whose vertex is represented by an observed GPS satellite. The spatial relation between the monitor stations and the satellites is analogous to the relation between the user and the satellites. Through the use of a continuous Kalman filter, the WAAS arrives at an ensemble solution for each satellite that is observed by its network of ground monitor stations.

The top-level physical architecture for WAAS allows for up to 70 monitor sites to be constructed and networked into four master control sites. As might be expected with any complex system of this nature, the nonrecurring engineering costs are daunting, and every effort is made to reduce them. Naturally, the FAA would not build all 70 ground sites and then determine if fewer could be used. Instead, the simulation tool is utilized to predict the system performance when specific combinations of components are synthesized together as a working system. Early results published by Lockheed Martin Federal Systems (LMFS) (prior to acquisition by Lockheed, LMFS was originally the Federal Systems Division of IBM) indicated that based upon their simulation results, a far smaller number of ground antennas would be necessary. The analysis used the LMFS Service Volume Model (SVM), a high-fidelity covariance-based simulation tool used to determine user-obtainable navigation accuracy and service availability.

In addition to these analysis results, LMFS undertook the development and fielding of a Wide-Area Differential Global Navigation Satellite System (GNSS) Testbed; see Figure 9.11 for the physical architecture block diagram. The purpose of the testbed, like the simulation, was to further develop knowledge about the allocated architecture and confirm the performance of the algorithms being considered for use on WAAS. A critical activity during the testbed's lifecycle was its deployment into an operational environment. For this task, the SVM simulation tool was used to determine optimal locations for the GPS receivers and ground antennas.

The top-level system objectives to be optimized for the testbed are easily expressed as (1) minimize user range error, (2) maximize the area of geographic coverage where the user range error is 7.6 meters or less, 99.9% of the time, and (3) minimize the cost (i.e., deployment and operational). The first two components require the use of the simulation while the third component is treated as a simple linear projection of the costs incurred from acquiring the testbed equipment, leasing test laboratory space, and paying periodic operational expenses (i.e., telephone, electrical, technical personnel, and miscellaneous). The results of the simulation were combined with cost data for the prospective sites and evaluated using a simple multi-attribute value analysis technique, which considered the top-level system objectives. Note that the deployment costs were determined to be roughly equal and, for purposes of the analysis, were considered to be equal among each set.

Many preliminary studies were undertaken to identify candidate locations for the ground monitor sites. Typically, these were in the eastern half of the United States and within close distances to one another to minimize travel time for deployment and maintenance. Of the many possible sites, several were conveniently collocated with an existing company facility. The site combinations were evaluated together as a location set. Different sets, or combinations of the sites, were evaluated using the SVM to determine if there would be a significant effect on the expected GNSS testbed performance. The multi-attribute value analysis of the combined simulation and cost results is summarized in Table 9.4. The table contains error values representing the SVM prediction for average vertical position accuracy (VPA) and average horizontal position accuracy (HPA) as well as the average user error, where user error is defined as the root-sum-square of VPA and HPA. Coverage represents the percent of evaluated grid points where the predicted accuracy at each point is less than the required threshold value of 7.6 meters.

Coverage for sets 2 and 3 were significantly worse than for sets 1 and 4, providing justification to eliminate sets 2 and 3 from consideration. Although set 4 meets the objectives of maximum coverage and minimum user error, the high operational cost of set 4 due to the usage of non-company property makes set 4 look inferior to set 1. Set 1 was preferred because it offered a reasonable geometry for determining wide-area corrections, had good coverage, and offered a smaller operational cost even though the average user error was the worst. The average user errors were all so close to each other that this objective was not very meaningful in discriminating among the alternatives.

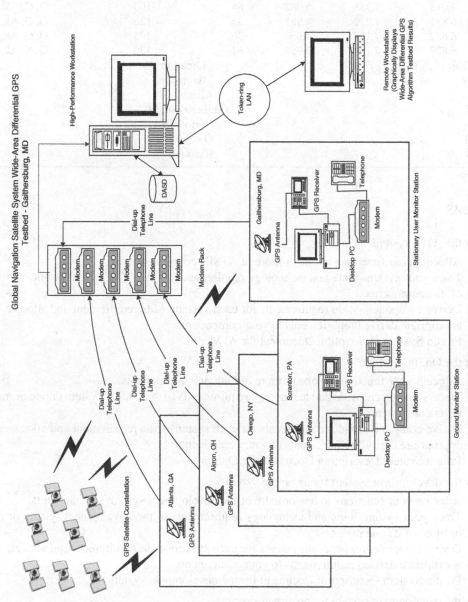

FIGURE 9.11 Physical architecture diagram for GNSS testbed.

TABLE 9.4   SVM Site Location Analysis Summary

Set	VPA A	HPA	User Error	Coverage (%)	Monthly Operational Cost	Sites
1	7.013	7.358	5.082	84	100	O, G, Ak, At
2	6.871	7.219	4.983	36	125	O, G, Ak, N
3	6.837	7.187	4.960	36	105	O, G, Ak, S
4	6.829	7.1	4.953	84	130	O, N, Ak, S

Site Key	Location
Ak	Akron, OH
At	Atlanta, GA
G	Gaithersburg, MD
N	Norfolk, VA
O	Owego, NY
S	Scranton, PA

PROBLEMS

9.1 For the ATM system:

    i. Allocate your functions to one or more of ATM's components.

    ii. Trace your system-wide and technology requirements to the ATM system or one or more of its components.

    iii. Derive component-wide requirements for each system-wide requirement and allocate the appropriate derived requirements to your components.

    iv. Print a System Description Document for ATM.

9.2 For the OnStar system:

    i. Allocate your functions to one or more of OnStar's components.

    ii. Trace your system-wide and technology requirements to the OnStar system or one or more of its components.

    iii. Derive component-wide requirements for each system-wide requirement and allocate the appropriate derived requirements to your components.

    iv. Print a System Description Document for OnStar.

9.3 For the development system for an airbag system:

    i. Allocate your functions to one or more of the development system's components.

    ii. Trace your system-wide and technology requirements to the development system or one or more of its components.

    iii. Derive component-wide requirements for each system-wide requirement and allocate the appropriate derived requirements to your components.

    iv. Print a System Description Document for the development system.

9.4 For the development system for an airbag system:

    i. Allocate your functions to one or more of the manufacturing system's components.

    ii. Trace your system-wide and technology requirements to the manufacturing system or one or more of its components.

iii. Derive component-wide requirements for each system-wide requirement and allocate the appropriate derived requirements to your components.

iv. Print a System Description Document for the manufacturing system.

9.5 A system that is available 90% of the time is said to have one "9" of availability. Of the 365 days in a year, such a system would be "down" about 36 days and 12 hours.

i. A system that is available 99% of the time has two "9's." How many days and hours per year is this system "down."

ii. How many days, hours, and minutes is a system with three "9's" of availability down?

iii. How many hours, minutes, and seconds is a system with four "9's" of availability down?

iv. How many minutes and seconds is a system with five "9's" of availability down?

v. How many minutes and seconds is a system with six "9's" of availability down?

vi. Where does the general class of personal computers fall in this spectrum of availability? Where do you think the air control system of the Federal Aviation Administration for a country should fall in this spectrum? Where does the telephone system fall? Where does your internet provider fall?

Chapter 10

Interface Design

10.1 INTRODUCTION

Interfaces are common failure points on systems. An interface is a connection resource for hooking to another system's interface (an external interface) or for hooking one system's component to another (an internal interface). The systems engineer's design problem includes identifying the interfaces, both external and internal, and allocating items (inputs and outputs) to the defined interfaces. Once these tasks are completed, the requirements for each interface must be derived from existing system-level requirements. Finally, in alternative interface architecture, alternatives must be examined, including the needed functions and the most cost-effective alternative chosen.

The interface requirements must address total system performance, the fidelity of the interface, and any system requirements meant to constrain interface design. Typical system performance requirements of concern in designing the interfaces are system throughput and response time. The fidelity of an interface is determined by the integrity of the items being transported, the guaranteed delivery of the items, and failure in detection and recovery within the interface. In other words, the interface should not change the items during the transmission process, should eventually deliver every item placed on the interface (and not create any items), and should detect faults early and recover gracefully (a hard but important word to define).

Section 10.2 discusses the process for developing the interface designs of the system. Generic architectures, introduced in Section 10.3, can be used as the architectural concept for any given interface. These generic architectures come from communication and computer systems. Section 10.4 discusses the important issue of standards, a major support in the definition and design of interfaces. Sections 10.5 and 10.6 address two major standards, one for communications systems and one for software architectures. The open systems interconnection (OSI) reference model serves as the basis for many standards related to telecommunications and computer networks. This reference model provides a rich basis for viewing interfaces. The common object request broker architecture (CORBA) is an industry standard for software systems integration. Section 10.7 addresses the design of an interface.

The Engineering Design of Systems: Models and Methods, Fourth Edition. Dennis M. Buede and William D. Miller
© 2024 John Wiley & Sons, Inc. Published 2024 by John Wiley & Sons, Inc.
Companion website: www.wiley.com/go/engineeringdesignofsystems4e

The generic interface architectures described in this chapter include message passing, shared memory, and network. Each of these architectures is described, followed by a discussion of strengths and weaknesses.

The OSI reference model and CORBA are introduced as well-conceived architectures for common interfaces. The discussion in this chapter is focused on the functions performed in these architectures so that the engineer of a system has samples of functions to draw from for designing any type of interface.

The *exit criterion* for completing the design of the system's interfaces is acceptance by the engineer responsible for the allocated architecture that the interface is consistent with the system's components and configuration items (CIs) as well as the performance objectives and requirements of the system.

10.2 OVERVIEW TO INTERFACE DEVELOPMENT

An *interface* is a connection for hooking to another system (an external interface) or for hooking one system component to another (an internal interface). The interface of a system contains both a logical element and a physical element (or link) that are responsible for carrying items (electromechanical energy or information) from one component or system to another. The interface must ensure that the item is delivered on time and in the same form as the item was received.

The development of the interface architecture is quite similar to the development of the allocated architecture of a system, as shown in Figure 10.1 [see Appendix B on the Companion website for the entire IDEF0 (Integrated Definition for Function Modeling) Model for Engineering a System]. The functions of defining requirements as well as the functional, physical, and allocated architectures are present. The only new function is the evaluation and selection of a high-level interface architecture; Section 10.3 defines and discusses the three major alternate interface architectures in use today in communication and computer systems. This high-level architecture for the interface is analogous to the concept selection for the system design. Before proceeding very far in the development of a system, high-level concepts, each having a different operational concept, are posited and evaluated.

This decomposition of functions for developing an interface architecture assumes that the functional process will be revisited several times in whole or in part. As interface changes arrive from the process responsible for the system's allocated architecture, the relevant functions for developing the interface architecture are triggered and set the whole process in motion to develop a revised interface architecture.

10.3 INTERFACE ARCHITECTURES

Most interfaces are communication systems or analogies of communication systems (e.g., a conveyer belt). *The principal communication architectures are message passing, shared memory, and networks.* An every-day example of each of these architectures is as follows:

Message Passing: mail delivery that predictably occurs once or twice a day and allows those receiving the mail to turn their attention to the mail immediately or wait until a more opportune time and permits messages of substantial volume. This approach is interrupt-driven communication, which is common to operating system designers.

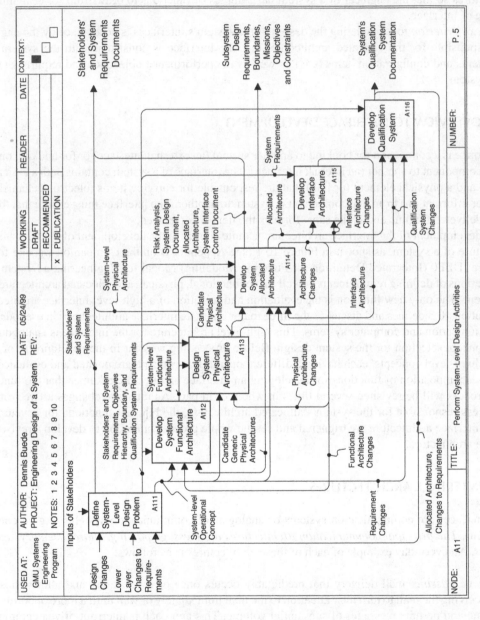

FIGURE 10.1 Development process for the interface architecture.

Shared Memory: a meeting or conference in which only one person speaks at a time and conveys relatively compact messages, all can hear what is said, but are restrained from other productive work during the meeting. This approach is a polling interface, which is natural to application coders.

Network: a telephone conversation that can involve messages of widely varying lengths and can be instigated at almost any time.

10.3.1 Message-Passing Architectures

The message-passing architecture involves multiprocessor systems, in which processors communicate without a global memory. Each processor has its own memory and communicates with the other processors using "send" and "receive" operations. A message is usually defined as an instruction, data, synchronization, or interrupt signal. In synchronous message passing, the sending processor stops executing until the receiving processor finishes processing the called function using the message and then receives the result from receiving processor. There are few system-level applications where this synchronous message-passing approach makes sense. Early synchronous message-passing systems used a bus architecture for communication among the processors. The message that is transmitted over the bus consists of a protocol and data segments. The protocol segment includes any information needed by the bus interchange unit to deliver the message; typically, this is information about the size of the message and address of the node to receive the message. For each transmitted message, the following communication process must he completed:

1. One node must win control of the communication channel through a priority scheme implemented by the system.
2. The winning node becomes the master and sends a protocol segment to the intended receiving node(s), called the slave(s).
3. The slave node(s) notifies the master that the protocol segment was successfully received.
4. The master sends (or receives) the data segment to (from) the slave(s).
5. The slave(s) notifies the master that the data segment transfer is complete.
6. The master surrenders control of the communication channel.

Asynchronous message-passing processors can store messages in buffers and perform send/receive operations at the same time as processing other tasks. Asynchronous message-passing architectures use a variety of interconnection structures: bus, hypercube, and mesh. With a bus interconnection, the bus system stores the sent message until the receiving processor is ready to receive the message. The receiving processor then processes the message and sends a response. Issues can occur here when the buffer on the bus is full. With hypercube and mesh interconnection systems, the processors will buffer the messages with their own cache memory until they are ready to process the message and return a result. Similarly, here cache memory size can become a design problem. Design options within the framework of a message-passing architecture include static versus dynamic and bus versus switch, both of which can be single or multiple.

Since message-passing architectures tend to be tightly coupled, critical design factors are the communication bandwidth (bits per second) and the communication latency (time delay). Two well-known modeling methods to support system designers of message-passing systems are Actor and Pi calculus [Milner, 1999; Sangiorgi and Walker, 2001; Agha, 1985; Hewitt and de Jong, 1983]. A key metric for these modeling methods is called process granularity, which is defined as computation time divided by communication time. Message passing is often employed for larger values of process granularity.

10.3.2   Shared Memory Architectures

Asynchronous communication requests of a byte to a few words in size that can be defined statistically are ideal for shared memory architectures. The shared memory architecture is a fast-access storage or global memory device, which is the interface among processors. The shared memory and interacting processors can either be part of the same hardware component or interface via global memory. Statistical predictions of message traffic are usually possible when message updates are within several clock cycles (e.g., nanoseconds).

The communication model for shared memory is:

A processor generates a read or write request for another address in shared memory.

The current owner of this variable is notified of the request.

The cache memory of the current owner is dumped to local memory.

The global variables of the current owner are dumped into shared memory.

The read or write request of the processor is completed with a data transfer.

Designs of shared memory architectures can be either bus or switch-based. In addition, the shared memory may be uniform (single memory device), nonuniform (shared memory associated with each processor), and cache-only (shared cache memory across the processors).

The performance of shared memory systems can be degraded substantially if a requesting processor needs information that is not in the cache memory of the shared memory interface. In this case, all activity is blocked until the shared memory can retrieve the variables needed. Besides the contention for memory issues, a second major issue for shared memory architectures is coherence (multiple copies of the same data that may drift out of agreement over time). Shared memory works best in highly parallel software applications in which the global data of each application must be accessed frequently by the application and infrequently or never by the other applications. But this approach has scalability issues compared to the others.

10.3.3   Network Architectures

Networks have become commonplace in the workplace with the local area network (LAN) products. These network architectures were initially developed in the late 1950s for radar and air defense systems. In many ways, the network architecture is a distributed collection of shared memory systems, in which each shared memory system has the ability to tap into the shared memory of the other systems on the network. The best analogy for communication is to a file server with access to slow storage devices; in this case, the communication of information is via a statistical block transfer process.

The transfer of information typically takes milliseconds to minutes, depending upon the size of dataset, and includes relatively large blocks of data. The scale of a network ranges from nanoscale, personal area, local area, home area, campus area, backbone, metropolitan area, wide area, enterprise area, virtual private area, and global area.

The main difference between the network and the message-passing architecture is that a network provides demand-based service, while message passing primarily uses scheduled transfers. Networks can service hundreds of nodes.

A network system typically includes the communication hardware and a software package, typically called a network operating system. There are many such commercial network operating systems. The software provides various priority-based queuing models, often with separate transmit and receive queues. The network provides extensive fault checking and does not suffer from the failure modes of message-passing architectures. A network interface controller and the Advanced

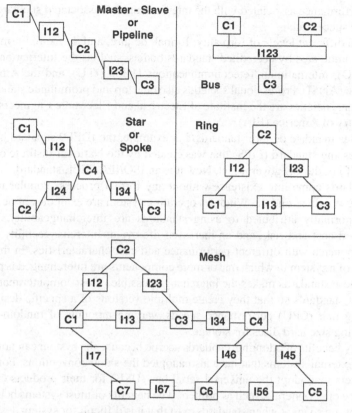

FIGURE 10.2 Network architectures.

Microcontroller Bus Architecture (AMBA) are examples. AMBA is defined by an open standard for the connection and management of functional processes using a microcontroller (system-on-a-chip). This AMBA technology is one of several technologies that have been used to enable multiprocessor systems.

There are many network architectures available. Five of the most common are shown in Figure 10.2. The pipeline architecture is a serial linkage of components that is most appropriate when the components only need to communicate with their neighbor in the network. The bus architecture is the most general; each component places its information on the bus, and the bus distributes the information to the appropriate sources. The bus architecture is most appropriate for many components. The spoke architecture isolates one component as the central processor that manages the communication process. The ring architecture is one of the most common architectures in office settings. The mesh architecture is an irregular connection of components that provides sufficient redundancy (pathways between any two nodes) for the system under consideration while stopping short of full interconnection. Duato et al. [1997] provide many examples of interconnection networks used within parallel computation devices and telephony systems.

10.4 STANDARDS

Standards help ensure that an interface will enable the connection of two components. Each component is required to meet a given standard, and the interface is designed to meet the same standard.

As long as the performance associated with the interface and the associated standard is satisfactory, the design will be successful.

Standards have different levels of formality: formal, de jure, and de facto. Formal standards are negotiated and promulgated by accredited standards bodies, such as the International Organization for Standards (ISO), International Telecommunications Union (ITU), and the American National Standards Institute (ANSI). Professional societies also develop and promulgate standards. Examples of such professional societies are the Institute of Electrical and Electronic Engineers (IEEE) and the Electronics Industry of America (EIA).

Legal authorities mandate de jure standards. For example, the IDEF0 standard became a federal information processing standard (FIPS) that was created by the National Institute of Standards and Technology (NIST) of the US government. Now it is an ISO/IEC/IEEE standard.

De facto standards come into existence without any formal process. Popular usage creates de facto standards. X Windows and the Windows operating system are examples of de facto standards.

The benefits normally attributed to using standards are interchangeability, interoperability, portability, reduced cost and risk, and an increased life cycle. Interchangeability is the ability to interchange components with different performance and cost characteristics. In this way, creating multiple versions of a system in which one or more components are interchanged is possible because the adoption of these standards makes the interchange possible. Most computer manufacturers have adopted sufficient standards so that they create multiple versions of a specific design with varying central processing unit (CPU) performance speeds, varying amounts of random-access memory (RAM), and varying size hard discs for storage.

Interoperability benefits of adopting standards accrue because the system can now operate with a wider variety of external systems that have also adopted the same conventions. For example, computer manufacturers that adopt the universal serial bus (USB) for their products can be interfaced with a wide variety of peripherals, such as printers. The benefits of most systems being interoperable with other systems are so great when standards exist that it is difficult for system designers to deviate from such standards. The answer to such deviations is limited performance by an aging technology. Predicting if and when a new technology will provide enough increased performance or decreased cost to justify changing a standard is often difficult.

Portability is a benefit for systems that operate on another system. Software systems obtain portability by adopting the standards necessary to run on multiple platforms with varying hardware or operating systems. Systems that require power obtain portability by having a power unit that permits power to be obtained from a standard wall socket. Systems like my laptop computer that require direct current (dc) current still need the portability to operate using power from alternating current (ac) sources and include a power unit that converts ac to dc power.

Adopting certain standards allows a system designer to buy modules that provide the needed performance characteristics at a reduced cost. Standards promote competition among vendors, competition that provides reduced cost, and reduced risk for equivalent performance.

An increased life cycle for the system is possible when long-lived standards are adopted. The system can use the interoperability of its components to upgrade its capabilities as new technologies come along, as long as these new technologies adopt the standards. Typically, the new technologies provide downward compatibility in the sense that the older products can be replaced by the new, but not vice versa.

10.5   OPEN SYSTEMS INTERCONNECTION ARCHITECTURE

In 1977, the ISO approved the initiation of work on a standard for the interconnection of computers comprised of different architectures and technologies [MacKinnon et al., 1990]. The first meeting, involving 40 experts, was held in March 1978. At the time, a few proprietary communications architectures were available (e.g., Digital Network Architecture [DNA] of Digital Equipment Corporation, Distributed Systems Architecture of Honeywell, and Systems Network Architecture [SNA] of IBM). In 1983, the ISO and the International Telephone and Telegraph Consultative Committee (CCITT) of the ITU approved the reference model for OSI [Schwartz, 1987]. This reference model defines a seven-layer architecture for network-based communication between end-user nodes in a telecommunications network. The OSI is a set of internationally accepted standards that revolve around this reference model; these standards were developed in international forums and have been accepted on an international basis for this reason. The OSI is also a set of products that conform to these standards.

The OSI reference model contains seven layers: physical, data link, network, transport, session, presentation, and application. The first four layers are known as the lower network layers. The last three layers are known as the higher layers; these higher layers plus the first four layers must be present in each end user or host node. On the other hand, intermediate nodes in the communications architecture must only possess the first three layers. Figure 10.3 presents a common representation of communication between two hosts using a communications network, such as a LAN or the Internet. Data is being transferred from an application on the left host node through the physical media and an intermediate node in the communications network to the host node on the right. The number of intermediate hosts depends not only on the communication network but also on the route selected through that communication network. In the communication network at the top of Figure 10.3 at least two intermediate nodes would be involved in communication between the two hosts shown; it is possible that all five nodes would be involved.

Some of the key definitions associated with OSI are [MacKinnon et al., 1990]:

System: an autonomous whole capable of performing information processing or information transfer.

Open System: a system that can create, transmit, receive, and act upon OSI messages.

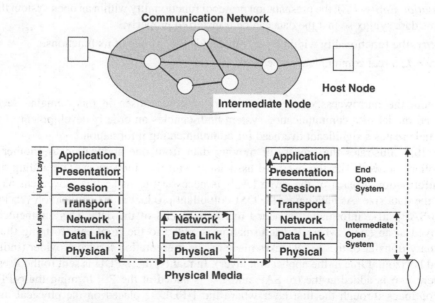

FIGURE 10.3   Communication in the OSI reference model.

Interconnection: ability to satisfy four types of activity – movement of digitized data over physical transmission media in a reliable manner, organization, and control of the paths between those open systems that are the sources and destinations of information, exchange of commands and data to manage the cooperation of the systems that desire to interwork to achieve a specified purpose, and provision of a variety of services and facilities that directly support the user applications.

Service Provider: the subsystem formed by a layer and all layers below it. This subsystem only serves the layer above it. So, the service provider formed by the transport layer includes the network, data link, and physical layers and serves the session layer.

Protocol: a complex multipart message that is passed between systems.

Protocol Control Information (P-N): information that is added at layer N to the front of a message received from the $(N+1)$ layer above; this information is used to control the transmission of the message among entities in layer N.

Protocol Data Unit (N-PDU): the message at layer N that contains the message from layer N 4-1 plus the protocol control information for layer N.

Interface Control Information (I-N): information that is added at layer N to the front (and possibly the end) of the protocol data unit of layer N to be sent to layer $N-1$.

Interface Data Unit (N-ID U): the message at layer N that contains the interface control information plus the protocol data unit of layer N and that will be sent to layer $N-1$ for transmission to the $N-1$ service provider.

Service Access Point [(N)-SAP]: the point of interaction between layers $N+1$ and N; the point at which I-N is added to the front (and possibly end) of the $N+1$-PDU being sent from layer $N+1$.

Application: a set of distributed tasks that satisfy some real-world information processing requirement.

Application Entity (AE): the portion of an application that is responsible for interconnecting via OSI.

Presentation Entity (PE): the presentation protocol functionality within an open system that transforms data syntax so that the data can be transferred properly.

(N)-entity: the functionality within layer N that adds P-N as one of its functions.

Subnetwork: a real communication network.

First, note the narrowness of the definition of system chosen in this domain. Second, the multilayered model of a communication system both enables an orderly development of standard products and creates a significant overhead for communicating information.

Figure 10.4 illustrates the process of moving data from one application to another over an OSI-compliant network and the overhead associated with that movement. The adding and stripping of information at each of the seven levels is necessary to make this movement happen but increases the data size. As data enters the OSI-compliant product at the application service access point [(7)-SAP], nI-7 information is added to the front end of the data. This augmented data is then received at an (AE), where P-7 information is added to the front end, forming the 7-PDU. An imaginary transfer of the 7-PDU takes place on the presentation service provider (indicated by the dashed horizontal line in the application layer). In reality, the 7-PDU is sent to the Presentation layer, where 1–6 is added at the (6)-SAP, and P-7 is added at the PE, forming the 6-PDU. This process continues through the first layer where the 1-PDU is placed on the physical media and transferred to the correct host. The process is repeated in reverse with the protocol and interface

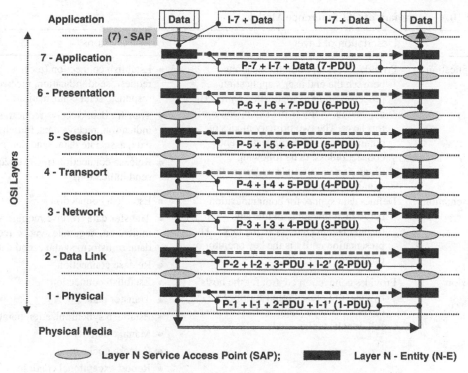

FIGURE 10.4 OSI process of adding and stripping PCIs and ICIs.

control information being stripped at successively higher layers until the original data is delivered to the application on the second host.

Table 10.1 provides a short description and the key functions of each layer [Levi and Agrawala, 1994; MacKinnon et al., 1990; Schwartz, 1987]. Each layer, except the first, is responsible for establishing a connection on the service provider below it, transferring the data to and from that service provider, and releasing the connection when finished or required. In addition, the layers conduct functions such as reporting exceptional conditions, providing error control, negotiating quality of service, and providing flow control.

While the OSI reference model has received a lot of attention as a standard, the world of products that incorporate communications systems has largely passed OSI by in favor of the de jure standard codified by the military: Transport Control Protocol/Internet Protocol (TCP/IP). This de jure standard has three layers above the physical layer: the network layer for which the LP is defined, the transport layer for which the TCP is defined, and the upper layers, which employ a variety of protocols.

10.6 COMMON OBJECT REQUEST BROKER ARCHITECTURE

From the inception of software applications, one of the most difficult problems for users is the communication of information among software applications developed by different organizations or programmers. Most software applications were designed to be a closed system, often involving proprietary code, algorithms, and interfaces. On occasion, several software applications were integrated vertically to address the problems in a single market. The Object Management Group (OMG)

TABLE 10.1   Summary of OSI Reference Model

Layer	Description of Layer	Layer Functions
(7) Application	Provides necessary communications between the end user's application processes and the application entity. The application entity is the key operator of this layer. The two primary modes of communication are connection and connectionless. (The following discussion in this table addresses the connection mode.)	• Establish connection (receive request, send indication, receive response, send confirmation) • Transfer data (receive request, send indication, receive data, initialize data, associate data, send data) • Release connection (receive request, send indication)
(6) Presentation	Defines data syntax for communication between application entities and maintains transparency to the hosts. The presentation entity is the key operator of this layer.	• Establish connection • Transfer data (receive request, send indication, negotiate syntax, receive data, transform syntax, send data) • Release connection
(5) Session	Provides connection control for the hosts by enabling presentation entities to organize the exchange of data in either full or half-duplex mode.	• Establish connection • Transfer data • Establish synchronization points • Manage activity • Release connection • Report exceptional conditions
(4) Transport	Establishes transparent and reliable end-to-end transmission of data between host nodes.	• Establish connection • Transfer data • Provide error detection and recovery • Release connection
(3) Network	Determines the establishment of connection without concern for the type of subnetwork and handles routing. Represents the interface between the communications carriers (layers 1–3) and the computer manufacturers (layers 4–7).	• Establish connection • Transfer data • Perform multiplexing • Provide error control • Provide sequencing and flow control • Release connection
(2) Data Link	Establishes reliable transmission on the physical layer.	• Establish connection • Negotiate quality of service (QOS) • Transfer data • Provide flow control • Reset connection • Release connection
(1) Physical	Defines how the physical network is accessed in order to provide bit transparent transmission on the physical media. Supports synchronous and asynchronous transmission; duplex, half-duplex, and simplex modes; and point-to-point and multipoint topologies.	• Determine presence of signaling pulses • Determine timing of signaling pulses

began operations in 1989 in response to this problem. The result is CORBA as a standard that would permit programmers to integrate software modules resident on the same network by treating each application as an object. The CORBA standard was developed via a set of requests for proposals developed by the OMG and subsequent development contracts issued to corporations such as Digital, HP, HyperDesk, and Sun.

The CORBA standard is all three standard types: formal, de jure, and de facto. Part of CORBA, the interface definition language (IDL), is a formal standard that has been adopted by the ISO and the European Computer Manufacturers Association (ECMA). The CORBA is a de jure standard in the United States and among several contractors and a de facto standard elsewhere in the world. The OMG and X/Open jointly publish CORBA.

The CORBA standard treats software applications as objects, and as such, sits at the application level of OSI's seven-layer architecture. See Figure 10.3. The CORBA is based on a client–server model for distributed computing. The IDL, a formal standard, is a universal notation for software interfaces defining a boundary between the client code (requests for services) and the software objects that implement those services. These software objects may be written to the standards defined by CORBA or may be legacy software that is "wrapped" by additional code that does adhere to CORBA standards. The IDL is both platform and language independent and has not changed significantly since first defined in 1991. In fact, IDL must remain stable or the associated standards inherent in CORBA will be broken. The IDL standard defines what is exposed in the interface between the service and its client(s); any other details and relationships are forbidden. For details on the IDL see Mowbray and Ruh [1997] or Mowbray and Zahavi [1995].

Although IDL is the key to making CORBA work from both a software development and architecture perspective, there are four additional categories of objects that comprise the CORBA architecture and are more important to this discussion of interfaces: the object request broker (ORB), CORBA services, CORBA facilities, and CORBA domains.

The first object category is the ORB, which is the core of CORBA and is an analogy to a bus network. The ORB is the interface between the client (software package requesting a service of another package) and the server (software package performing the service requested). So, in fact, the ORB can be viewed (Fig. 10.5) as a bus architecture that operates in the application layer of the OSI network communication model. The main role of the ORB is to standardize access between software applications, enabling CORBA to hide the programming, platform, and location peculiarities of client and server software objects. Each software object registers its interface characteristics with the ORB. The ORB receives all requests for service by another software application and knows which application to task with the request, where that application is, and how the request has to be translated so that the application will understand the request. The ORB requires that each software application

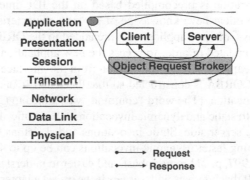

FIGURE 10.5 CORBA overlaid on OSI seven-layer model.

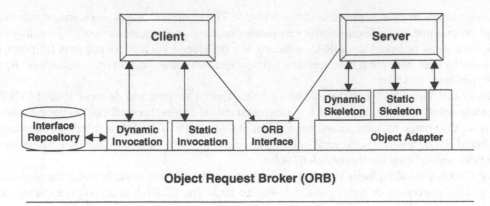

Object Request Broker (ORB)

FIGURE 10.6 ORB interactions with clients and servers.

be written in accordance with CORBA standards as defined by the IDL or wrapped in a software application (wrapper) that adheres to IDL and interfaces with the non-IDL software application. This bus architecture is the reason that CORBA can be efficient in interfacing software applications. Without an ORB-like network, each application must be able to interface with every other application; if there were N applications and a new one is added, the new application must have N new interfaces developed. With CORBA, each new application requires either an IDL wrapper to connect it to the ORB or the adherence to the IDL architecture.

Parts of the ORB are exposed to the applications (clients and servers), as shown in Figure 10.6. The dynamic invocation, the ORB interface, and the dynamic skeleton are defined as part of the CORBA specification and provided by all ORB environments. The ORB interface contains several general-purpose methods.

The dynamic invocation interface allows the client to request a service without requiring that precompiled stubs be part of the ORB. Dynamic invocation means that interface-related information about the server is acquired at the time of the invocation, providing great freedom and flexibility. The dynamic skeleton associated with the server's interaction with the ORB provides a dynamic bundling of the information in the request from the client into input parameters for the server and a dynamic bundling of the results obtained by the server for return to the client. The combination of dynamic invocation and dynamic skeletons enable users to create implementations of objects that form a gateway to often-used applications such as word processing and databases.

Static invocations (sometimes called stubs) and static skeletons are also available as extensions of the ORB. A static invocation is precompiled based on the IDL interface of the client to the ORB and requires that the client have knowledge of server's characteristics before the request is made. As additional objects (software applications) are added to the ORB, a client relying on static invocation will have to be updated in order to access the new applications. A client using dynamic invocation will be able to learn the needed information from the interface repository while building the request. Interestingly, CORBA is constructed so that the server is unaware of the nature (static versus dynamic) of the invocation. (The word "common" was added to CORBA when the decision was made to implement both static and dynamic invocations.) The static skeleton is analogous to the static invocation but on the server side. Static invocations and skeletons have the benefits of being easier to program, performing faster (dynamic invocations can be up to 40 times slower than static invocations [Orfali et al., 1997, p. 71] more robust, and easier to understand.

The final part of the ORB that interacts with servers is the object adapter. The major function of the object adapter is to define how an object is activated. One software application that can satisfy many types of requests could use a different object adapter for each request type. The CORBA standard requires that a basic object adapter be available in every ORB; this basic object adapter is sufficient for most applications. The basic object adapter performs the following functions: installation and registration of an object implementation (implementation repository), generation and interpretation of object references, activating and deactivating object implementations, invoking methods, and passing method parameters.

CORBA services include the types of services that are part of operating systems and are globally applicable. Examples of these services are lifecycle, relationship, naming, externalization, event, object query, object properties, transaction, concurrency, licensing, security, trader, and object collection. These services are packaged as objects with IDL interfaces and are augmentations of the ORB. These services enhance the effectiveness, efficiency, and security of the ORB and were proposed by platform and ORB vendors. Each service is implemented as an object so that it can be used by any application.

CORBA facilities are objects that provide services to application objects for application developers and are keyed to interoperability issues of the applications. Examples of CORBA facilities are Internet Facility, Distributed Document Facility, Print Facility, Workflow Management Facility, and Calendar Facility. The initial architecture for CORBA facilities is divided into user interface management, information management, system management, and task management. Note, these are the same elements as CORBA services except that user interface management in CORBA facilities replaces infrastructure services and elements in CORBA services. The applications in CORBA facilities are likely to change the way the user views computing and to enable the ORB to distribute the computing associated with a user's need across the platforms associated with the ORB in the most efficient manner.

CORBA domains, the final category of the CORBA, are domain-specific or line-of-business standard groups within the OMG, examples include electronic commerce, finance, telecommunications, life sciences, medical, manufacturing, and business objects.

CORBA is not unique in its efforts to enable integration of software applications for users. Other attempts to integrate applications are the distributed computing environment (DCE) of the Open Software Foundation (OSF) and Microsoft's distributed component object model (DCOM). In fact, these three approaches compete with and complement each other. The remote method invocation of JAVA is also related to these three approaches, see Mowbray and Ruh [1997] for a comparison of these approaches.

10.7   INTERFACE DESIGN PROCESS

Interface design is central to the success of the systems engineering process. The need for specific interfaces is determined when the system's components are defined and allocated functions *in* the process of defining the allocated architecture. Engineers of the system who are performing these tasks identify those items (inputs and outputs) that pass between components. The transportation of these items must be allocated to some physical entity; additional low-level functions must be defined that make the transition across this transportation entity possible. The IDEF0 diagram in Figure 10.1 shows the design process of the system-level interfaces. As discussed earlier, this design process has all of the elements of the system's design process.

Design of the interface must pay special attention to the system performance issues associated with the interfaces outputs. Concerns about the timeliness, accuracy, and reliability of the outputs of the interface need to be considered carefully. The fidelity of the interface is defined as the insurance of the integrity and delivery of items being transferred; that is, the item being sent is the same as the item being delivered, and the item is delivered in a reasonable amount of time. Clearly, the interface

- Define Interface Requirements
  - Identify the Items to Be Transported by the Interface
  - Define the Operational Concept
  - Bound the Problem with an External Systems Diagram
  - Define the Objectives Hierarchy
  - Write the Requirements
- Select a High-Level Interface Architecture
  - Identify Several Candidate Architectures
  - Define Trial Interfaces for Each Candidate
  - Evaluate Alternatives against Requirements
  - Choose High-Level Interface Architecture
- Develop Functional Interface Architecture
  - Specify Functional Decomposition
  - Add Inputs and Outputs
  - Add Fault Detection and Recovery Functions
- Develop Physical Interface Architecture
  - Identify Candidates based upon High-Level Architecture
  - Eliminate Infeasible Candidates
- Develop Allocated Interface Architecture
  - Allocate Functions to Components of the Interface
  - Analyze Behavior and Performance of Alternatives
  - Select Alternative
  - Document Design and Obtain Approval

FIGURE 10.7   Interface design process.

needs to be sized to handle some determinable quantity of items. Finally, there must usually be extensive failure detection and recovery algorithms to address the integrity and delivery of items.

The design process for an interface includes the steps shown in Figure 10.7. First, defining the components to be addressed, the items that are transferred between them, and any interfaces that have already been specified should bound the interface design problem. Next, we must identify those items that are to be included in the interface for which we are concerned. Before getting into the design, we must derive the requirements for this interface from the current requirements specification. Included in these requirements are the performance, cost, and trade-off requirement that will be instrumental in selecting the interface.

The design steps are to choose an interface architecture (e.g., shared memory, message passing, and bus network); define specific trial interface alternatives (e.g., various bus network alternatives); evaluate these alternatives against the requirements, specifically the performance, cost, and trade-off requirements; and finally choose a specific interface alternative.

Once the interface has been chosen, the behavior of the interface must be detailed and added to the functional architecture. Next, functional behavior is allocated to the existing components and the new interface. Finally, the performance of the segment containing the components and interface must be evaluated, and critical fault detection and recovery behavior must be added to the functional architecture and then allocated to the components and interface.

Figure 10.8 provides a sample result of the above interface design for an elevator. The interface is an external one between the elevator and the building for the purpose of transferring emergency communications between passengers in the elevator and appropriate emergency response unit

---

- Define Interface Requirements
  - *Identify the Items to Be Transported*: Emergency communications from the elevator to the building (and onto the emergency response team)
  - *Define the Operational Concept*: Passenger encounters emergency and requests ability to make emergency known to emergency response team; elevator provides resource for passenger to use; passenger communicates
  - *Bound the Problem with an External Systems Diagram*: (skipped)
  - *Define the Objectives Hierarchy*: Objectives are (1) availability of interface, (2) fidelity of the communicated message, (3) operational cost (monthly), and (4) deployment cost.
  - *Write the Requirements*: (skipped)
- Select a High-Level Interface Architecture
  - *Identify Several Candidate Architectures*: (1) Telephone connection to building, (2) dedicated communication system network to emergency response team
  - Define Trial Interfaces for Each Candidate (skipped)
  - *Evaluate Alternatives against Requirements*: Dedicated network is too expensive to install
  - *Choose High-Level Interface Architecture*: Telephone connection is chosen
- *Develop Functional Interface Architecture*: Not needed because interface is standard
- *Develop Physical Interface Architecture*: Not needed because interface is standard
- *Develop Allocated Interface Architecture*: Not needed because interface is standard

---

FIGURE 10.8   Sample interface design between elevator and the building housing the elevator.

(e.g., police). In this case, a standard interface item, a commercial telephone system, is chosen, making most of the interface design process unnecessary. Commercial standards are often chosen as interfaces for this reason.

10.8   SUMMARY

Interfaces are the primary responsibility of the systems engineer and are the most common failure point on systems. Designing the interfaces of a system begins with identifying the interfaces, both external and internal, and allocating items (inputs and outputs) to the defined interfaces. Next, the requirements for each interface must be derived from existing system-level requirements. As part of the system's requirements, interface requirements will be derived that define the performance and fidelity of the interface. System throughput and response time are the common performance issues that are relevant to designing the interfaces. The fidelity of an interface means ensuring the quality of the items being carried.

As part of the design process alternative interface architecture options must be examined and the most cost-effective chosen. These alternatives can be based on message passing, shared memory, or network architectures, depending upon the characteristics of the items being transported and the performance issues associated with the system.

Standards play a major role in the design or selection of an interface. If a standard can be selected as an interface, then the design information that needs to be communicated in any component or CI specification is readily available and probably well understood. Standards can be formal (adopted by a recognized standards-setting body), de jure (mandated by legal authorities), and de facto (adopted via popular usage by many commercial concerns).

Two major standards, the OSI reference model and the CORBA were presented in this chapter. These two standards demonstrate the complexity associated with most significant interfaces in terms of design issues and functionality.

CASE STUDY: PATHFINDER COMMUNICATIONS FAILURE

The *Pathfinder* system that was deployed to the surface of Mars for a landing on July 4, 1997, was truly a success in many ways. Unique system design features included a landing on air bags and the small but effective Sojourner rover.

However, a few days into the mission operators on the ground noticed that infrequent total system resets were occurring that were causing the loss of data. The *Pathfinder's* information system contained an interface described as a "bus or shared memory area" [Jones, 1997]. A priority system had been established for giving various system activities access to this interface. A bus management task had high priority and ran frequently to accept specific data elements into the shared memory area and then distribute them to their proper locations. A task for gathering and publishing meteorological data had low priority. A particularly lengthy communications activity employed by the spacecraft had a medium priority. Mutual exclusion locks were employed to give an activity access to the interface. A mutual exclusion (mutex) lock is given to an activity and grants that activity control of the communications interface until it releases control back to the interface. VxWorks is the commercial package used on *Pathfinder* to handle these scheduling activities on the interface. Wilner [1997] described the problem causing the system resets and the process used to diagnose and fix this problem.

The meteorological data gathering activity was an infrequent user of the communications interface and involved in the publishing of a substantial amount of data. This data was so voluminous that the meteorological data activity would have to obtain and release mutexes several times before it was finished. The meteorological activity was broken into short enough segments that the high-priority bus management task could gain control for its important functions during the meteorological activity. However, the long-running, medium-priority communications activity would infrequently interrupt the meteorological activity during one of its pauses and gain control of the interface. The durations of this medium priority communications task and the previous segments of the meteorological task were sufficiently long to invoke a watchdog timer that was employed to ensure that the high-priority bus management task was executing appropriately. In these rare cases, the watchdog timer would invoke a total system reset as a hedge against the system being in a deadlock or failure mode. Whenever the reset occurred, the data in the interface would be lost.

Fortunately, VxWorks had a feature for recording a total trace of system events. Jet Propulsion Lab (JPL) engineers ran the *Pathfinder* replica on Earth in their lab until the reset situation was replicated. They found that VxWorks had been programmed to run without a feature called priority inheritance. Enabling this priority inheritance feature would solve this problem by keeping the medium-priority communications task from slipping into the middle of the meteorological publishing task. The JPL engineers uploaded a short C program that enabled the priority inheritance feature. *Pathfinder* experienced no more system resets or loss of data.

PROBLEMS

10.1 Develop a functional, physical, and allocated architecture for an OSI-compliant communication system, using the material presented in this chapter for the OSI reference model. Note the physical architecture of the communication system will include the physical communication network as well as the layers of the OSI reference model.

10.2   Develop a functional, physical, and allocated model for a CORBA-compliant software system. Use a physical architecture comprised of the IDL, ORB, CORBA services, CORBA facilities, and CORBA domains.

10.3   Select several items for your OnStar project from previous chapters and design an interface for those items.

10.4   Select several items for your ATM project from previous chapters and design an interface for those items.

Chapter 11

Integration and Qualification

11.1 INTRODUCTION

Integration is the process of assembling the system from its components, which must be assembled from their configuration items (CIs). *Qualification* is the process of verifying and validating the system design and then obtaining the stakeholders' acceptance of the design. Recall that verification is the determination that the system was built right; while validation determines that the right system was built. Both activities are conducted by the systems engineering team as part of the development process, primarily during integration. Validation has critical early elements (conceptual, design requirements, and validity) that are completed during the design phase. The system that is used to qualify the system being designed must be built for that purpose. So, while the operational system is being designed, the qualification system for the operational system is also being designed and integrated. The operational phase for this qualification system is during integration and qualification. Also keep in mind that other systems are being developed concurrently with the operational system, namely, some or all the manufacturing, deployment, training, refinement, and retirement systems. Each of these also has a qualification system.

The terms *testing* and *qualification* are used interchangeably in parts of this chapter. The word *testing* is associated with the keywords of *acceptance, validation, and verification* by most systems engineers. However, the process of acceptance, validation, and verification comprise what is being called qualification in this chapter. The confusing usage arises when an instrumented test is mentioned as one of four methods that comprise qualification (testing), and the other three methods do not contain the word test: inspection, demonstration, and analysis and simulation. In fact, these three methods are forms of test. The word *qualification* is used in this chapter as often as possible to mean the process that comprises acceptance, validation, and verification testing. The word *testing* will be used with these three terms but is meant to be associated with the methods used in the qualification process during integration.

This chapter begins by providing a detailed definition of the elements of qualification: acceptance testing, validation, and verification. Section 11.3 discusses the concept of integration since qualification takes place as integration is progressing; alternate processes for integration are discussed in

The Engineering Design of Systems: Models and Methods, Fourth Edition. Dennis M. Buede and William D. Miller
© 2024 John Wiley & Sons, Inc. Published 2024 by John Wiley & Sons, Inc.
Companion website: www.wiley.com/go/engineeringdesignofsystems4e

Section 11.4. Then qualification is described in detail, beginning with planning and proceeding to a detailed discussion of qualification methods. Qualification terminology is presented in Section 11.5. Defining the qualification system is presented in Section 11.6. Qualification methods are covered in Section 11.7. Special topics in acceptance testing focused on usability are described in Section 11.8.

The *exit criterion* for integration and qualification is acceptance of the design by the stakeholders. This is often done conditionally; that is, with the provision that certain system elements be revised to enable greater cost-effectiveness during operation.

11.2 DISTINCTIONS AMONG ACCEPTANCE, VALIDATION, AND VERIFICATION TESTING

In Chapter 1, the concepts of verification, validation, and acceptance were introduced. (Grady [1997] provides additional detail on the distinctions being discussed here.) *Acceptance* is a stakeholder function for agreeing that the designed system, as tested or otherwise evaluated by the stakeholders, is acceptable. As such, acceptance is driven by the stakeholders, with the knowledge of the results of validation and verification activities that have preceded it, see Figure 11.1.

Validation is the process of determining that the systems engineering process has produced the *right system,* based upon the needs expressed by the stakeholder. Validation is carried out by the systems engineers, based upon what they believe the stakeholders' needs to be. The most reliable and early statement of the stakeholders' needs is the operational concept. Therefore, *operational validity* is the matching of the capabilities of the designed system to the operational concept; this naturally occurs late in the integration phase after the designed system has been verified. However, conceptual

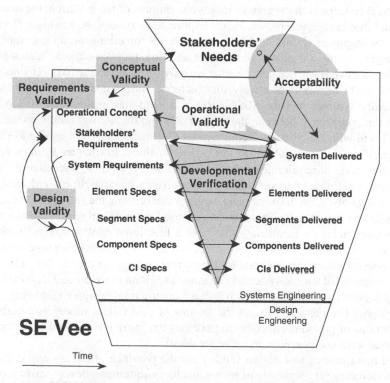

FIGURE 11.1 Verification, validation, and acceptance.

validity, requirements validity, and design validity are important aspects of validity and need to be addressed early in the design phase. Conceptual, requirements, and design validity are called *early validation*, the determination that the right problem is being defined at the current level of abstraction, given the validity of the problem definition at a higher level of abstraction.

Conceptual validity is the correspondence between the stakeholders' needs and the operational concept. Conceptual validity needs to be established at the outset of the design process via interactions among the systems engineers and the stakeholders; however, the systems engineer cannot assume that once established there is no more work to be done. Stakeholders' needs change and the operational concept must change with those needs. Note operational validity only makes sense if conceptual validity has been established. If both conceptual and operational validity are solid, then the stakeholders' acceptance should be nearly guaranteed.

Requirements validity is the correspondence between the operational concept and the stakeholders' requirements. In requirements validity, the operational concept is assumed to be an accurate reflection of the stakeholders' needs; the validation occurs by establishing that the stakeholders' requirements have neither introduced new issues nor left issues out of the operational concept, thus causing the design of a different system than envisioned in the operational concept. But recall that the operational concept and stakeholders' requirements should be stated in design independent terms, making this task of requirements validity quite difficult. Elements of requirements validity are ensuring: There are input/output *requirements* for all of the inputs and outputs *in* the operational concept, that every objective in the objectives hierarchy has a performance requirement in the Stakeholders' Requirements Document (StkhldrsRD), that every external interface to the system has been considered for an external interface requirement, and so forth. The ecosystem model and objectives hierarchy (discussed in Chapter 6) are key tools for establishing this requirements validity. In addition, intermediate products such as a data model that relates the inputs to and outputs from the system in the operational concept to the aggregate inputs and outputs of the system in the external systems diagram can and should be developed to support requirements validation. At a higher level of abstraction, the systems engineers should be asking "Can we get something we do not want even though these requirements stating our needs are met?" In addition, they should ask "Can we get what we want (the problem solved) without getting what we have asked for in the requirements?" If either of these questions can be answered positively, there is more work to do on the requirements.

Design validity assumes that the StkhldrsRD is a valid statement of the stakeholders' needs and addresses the congruence between the StkhldrsRD and the derived requirements. The derived requirements begin with the Systems Requirements Document (SysRD), evolve to subsystem and component specifications, and culminate in CI specifications. In Chapter 9, three techniques for flowdown or derivation of requirements are discussed: apportionment, equivalence, and synthesis. Establishing design validity for apportionment and equivalence is straightforward. Design validation when synthesis is involved, on the other hand, requires establishing the validity of the models used to complete flowdown via synthesis. These models are used to transform requirements on one or more variables to requirements on parameters that have a functional relationship with these variables. A common cause for failure in this synthesis process is that the models being used were valid in previous engineering efforts but are not valid for the current system; yet the validity of the models from previous developments of similar systems is assumed to pertain to the current development. Petroski [1994] provides extensive evidence of such failures in structural design engineering; he highlights failures of bridges. The designers forgot the lessons of past failure modes and built bridges that were extrapolations of previous efforts, extrapolations that were not justified based upon modeling assumptions that were not examined in sufficient detail.

Conceptual requirements and design validity are the province of the systems engineering team and must be undertaken very seriously to ensure that the requirements development process does not redefine the problem being solved. There are two chains that must be strong; see Figure 11.2. The first

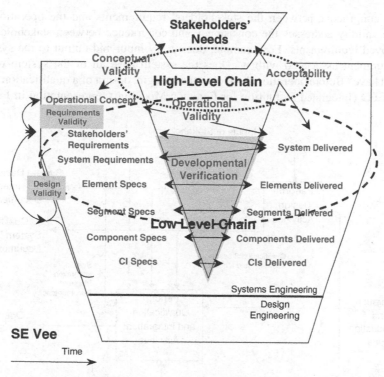

FIGURE 11.2 Two qualification chains. The high-level chain consists of conceptual validity, operational validity, and acceptability. The low-level chain consists of design validity, requirements validity, developmental verification, and operational validity.

chain consists of conceptual validity, operational validity, and acceptance testing. Requirements validity, design validity, verification, and operational validity comprise the second chain. Each of these chains is only as strong as the weakest link.

Verification is the matching of CIs, components, subsystems, and the system to their corresponding requirements to ensure that each has been *built right*. This process of design verification is also carried out by the systems engineering team to ensure that the design problem defined in conjunction with the stakeholders is being solved appropriately. For verification to be successful, the originating and derived requirements must be testable; that is, the requirements must be single statements that are unambiguous, understandable, and verifiable (see Chapter 6).

Verification begins in the design phase with the definition of the derived requirements and becomes the focus of activity early in the integration phase when the systems engineers can match the derived requirements to the capabilities of the CIs and the components. However, the design of the test system to achieve this verification must occur in the design phase of the system or be revalidated if an existing test system from a precedent system development is reused for the follow-on system development. This second case is more common in system development of a product line based on a platform architecture and where there is a pipeline of follow-on systems in different stages of being developed.

It is a misconception to picture verification as beginning and ending before validation, which begins and ends before acceptance testing. In fact, as can be seen in Figure 11.1, validation must begin with the definition of requirements to ensure that there is conceptual validity between the operational concept and the stakeholders' needs. Requirements validity also begins almost immediately to address the congruence between the stakeholders' requirements and the operational concept. Finally, design validity addresses the consistency and congruence between stakeholders' requirements and derived requirements. For example, does every input and output to the system have at least one requirement associated with it? Does the system have all of the system-wide requirements it should have? Before operational validation can begin, design of a qualification system must occur. The IDEFO (Integrated Definition for Function Modeling) representation in Figure 11.3 of

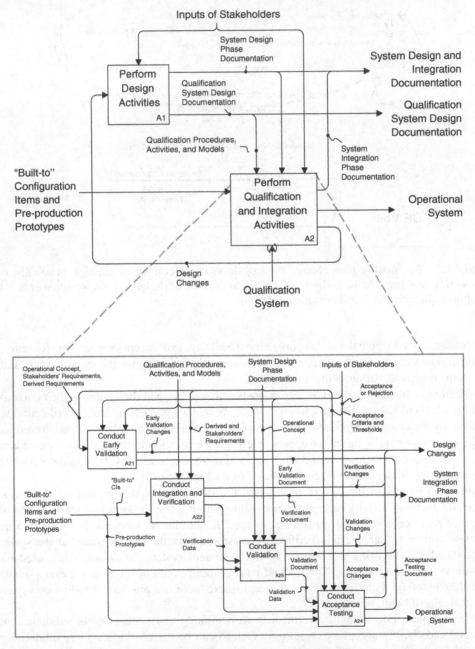

FIGURE 11.3   Bottom-up integration process.

early validation, verification, operational validation, and acceptance testing suggests the most likely sequential ordering. In practice, though, there is substantial concurrency involving these processes, making the results even more difficult to get right.

Finally, for the acceptance test to be successful, there must be clear agreement between the acceptance thresholds and the early design documents of the operational concept and stakeholders' requirements. Therefore, design of the acceptance test must begin early enough to enable both conceptual and design validity.

Successful integration relies critically on the complete and consistent development of stakeholders' requirements, the proper flowdown of stakeholders' requirements into derived requirements and tracing of requirements to functions and components/CIs, and the analysis of system performance and cost considering the stakeholders' fundamental objectives. These are design activities associated with the system. The development of test requirements, including the verification, validation, and acceptance test plans, initializes integration and helps formalize the design process.

11.3 OVERVIEW OF INTEGRATION

Textbook integration is a bottom-up process (see the top half of Fig. 11.4) that combines multiple CIs into components, multiple components into subsystems, and multiple subsystems into the system. At each level of integration, the appropriate interfaces and models of the external systems, components, and CIs must exist for this subset of the system. These interfaces and models are stimulated by defined sets of inputs and tested to determine if the appropriate outputs are obtained. In addition, the physical combination of the CIs, components, or subsystems is examined to determine that the fit of these system elements is acceptable. This is not to say that integration can only be bottom up and must wait for the last available CI before proceeding to the component level. In fact, design stubs (shells or model replicas) for specific CIs, components, or even subsystems can he developed as part of the integration process to reduce risk, speed up integration, and enhance the testing effort. Alternate integration processes are discussed later.

Figures 11.4–11.6 show three different representations of the major integration functions. The bottom half of Figure 11.4 shows this information as an IDEFO diagram with the functions and flow of data among the functions; the major functions are (1) inspect and test the CI (component or subsystem), (2) identify and fix any correctable deficiencies found in the first function, (3) assess the impact of any uncorrectable deficiencies found in the first function, (4) redesign the CI (component or subsystem) to address unacceptable impacts of any uncorrectable deficiencies as identified in the third function, (5) modify the baseline of the design to account for any fixes (function 2) or acceptable impacts (requirements changes from function 3), and (6) integrate with the next CI (component or subsystem) and repeat until all CIs (components or subsystems) have been integrated. Figure 11.4 addresses component integration but has the identical structure for the higher-level integration at the subsystem and system levels.

Figure 11.5 shows logic structure of integration at the subsystem level, that is, integrating every subsystem of the system until all subsystems have been integrated. First, a selected subsystem is inspected and tested to determine if it meets the requirements defined In the specification for that subsystem; this is verification. If the subsystem is not deficient, the next subsystem begins the verification process. If the subsystem is deficient, modifications and fixes are made if possible, and the design baseline is modified accordingly. However, if there are remaining deficiencies, the impact of these deficiencies must be assessed. If the deficiencies are acceptable, no redesign is necessary, and the requirements baseline is modified. However, if the deficiencies are unacceptable, the subsystem must be redesigned, usually at great cost and delay in time. If any changes are made

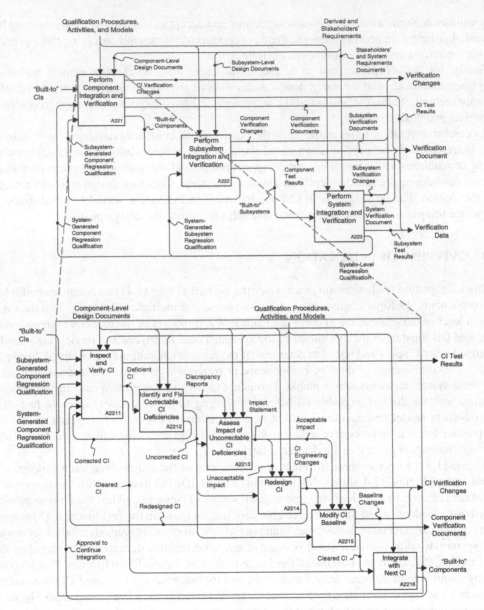

FIGURE 11.4   Major integration functions for component integration.

at all, the subsystem must be retested (called *regression testing*) in case any new problems were introduced.

These six functions cannot flow in serial sequence. In fact, some functions may not be executed at all. If there are no deficiencies, functions 2–5 are never executed. If all deficiencies are correctable, functions 3 and 4 are not executed. Figure 11.6 shows the control structure needed to make these function work as an activity diagram (act). Figure 11.6 shows the functions at the subsystem level of integration, but again this structure applies equally at the component and system levels.

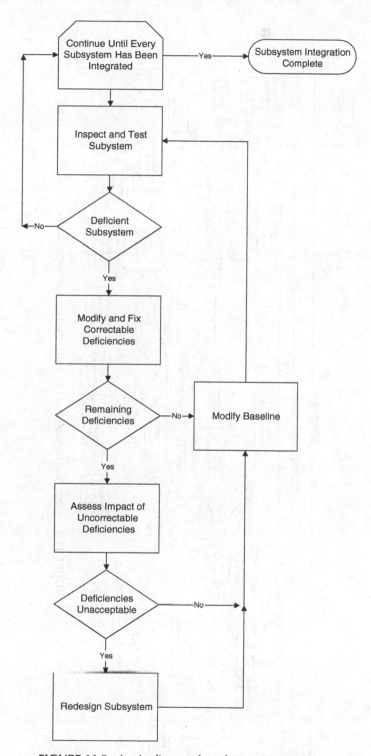

FIGURE 11.5   Logic diagram for subsystem integration.

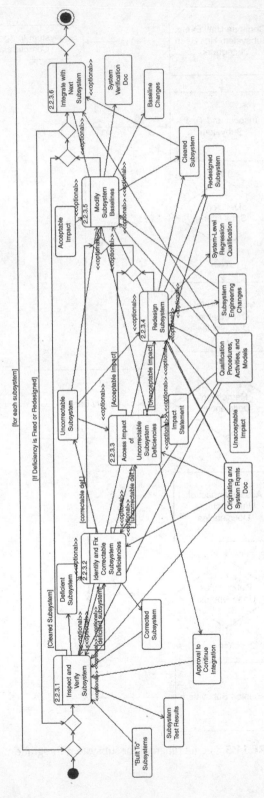

FIGURE 11.6 Integration control structure and data/material/energy flows.

11.4 ALTERNATE INTEGRATION PROCESSES

As discussed, earlier bottom-up integration is commonly discussed in textbooks as the desired approach. In fact, in Chapter 1, the Vee model of systems engineering represented the bottom-up integration process as the appropriate one. However, there are alternate integration processes (described in Table 11.1) that are appropriate to systems engineering; these alternate approaches have been investigated and described by the software engineering community [Perry, 1988]. The top-down integration process was commonly used in software engineering as part of top-down software design and development. The most commonly used integration process in the software industry [Perry, 1988] is "big bang" integration, in which CIs are combined as they become available and have completed testing. Middle-out integration recognizes that many system developments are not unprecedented and reuse elements from existing systems.

Top-down integration begins by examining the top-level core of the system, which is followed by adding major components to this core and testing and ends by adding the individual CIs to the cores of the components and testing. Top-down integration is very difficult to accomplish for systems with hardware, people, and facilities that are designed from scratch. It is difficult to define a system core that is hardware, people, and facilities unless a large part of the system already exists, commonly referred to as "commercial off-the-shelf" (COTS) components or CIs. However, as more and more new systems are made up of larger and larger amounts of COTS components, top-down integration has greater usefulness in systems engineering.

Both the bottom-up and top-down integration processes can proceed for the entire system by adding or peeling a layer of the system as one would an onion; this is referred to as phase integration. For bottom-up integration, this means that all the CIs are integrated into their respective components before any components are integrated. However, it is commonly counterproductive from schedule and cost perspectives to delay the integration of some of the components until all the CIs are ready.

At the other extreme is incremental integration *in* which one subsystem at a time is integrated from the CIs up through its components before the integration of any other subsystem is begun. Just as phase integration is impractical, so too is pure incremental integration. A major element of test planning is the creation of a realistic schedule for when each CI will be ready so that integration can proceed at an orderly pace and test system devices and models can be ready when needed. This typically involves a mixture of phase and incremental integration.

Middle-out integration begins with an existing system. Select subsystems and components may be removed from the system. Newly developed subsystems and components are integrated into the system.

Finally, big-bang integration is a relatively undisciplined, but much used, approach to integration. At the worst extreme, this approach begins assembling CIs as they become available and undertakes testing as an afterthought. Since there is no serious planning for testing sequences, fault detection and fault localization and diagnosis become very difficult. At its best, this approach combines bottom-up and top-down integration in a disciplined and rigorous manner. When done well, this approach often takes more planning and development of test rigs but can be accomplished more quickly.

Another major element of the development of the qualification system and qualification planning is the creation of the appropriate test stubs and scaffolds with drivers for the relevant qualification scenarios. Each CI, component, and the system as a whole must be stimulated by a given set of inputs for each qualification case. In addition, test equipment must be put in place to capture the outputs of these CIs, components, and the system. The qualification plan ensures that these qualification system elements will be in place at the right time to enable the planned integration sequence of CIs and components. The plan typically breaks down when planned tests are failed by specific CIs, components, or the system. A well-designed qualification plan will address schedule adjustments for possible qualification failures as part of risk mitigation.

TABLE 11.1   Principal Integration Processes.

Top-Down	• Integration begins with a major or top-level module. • All modules are called from the top-level module are simulated by "stubs" (shell or model replica). • Once the top-level module is qualified, actual modules replace the stubs until the entire system has been qualified. • This is most useful for systems using large amounts of COTS components.
	Phase Integration: Integration is done from the top down to the lowest level; one peel of the onion at a time. **Incremental Integration**: Integration is done for a specific module from top to bottom; one slice of the system at a time.

	Advantage:
	    Early demonstration of the system is allowed.
	    Representation of the test cases is easier.
	    This is more productive if major flaws occur toward the top of the system.

	Disadvantage:
	    Stubs have to be developed.
	    Representation of test cases in the stubs may be difficult.
	    Observation of test output may be artificial and difficult.

Bottom-Up	• Integration begins with the elementary pieces (or CIs) that comprise the system. • After each CI is tested, components comprising multiple CIs are tested. • This process continues until the entire system is assembled and tested. • This is the traditional systems engineering integration approach.
	Phase Integration: At any point in the integration, all of the subsystems are at the same stage of integration testing. **Incremental Integration**: Integration proceeds one slice of the system at a time.

	Advantage:
	    It is easier to detect flaws in the tiniest pieces of the system.
	    Test conditions are easier to create.
	    Observation of the test results is easier.

	Disadvantage:
	    "Scaffold" systems must be produced to support the pieces as they are integrated.
	    System's control structure cannot be tested until the end.
	    Major errors in the system design are typically not caught until the end.
	    System does not exist until the last integration test is completed.

Middle-Out	• Integration begins at an intermediate (i.e., component) level of the system. • Integration at levels above and below the component level performed concurrently. • Integration above continues until it is completed at the top-level module. • Integration below continues until it is completed for the underlying CIs.
	Advantage: Combines advantages of both top-down and bottom-up integration.

	Disadvantage:
	    "Scaffold" systems must be produced to support the pieces as they are integrated.
	    More complex approach than top-down and bottom-up.
	    Requires more resources/cost per time interval.

TABLE 11.1 (*Continued*)

Big Bang	• Untested CIs are assembled and the combination is tested.
	• This is a commonly used and maligned approach.

        Advantage:
            Immediate feedback on the status of system elements is provided.
            Little or no pretest planning is required.
            Little or no training is required.

        Disadvantage:
            Source of errors is difficult to trace.
            Many errors are never detected.

11.5 SOME QUALIFICATION TERMINOLOGY

The purpose of qualification is not only to find faults and failures but also to prevent them and to provide comprehensible diagnoses about their location and cause. Recall the following definitions from Chapter 7:

Failure: deviation in behavior between the system and its requirements. Since the system does not maintain a copy of its requirements, a failure is not observable by the system.

Error: a subset of the system state, which may lead to a failure. The system can monitor its own state, so errors are observable in principle. Failures are inferred when errors are observed. Since a system is usually not able to monitor its entire state continuously, not all errors are observable. As a result, not all failures are going to be detected (inferred).

Fault: defects in the system that can cause an error. Faults can be permanent (e.g., a failure of system component that requires replacement) or temporary due to either an internal malfunction or external transient. Temporary faults may not cause a sufficiently noticeable error or may cause a permanent fault in addition to a temporary error.

The qualification designer should realize that the design of the qualification system is not only important in terms of finding and defining faults and errors but also in guiding designers to preclude them from introducing faults in the first place. In addition, the qualification designer must realize that no qualification procedure is perfect. As Glegg [1981] points out, no procedure can answer all questions of interest. Some procedures do well at capturing what happened; others do much better at explaining why these things happened. As a result, a number of complementary procedures must he employed for success. When complete, the qualification design must document the qualification procedures in detail and the expected qualification results (requirements) for each procedure. In fact, recall that the qualification process is being conducted by a qualification system; the qualification design should be tested just as any system would be.

To design the qualification system, some basic knowledge of faults is needed, and some modeling of fault importance should be completed. The software community [Beizer, 1990] has written much more extensively on these topics than has the systems engineering community. Beizer [1990] presents three laws of software testing that are directly relevant to systems:

First Law: The Pesticide Paradox – Every method you use to prevent or find bugs leaves a residue of subtler bugs against which those methods are ineffectual.

Corollary to the First Law – Test suites wear out.

Second Law: The Complexity Barrier – *Software* complexity (and therefore that of bugs) grows to the limits of our ability to manage that complexity.

Third Law – *Code* migrates to data.

For systems, replace the word bug with fault. The third law becomes "hardware and people migrate to software which eventually migrate to data." Theoretically, Manna and Waldinger [1978, p. 208] summarized the barriers to verification (the easy part of qualification) as

- "We can never be sure that the specifications are correct."
- "No verification system can verify every correct program."
- "We can never be certain that a verification system is correct."

These barriers generalize to validation.

Beizer [1990] also provides a taxonomy of bug (fault) consequences:

Mild: The symptom offends us esthetically, for example, misspelling or poor formatting.

Moderate: Outputs are misleading or redundant, affecting system performance.

Annoying: The system's behavior is dehumanizing, for example, names are truncated, bills for $0.00 are sent, operators must resort to unnatural command actions to obtain the desired response.

Disturbing: The system refuses to handle legitimate functions.

Serious: The system loses track of functions and gobbles unique inputs, for example, your deposit is lost.

Very Serious: The system mixes input and output streams, for example, your deposit is credited to another account.

Extreme: The problems are not limited to a few situations but occur on a frequent basis.

Intolerable: The system causes long term, unrecoverable corruption of the database, and this corruption is not easily detected.

Catastrophic: The system decides on its own to shut down, causing unrecoverable corruption of the database.

Infectious: The system completes its own functions, but in so doing, it corrupts the functioning of other systems.

This type of fault categorization is the first step in defining the importance of faults; these categories define distinctions among the consequences of faults. The other key element of fault importance is the frequency with which the fault occurs. (Note Beizer's extreme category is a variation of very serious that increases the frequency. In a taxonomy on consequences, extreme should be removed.) Consider the set of scenarios ($j = 1, 2, \ldots, J$) in the operational concept (or preferably some aggregation of these scenarios). Develop the following two metrics for each scenario and each fault category ($i = 1, 2, \ldots, I$):

p_{ij} = probability of fault i in scenario j;

c_{ij} = dollar (or some other value measure) consequence of fault i in scenario j.

The measure of the importance of the fault types I_i is

$$I_i = \sum_{j=1}^{J} V_j p_{ij} c_{ij}$$

where V_j is the relative measure of the importance of each scenario. (Note if c_{ij} is in dollars, the term V_i can be set to 1.0; however, if c_{ij} is in non-dollar units, V_i will be needed to calibrate across scenarios.) This measure works well if the likelihood of each fault type in each scenario is relatively rare. If some fault types may occur multiple times in a scenario, then a more complex measure should be used.

Beizer [1990] also presents a taxonomy of "bugs" (software faults) for software programs based upon the cause or source of introduction of the bug. This taxonomy includes requirements, features and functionality, structure, data, implementation and coding, integration, system and software architecture, and testing. Beizer [1990] provides detailed summary statistics for the frequency of these types of bugs.

11.6 DEFINING THE QUALIFICATION SYSTEM

There are four major levels of qualification planning: Plan the qualification process, plan the qualification approaches, plan qualification activities, and plan specific tests. The first three qualification planning functions are conducted for verification, validation, and acceptance testing. The fourth planning function is conducted for every specific qualification activity identified in the three prior planning functions. These final plans should stipulate that every requirement be tested individually. Table 11.2 shows the elements of each of the four qualification planning functions. Recent research has been conducted in this area by Meisenzahl et al. [2006], Levardy et al. [2004], and Hoppe et al. [2003].

The system's objectives discussed in Chapter 6 become key for the initial activity of planning the qualification process. These objectives of the system drive the qualification objectives. A key part of the qualification objectives is determining whether the test was passed by the system design or not. Defining the threshold for passing the test is a difficult balancing act; the threshold cannot be too low or there is no reason to conduct the test. At the same time, the threshold cannot be too high or there is too great a chance that development money will be wasted fixing deficiencies that were not worth fixing and delaying the production and delivery of a system that is badly needed by the stakeholders, especially when competitive advantage is involved. The qualification objectives must be focused on determining whether the system passes or fails the threshold criteria. This focus on qualification objectives and pass/fail thresholds is the identification of alternate concepts for the qualification system, culminating in the selection of that concept that is deemed most appropriate. This concept selection decision must trace back to the original system concept selection.

Once the qualification objectives have been established, the operational concept for qualification (including key scenarios) can be defined. This operational concept will produce a definition of all high-level inputs and outputs of the tests. The definition of the qualification scenarios in consideration of the qualification objectives is establishing at a high level what should be tested and to what precision of confidence. The qualification requirements, based upon the threshold criteria for passing, determine how well the test should be conducted in each area. Each specific test should be considered a system; the major test functions are needed to help define the resources needed for the test. These qualification functions enable the development of qualification requirements; both input/output requirements and qualification-wide/technology requirements. The qualification requirements in this case involve the examination of the qualification system design to ensure that it satisfies the requirements involved in meeting the qualification objectives. Qualification coverage matrices involve comparisons of the qualification requirements to the qualification activities; these matrices enable the management of qualification requirements to ensure that every requirement is being met by some activity. Even more so than with most systems, there may be risks that the testing process will not be completed in a timely manner; test failures at certain points may cause delays in fixing deficiencies or replacing test items. Therefore, extra effort should be expended to identify

risks to meeting qualification-wide requirements such as schedule and time and develop risk mitigation strategies for dealing with such risks. Finally, the plan for the qualification process should be
documented in a master qualification plan.

The second major qualification planning function of Table 11.2, plan the qualification approach,
involves creating specific test activities (subfunctions), as well as the physical and allocated architectures for the qualification system. The physical architecture for a test includes test equipment
and facilities, as well as the organizations (people) that will conduct a specific test. After one or
more generic qualification architectures have been devised and several instantiated qualification

TABLE 11.2 Qualification Planning Functions.

Plan the qualification process	• Review system objectives
Acceptance test Validation test Verification test	• Identify qualification system objectives • Identify pass/fail thresholds • Define qualification operational concept • Define qualification requirements • Define qualification functional architecture • Define qualification generic physical architecture • Generate qualification coverage matrices (allocate requirements to functional architecture and functions to the generic physical architecture) • Identify risks and mitigation strategies • Create master qualification plan
Plan the qualification approaches	• Define subfunctions (or test activities) for the functional architecture
Acceptance test Validation test Verification test	• Define qualification resources and organizations (instantiated physical architecture) • Assign qualification activities to organizations • Allocate qualification activities to resources • Develop qualification schedules consistent with development schedule
Plan qualification activities	• Develop detailed derived qualification requirements for the test activities
Acceptance test Validation test Verification test	• Develop functional architectures for fulfilling the test activities • Define detailed component architectures for the test resources (identifying what special test fixtures and test stubs are needed) • Generate coverage matrices (allocate derived requirements to functional architectures and functions to physical architectures) • Write activity-level qualification plans for each qualification component • Assign qualification responsibilities
Plan specific tests	• Create test scenarios
Acceptance test Validation test Verification test	• Identify required stimulation data for each activity • Write test procedures • Write analysis procedures • Define test and analysis schedules

architectures are identified, decisions can be made about the most cost-effective means for achieving the qualification objectives with a reasonable risk. As part of this process for selecting an allocated qualification architecture, the allocation of qualification activities to equipment, facilities, and organizations must be considered. Planned previous qualification data must also be considered so that each test does not retest or overtest certain requirements. Finally, these qualification activities can now be planned in time so that the qualification resources are used efficiently, and development schedule requirements are met.

The last two qualification planning functions in Table 11.2 define the qualification activities in greater detail, that is, at the component and CI levels. Planning the qualification activities decomposes each activity to two or three levels of detail and matches these subactivities to requirements and resources. Planning the specific tests takes each test activity and creates detailed scenario and data specifications of the activity. Test procedures for handling the test equipment and test data are also produced. Finally, detailed schedules are produced.

Figure 11.7 depicts the design process of the qualification system as an IDEF0 diagram. Note this is essentially the same process discussed in Chapters 6–10 for any system. However, a final activity is added to address the development of all the models needed for qualification.

11.7 QUALIFICATION METHODS

Four categories of *qualification methods* are inspection, analysis and simulation, instrumented test, and demonstration. Table 11.3 summarizes each of these methods by describing each, discussing when each is used and when each is most effective.

Inspection is used for physical, human verification of specific requirements. As automation has come to replace humans in the performance of certain activities, more and more of inspection can be accomplished by computers, which falls under instrumented test. A major example of this migration from inspection to instrumented test is the examination of software code for key features or the lack of key features. Finally, qualitative models that are now available with systems engineering tools allow for extensive inspection opportunities related to design validity and design verification.

Analysis and simulation involve the use of models to test key aspects of the system. Models have always been used in engineering; see Chapter 3 and the discussion of mental models. The most common use of models is to examine the performance of the system in a range of environmental conditions. Initially, these models support the design process by enabling the comparison of alternate physical architectures. However, as verification and validation begin, these same models can be used to augment instrumented test and demonstration. Initially, the results of instrumented test are fed back to the models and used to refine parameters embedded in the model. Later, the models can be used to predict the results of instrumented tests and demonstrations. As confidence in a specific model increases, the model can be used to replace some of the instrumented tests and demonstrations. An important example of this interplay between models and instrumented tests is the development of estimates for such parameters as reliability, availability, and durability [see Holmberg and Folkeson, 1991]. Lee and Yannakakis [1996] provide a detailed survey of the use of one class of models (finite-state machines) in testing. Additional advances are being made in the verification of models that directly relates to verifying systems, see Baier and Katoen [2008].

Table 11.4 describes testing methods that can be used at the system level and lower. These functional and structural testing methods are used in conjunction with top-down, bottom-up, middle-out, and big-bang integration. Functional testing examines the system at the level of inputs and outputs under mostly nominal conditions. Structural testing deals with specific characteristics of the outputs as well as the system-wide properties such as safety, availability, and recovery. Structural testing pays particular attention to the most extreme environments that the system will experience.

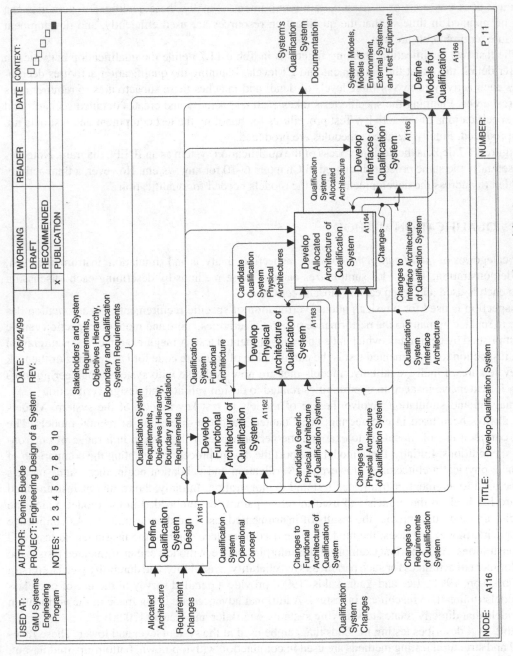

FIGURE 11.7 The process for developing the qualification system.

TABLE 11.3 **Qualification Methods.**

Method	Description	Used During	Most Effective When
Inspection (Static Test)	Compare system attributes to requirements.	During all segments of verification, validation, and acceptance testing for requirements that can be addressed by human examination.	Success or failure can be judged by humans; examples include inspection of physical attributes, code walk-throughs, and evaluation of user's manuals.
Analysis and Simulation	Use models that represent some aspect of the system. Examples of models might address system's environment, system process, system failures.	Used throughout qualification, but emphasis is early in verification and during acceptance. Often used in conjunction with demonstration.	Physical elements are not yet available. Expense prohibits instrumented test, and demonstration is not sufficient. Issue involves all or most of the system's life span. Issue cannot be tested (e.g., survive nuclear blast).
Instrumented Test	Use calibrated instruments to measure system's outputs. Examples of calibrated instruments are oscilloscope, voltmeter, LAN analyzer.	Verification testing.	Engineering test models through system elements are available. Detailed information is required to understand and trace failures. Life and reliability data is needed for analysis and simulation.
Demonstration or Field Test	Exercise system in front of unbiased reviewers in expected system environment.	Primarily used for validation and acceptance testing.	Complete instrumented test is too expensive. High-level data/information is needed to corroborate results from analysis and simulation or instrumented test.

Samson [1993] postulates four facets for any qualification activity: structural (relation to system implementation), function (relation to system functions), environment (relation to environmental conditions), and conditions (relation to requirement characteristic). The first two of these facets are mutually exclusive and are described in Table 11.4. The second two need to be added to each specific structural or functional test to make it complete. In other words, there has to be an environmental facet and a conditional facet for each functional test and each structural test. Table 11.5 shows Samson's examples of these facets.

Today's systems are often computer and software-centric. Black-box, white-box, and gray-box (or grey-box) testing methods (Table 11.6) are commonly employed in software testing. For each method, test cases must be specified, and test data generated as inputs. These inputs are then injected into both the system prototype (which is essentially a model of the eventual system) and a model of the system. The outputs of the system and the model are compared; any discrepancies are checked to determine whether the system or the model is incorrect [see Chusho, 1987; Richardson and Clarke, 1985; Voges and Taylor, 1985].

TABLE 11.4   Testing Methods.

Functional testing	Test conditions are set up to ensure that the correct outputs are produced, based upon the inputs of the test conditions. Focus is on whether the outputs are correct given the inputs (also called black-box testing).
Structural testing	Examines the structure of the system and its proper functioning. Includes such elements as performance, recovery, stress, security, safety, and availability. Some of the key elements are described later.
Performance	Examination of the system performance under a range of nominal conditions ensures system is operational as well.
Recovery	Various failure modes are created and the system's ability to return to an operational mode is determined.
Interface	Examination of all interface conditions associated with the system's reception of inputs and sending of outputs.
Stress testing	Above-normal loads are placed on the system to ensure that the system can handle them; these above-normal loads are increased to determine the system's breaking point; these tests may proceed for a long period of time in an environment as close to real as possible.

TABLE 11.5   Examples of Testing Facets.

Structural Facet	Functional Facet	Environmental Facet	Conditional Facet
Compliance	Algorithm analysis	Computer supported	Accuracy
Execution	Control	Live	Adequacy
External	Error handling	Manual	Boundary
Inspection	Intersystem	Prototype	Compliance
Operations	Parallel	Simulator	Existence
Path	Regression	Testbed	Load
Recovery	Requirements		Location
Security			Logic
			Quality
			Sequence
			Size
			Timing
			Typing
			Utilization

11.8   ACCEPTANCE TESTING

Acceptance testing is the final step in qualification and is separated from validation because acceptance testing is conducted by the stakeholders, whereas verification and validation have been conducted by the development team of systems engineers. In order for the development process to proceed efficiently and effectively, the thresholds for acceptance need to be defined early in the requirements development process by the stakeholders with the help of the systems engineering team. In fact, in Chapter 6, the agreement on the acceptance criteria was defined to be the exit criterion for the requirements development.

The acceptance test determines whether the stakeholders, especially the bill-payer, are willing to: accept the system as it is, accept it subject to certain changes, not accept it, or accept it after

TABLE 11.6   Black-Box, White-Box, and Gray-Box (or Grey-Box) Testing.

Black-box testing	The intended inputs and outputs of the system and its modules are known, and unintended inputs and undesired outputs and specified mitigations are characterized. Outputs are determined correct or incorrect based upon inputs; inner workings of the module are ignored. Both positive and negative testing have to be employed. This approach is scalable to system-level testing. • Positive testing pulls the test data and sequences from the requirements documents. • Negative testing attempts to find input sequences missed in the requirements documents and then determine how the module reacts. Crash testing is an example.
White-box testing	Complete inner workings of the module are known and examined as part of the testing to ensure proper functioning. Usually used at the CI level of testing; this method becomes impractical at the system level. • Path testing addresses each possible simple functionality and is based upon a prescribed set of inputs. • Path domain testing partitions the input space and then examines the outputs for each partition of the input space. • Mutation analysis injects predefined errors and tests the error detection and recovery functionalities.
Gray box (or grey-box) testing	The intended inputs and outputs of the system and its modules are known, and unintended inputs and undesired outputs and specified mitigations are characterized. There is limited knowledge of the internals of the system and its modules such as interface definitions, functional specifications, state-based models, or architecture diagrams of the target system and its modules, but not the software code or binaries. • Testing is based on requirement test-case generation with traceability to the concept of operations. • Gray-box testing techniques include matrix testing, regression testing, pattern testing, and (black-box) statistical orthogonal array testing.

certain changes have been made. Acceptance testing focuses on the use of the system by true users, typically a small, but representative sample of users. (During verification and validation, members of the systems engineering team and discipline engineers conducted the use of the system.) As a result, usability characteristics of the system are a major focus. Another characteristic of acceptance testing is the lack of time and money to conduct thorough, controlled tests of the system with users from which inferences, based on classical statistics, can be drawn. The two big issues in acceptance testing are what to test and how to test the usability of the system.

11.8.1   Deciding What to Test

Common wisdom says that everything possible, including all functionalities or paths, should be tested. The case study about the *Ariane 5* failure is one of many examples that support this wisdom. In fact, during verification and validation the key question is not "what should be tested?" but "what have we forgot to test?" The more systematic the design process the more likely it is that key issues for testing will arise. Nonetheless, it is imperative that everyone involved in the design and integration process constantly question where problems might arise. If only someone on the *Ariane 5* development team had insisted on running the new flight envelope through the software of the inertial reference system, the design flaw would have surfaced. This is an area in which the brainstorming techniques discussed in Chapter 9 can be useful to generate potential test issues, not all of which will be meaningful, but some of which may save the system from the disasters of *Ariane 5* and Hubble.

The question of "what should he tested?" becomes very relevant during acceptance testing. Acceptance testing substitutes developmental testers with real users but must rely on all of the previous testing activities. Exhaustive repetition of verification and validation is not feasible during acceptance testing due to the limits of time and money. The focus of acceptance testing is whether the system, as currently defined, is acceptable or not; and if not, why. But what does it mean to say that the system is acceptable? Can we distinguish only between acceptable and unacceptable? Acceptability is defined here to mean the stakeholders want to deploy the system as it is as soon as practically possible, with whatever flaws there are. More flaws are acceptable to stakeholders when the current system's deficiencies are causing severe problems for the users in accomplishing their goals, for the buyers in maintaining market share, or with the victims in suffering too many losses. However, the stakeholders may be willing to accept the system, yet still demand major changes quickly. The system is unacceptable when it will cause more problems than the current system. Similarly, the system can be totally unacceptable beyond the possibility of improvement or unacceptable until certain changes are made.

The acceptance test can either be designed under the assumption that the system is acceptable or that it is not. If the assumption that the system is acceptable is chosen, the test should be designed to prove it is not. A test designed to try to prove that the system is not acceptable would probably include a relatively small set of challenging activities that are key to the system's performance. If the system cannot perform some of these challenging activities, then it can be failed. On the other hand, if the test design assumption is that the system is not acceptable, then a reasonable amount of standard activity would be included in the test for the test to prove that the system is acceptable. If the system can pass most of these standard activities, then it can be accepted. Recall that a statement cannot be proven true by example, but it can be proven false by example. This latter approach is the more common in acceptance tests but not the more defensible.

Decision analysis (see Chapter 14) provides a rational, defensible way to analyze alternate acceptance test designs, including a seldom used option of no acceptance test or accept the system after verification and validation. The decision is whether to accept or not accept the system; the other options of accept but fix and do not accept until fixed should also be included. Now test designs are ways to gather information about system parameters about which uncertainty exists. This increased information, when collected during the test, may update this uncertainty in ways that are sufficient to justify accepting or not accepting the system.

11.8.2 Usability

In Chapter 6, usability testing with prototypes is discussed as a method of generating requirements. In qualification, usability testing is again used as part of acceptance testing to determine the success with which the requirements have been met.

In fact, usability testing is also used as part of verification testing when an iterative or evolutionary design process is employed. Limited experimental results for evaluating the effectiveness of evolutionary design are reported by Nielsen [1993, p. 107]. The median improvement over four projects was 38% per iteration, but with a high degree of variability. As a result, at least three iterations are recommended.

Recall from Chapter 6 that usability concerns five aspects of a user's interaction with a system: learnability, efficiency or ease of use, memorability, error rate, and satisfaction. These characteristics should be part of the system-wide requirements for most systems. These characteristics can typically not be tested adequately until the entire system has been assembled or simulated. During validation, the characteristics are tested by specially defined sets of users. Larger samples of users, often uncontrolled sets of users called beta testers, address these five aspects during acceptance testing.

When designing any test queries, there are two central issues: Is the query reliable and is the query valid. Reliable queries are queries that will result in the same response when repeated. Reliability is a major problem that cannot be solved completely due to the large individual differences among users. Segmenting the users into relatively homogeneous groups along the dimensions of domain experience, computer experience, and experience with the system under development helps significantly to obtain a reasonable chance of repeatability. To obtain sample users in this last of the three dimensions, there must be a sustained effort to train selected users to become very experienced users. Care must be used in defining homogeneous segments of users. If each of the three dimensions is categorized at two extremes, there are 8 (2^3) different combinations. Not all these combinations may be that interesting for the system in question. There may be some interest in user groups that are midrange in one or more of these dimensions; for most systems, the predominant number of users will be neither I nor expert along any of these three dimensions. However, there are some systems for which all users will be trained extensively before even being allowed access to the system, for example, air traffic control systems, and aircraft. However, for these systems, the memorability factor of usability may be critical.

Valid queries are those that are measuring the right or appropriate aspect of the system. For usability, this will refer to the five concerns outlined earlier, see the metrics in Table 6.5.

The best way to achieve reliability and validity of test measures is to set up relevant tasks on which tests will be conducted and measures taken. These tasks should be drawn from the operational concept; each task may be a complete scenario or a small segment of a scenario, depending on where in the qualification process the test is being used. Complete scenarios should be used during acceptance and the latter stages of validation. Segments can be used during prototyping and the early stages of validation. Each task must define a realistic setting for the user in terms of the system and its context, a specified set of circumstances in which to be performing the task, a well-defined outcome that the user is expected to achieve, and a realistic time interval in which to complete the task.

Cox et al. [1994] state the most serious obstacles to successful usability tests are

- Obtaining test participants that represent the real users of the system
- Securing a representative sample that will be predictive of how the total population will evaluate the system
- Selecting the tasks that are most critical to the usability needs of real users
- Writing test scenarios that accurately represent real task situations that a user will encounter in the system's environment
- Predicting which of the user interface characteristics are most critical or most often used

Yet, these obstacles must be overcome for usability testing to be successful.

11.9   SUMMARY

Integration begins when assemblies of CIs and components are evaluated in terms of the derived requirements. This process is part of verification, determining that the system was built right. There are several approaches to integration; bottom up being the most common one to systems engineering. Top-down and big-bang integration are more common in software engineering. Middle-out integration is only recently recognized as a distinct approach. Integration and verification end at the system level.

Qualification consists of verification, validation, and acceptance testing. Verification addresses the comparison of the specifications for the system's CIs, components, subsystems. and the system to the actual designs to make sure the designs are right, that is, meet the specifications.

Validation consists of early validation and operational validation. Early validation (conceptual, requirements, and design validity) proceeds during design to ensure that the design process is valid. Conceptual validity addresses the congruence between the stakeholders' needs and the operational concept. This is the hardest element of validation to complete successfully. Requirements validity applies to the conformity between the operational concept and the stakeholders' requirements. Design validity addresses the coherence between the stakeholders' requirements and the layers of derived requirements associated with the system, components, and CIs.

Operational validity may begin before verification is complete but ends after verification is complete and addresses the conformance of the system as it has been built with the operational concept. This is the last phase of the development process under the complete control of the systems engineering team.

Acceptance testing is controlled by the stakeholders and provides the stakeholders the final opportunity to review the design and verify that it meets their needs. Acceptance testing should fully utilize all the data and analyses that have been part of verification and validation. At the same time, though acceptance testing is focused on the use of the system by representatives of the stakeholders' community, whereas verification and validation employ highly qualified users (i.e., engineers) as stakeholders for the most part. As a result, the system's usability is a major focus during acceptance testing.

There are two critical chains whose links are checked during qualification. The top-level chain consists of the links: conceptual validity, operational validity, and acceptability. The first link is validated early in the design phase; the last two links are addressed at the end of integration. The second chain consists of requirements validity, design validity, verification, and operational validity. Note that operational validity is common to both chains, and recall that the chain is as strong as its weakest length. Therefore, it is a mistake to assert that any one of the links is more important than any of the others.

The development of the qualification system should be approached just as the development of any system, as described in Chapters 6–10. The operational concept, external systems diagram, objectives hierarchy, requirements, and architectures (functional, physical, and allocated) are all critical elements of the development of the qualification system. Besides addressing verification, validation, and acceptance, the qualification system is often broken into four methods: inspection, analysis and simulation, instrumented test, and demonstration.

While it is common to visualize the qualification system as the system that will detect and isolate faults in the product system's design, design of the qualification system, when done right, also reinforces the design process and reduces the introduction of faults into the design of the system.

In summary for Section 2 of this book, the traditional, top-down systems engineering (TTDSE) process has been described in some detail. Figure 11.8 integrates Figures 1.6 and 1.19 to bring the major elements of Chapters 6–11 together into a single picture. The point of this figure is that the

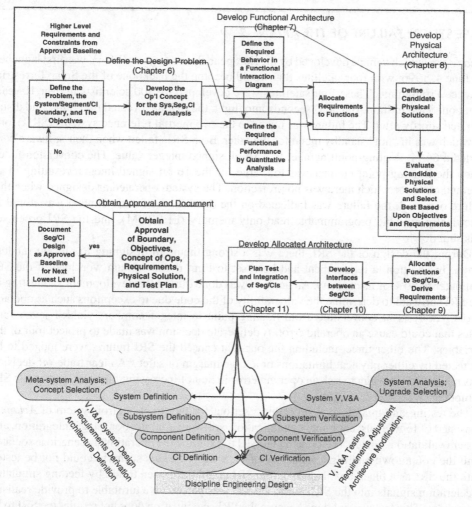

FIGURE 11.8 Repeated application of TTDSE to the layers of the system's design.

process described in Chapters 6–11 is repeatedly applied to the process of "peeling the onion" of the layers of the system. Each preceding layer provides the starting information for the layer before it. The major difficulty is getting started when very little needed information is available.

CASE STUDY: THERAC-25

The Therac-25 was a computer-controlled machine that provided radiation therapy in the late 1980s. Three patients were killed, and one seriously injured by radiation overdoses in the 1985–1987 time frame when four different operators entered an acceptable, but infrequently used, sequence of commands. While this tragedy can be traced to requirements and design errors, the qualification process should be focused on catching just this sort of flaw. This was clearly a case in which all possible data entry sequences should have been tested [Jacky, 1990].

CASE STUDY: FAILURE OF THE *ARIANE 5*

Ariane 5, a launch vehicle developed by the European Space Agency (ESA), was first launched on June 4, 1996, with four satellites that would become the backbone of the Solar Terrestrial Science Programme. These four satellites were developed by 500 scientists in over 10 years for about $500 million. But at 37 seconds into flight, the *Ariane 5* veered off course and disintegrated shortly after. The failure was traced to the two inertial reference systems (SRIs), one of which was in "hot" standby mode for the other. Both SRIs failed when their software converted a 64-bit floating-point number to a 16-bit signed integer value. The conversion failed when the floating-point number was too large for the 16-bit signed integer, resulting in an operand error for which there was no protection. The system operated as designed when this failure occurred: the failure was indicated on the databus, the failure context was stored in electrically erasable programmable read-only memory (EEPROM), and the SRI processor was shut down.

During the design of the SRI, there was a strong theme of designing to prevent random errors. In addition, a requirement had been set to limit the maximum workload of the SRI computer to 80% of its capacity. An analysis was done during the development and testing of the SRI software to determine the vulnerability of the code due to exceptions such as operand errors. Analysis of conversions from floating point to integer numbers yielded seven variables that could cause an operand error. A deliberate decision was made to protect four of the variables. The other three, including the one that caused the SRI failure, were judged to be protected by either physical limitations or a large margin of safety. A clear trade-off decision was made in this design to risk an operand error in lieu of increasing the workload on the SRI computer.

The testing and qualification procedures set out for the flight control system of *Ariane 5* consisted of four levels: equipment qualification, software qualification, stage integration, and system validation. No test was done on the SRI to examine the operational scenario associated with the countdown and flight trajectory of the *Ariane 5*. This scenario could not be tested with the SRI as a black box. However, the SRI could have been tested by feeding simulated acceleration signals into the SRI, while the SRI was placed on a turntable to provide realistic movement. This test was not done because the SRI specification does not require the SRI to be operational after launch. The purpose of the SRI was to provide inertial reference data prior to launch. Even though the SRI served no useful purpose after launch, its operation after launch was sufficient to cause the destruction of *Ariane 5* 37 seconds into the flight.

Much of the *Ariane 5* requirements and software were inherited from earlier versions of *Ariane*. Ten years earlier requirements had been established that the SRI operate 50 seconds beyond the initiation of flight mode. Flight mode started at 9 seconds for *Ariane-4;* this allowed restarting the countdown without waiting for a normal alignment of the spacecraft, which takes about 45 minutes. However, *Ariane 5* had a different initiation sequence that did not require the SRI being active during flight. *This is one case in which the old adage "if it ain't broke, don't fix it" caused a failure.*

The final stage at which this error could have been detected was at the functional simulation facility (ISF) which tests (1) guidance, navigation, and control performance in the whole flight envelope, (2) sensor redundancy operation, and (3) flight software compliance with all equipment of the flight control electrical system. "Technically valid arguments" [Lions, 1996] were presented for not having the SRIs in the loop for the tests conducted at the ISF. As a result, the SRIs were never tested for the *Ariane 5* launch. The trajectory profile of *Ariane 5* was sufficiently different than the profiles of previous *Ariane* launches that this operand error would always occur; a major requirements' failure followed by a failure of test design [Lions, 1996].

PROBLEMS

11.1   Describe a process of establishing conceptual validity that identifies the elements of conceptual validity and links between pairs of these elements. This process should then establish characteristics such as completeness, consistency, and correctness.

11.2   Describe a process that could he used to establish requirements validity. This process should identify the elements of moving from the operational concept to the stakeholders' requirements, as discussed in Chapter 6. Additional products beyond those discussed in Chapter 6 should be identified that would enable the validation of such characteristics as completeness, consistency, and correctness when comparing the operational concept to the stakeholders' requirements. Examples of comparisons that should be involved are

- Matching of operational concept elements to elements of the external systems diagram
- Matching of operational concept elements to input/output requirements
- Matching the objectives hierarchy to elements of the external systems diagram
- Matching the objectives hierarchy to input/output requirements
- Matching elements of the external systems diagram to input/output requirements
- Tracing input/output requirements to external items
- Matching the objectives hierarchy to system-wide requirements

11.3   Describe the types of activities (similar to those in Problem 11.2) that could be used to establish design validity. Identify intermediate products that could be used for establishing design validity. In particular, focus on developing the best definition of completeness for requirements that you can.

11.4   Develop an operational concept, external systems model, objectives hierarchy, and requirements for the qualification system for a traffic light system.

11.5   Develop the functional, physical, allocated, and interface architectures for the qualification system for a traffic light system.

11.6   Develop the integration and test functional, physical, allocated, and interface architectures for the development for a traffic light system.

Chapter 12

A Complete Exercise of the Systems Engineering Process

12.1 INTRODUCTION

This chapter serves as a review and a summary of Chapters 6–11. In this chapter, we will conduct all the system-level systems engineering steps for an automated soda machine and then discuss how we go about repeating these steps for the subsystem layer (or onion peel). The purpose of this chapter is to drive home the systems engineering process and how it is repeated as often as necessary for the subsystems and components that need further decomposition.

12.2 OPERATIONAL CONCEPT

Here are nine usage scenarios for the automated Soda Machine that were developed to cover a range of interactions with a "thirsty patron," supplier, and maintenance person. Notice that none of these usage scenarios address components of the Soda Machine.

(1) Thirsty patron stops by the Soda Machine, receives information about the price of a cold soda, inserts the correct amount of change, and selects one of five different sodas; the Soda Machine provides the selected soda chilled to be determined (TBD) degrees.

(2) Thirsty patron inserts coins in excess of that required for a soda, and the Soda Machine provides a change in coins with the selected cold soda.

(3) Thirsty patron inserts more than enough money in dollar bills and selects one of the soda types; the Soda Machine provides change for the money with the selected cold soda.

(4) Thirsty patron stops by the Soda Machine and sees that it is not available.

(5) Thirsty patron arrives and sees that the Soda Machine is out of the sodas she desires.

(6) Thirsty patron arrives and sees that the Soda Machine is out of change.

(7) The Soda Machine sends status reports on change and soda availability over a phone line installed in the building. The building provides electricity to the Soda Machine.

The Engineering Design of Systems: Models and Methods, Fourth Edition. Dennis M. Buede and William D. Miller
© 2024 John Wiley & Sons, Inc. Published 2024 by John Wiley & Sons, Inc.
Companion website: www.wiley.com/go/engineeringdesignofsystems4e

(8) Supplier for the Soda Machine empties the Soda Machine of deposited money and resupplies it with warm sodas and change.

(9) Maintenance person arrives and performs diagnostic tests to determine operational problems; the Soda Machine provides test results on the basis of the queries.

12.3 EXTERNAL SYSTEMS MODEL

First, we create the external systems diagram from the above usage scenarios. The first usage scenario deals with the thirsty patron and the Soda Machine, so we need to create functions for each of these two systems. Let us call the function for the thirsty patron "Request and Receive Cold Sodas." The function of the Soda Machine will be "Provide Cold Sodas."

Now, let us identify the elements of the first usage scenario that are inputs to the Soda Machine and outputs of the Soda Machine. The "Thirsty Patron" performs the "Request & Receive Cold Sodas" function and provides inputs of "correct change" and "selection" to the "Soda Machine" (performing the "Provide Cold Sodas" function). In return, the Soda Machine provides information about the soda selections and prices to the Thirsty Patron as well as providing a chilled soda after the correct change and selection have been entered. Figure 12.1 shows a modified N^2 chart for this first usage scenario. This N^2 chart is modified because the systems for the functions have been added at the bottom for clarity. The functions are shown with a gray background while the systems performing those functions are shown with a dotted background. The items flowing between functions are shown with a white background.

After all of the usage scenarios have been added to the N^2 chart, we see Figure 12.2. Note that some of the cells in the N^2 chart are blank – always ask if this makes sense because this is a way of discovering important missing usage scenarios. In our make-believe example, we will assume the blank cells do make sense.

Also, note that there are a lot of entries in some of the cells. For real systems, this overabundance of items requires us to consolidate some of the items into more generic categories.

As we translate the N^2 chart in Figure 12.2 into a SysML activity diagram, shown in Figure 12.3, we will consolidate (1) soda prices, availability of the machine, availability of sodas, and availability

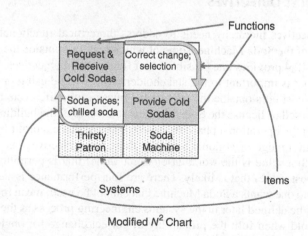

FIGURE 12.1 Modified N^2 chart for the first usage scenario.

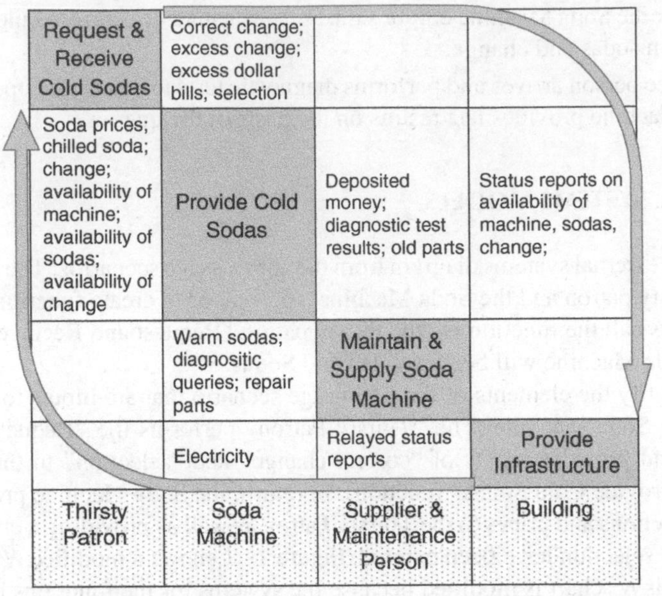

Request & Receive Cold Sodas	Correct change; excess change; excess dollar bills; selection		
Soda prices; chilled soda; change; availability of machine; availability of sodas; availability of change	Provide Cold Sodas	Deposited money; diagnostic test results; old parts	Status reports on availability of machine, sodas, change;
	Warm sodas; diagnositic queries; repair parts	Maintain & Supply Soda Machine	
	Electricity	Relayed status reports	Provide Infrastructure
Thirsty Patron	Soda Machine	Supplier & Maintenance Person	Building

FIGURE 12.2   Modified N^2 chart for all usage scenarios.

of change into "customer feedback" and (2) correct change, excess change, and excess dollar bills into "money." While it greatly improves communication with the activity diagrams to create these consolidations, it is also very important when writing the input/output requirements later to keep track of these consolidations because the input/output requirements need to address the detailed inputs and outputs.

One external input has also been added to make sure it is clear what the source of the electricity for the Soda Machine is.

12.4   FUNDAMENTAL OBJECTIVES

The fundamental objectives hierarchy needs to address the critical functional and non-functional performance aspects of the Soda Machine, some of which may fall outside the operational phase of the life cycle. Figure 12.4 provides one possible set of fundamental objectives. These are developed by thinking about what is important to the stakeholders who will be using or paying for the Soda Machine. Production cost falls outside the operational phase of the life cycle but is very important when it comes time to sell or license the Soda Machine to owners of the buildings. Operating cost is very important during the operational phase to those who have made a profit by installing the Soda Machine. Note we put a range of notional $ next to these two objectives (as we have done for all the objectives). The first value is the worst case (worst value) that is permitted. The second value is the highest case (best value) that is likely. There are four performance issues: availability of the Soda Machine since no one wants a Soda Machine that cannot be used; mean time between supplies, which will ultimately be defined later in the systems engineering process as the number of sodas the Soda Machine can hold when full; the probability of correct change for obvious reasons; and the probability of the correct soda being dispensed, also for obvious reasons.

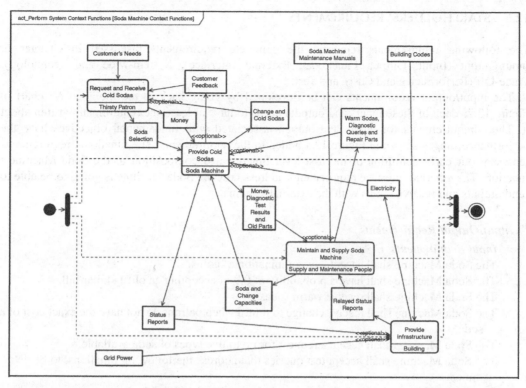

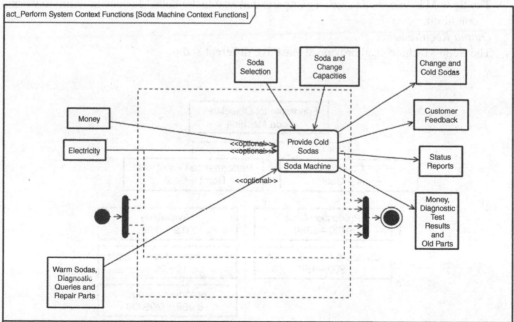

FIGURE 12.3 SysML activity diagrams of the external systems and context.

12.5 STAKEHOLDERS' REQUIREMENTS

The following requirements represent the complete requirements discussed in Chapter 6: Input/Output (Input, Output, Functional, External Interface), System-wide and Technology, Trade-Off (Performance and Cost), and Test.

The input/output requirements can be developed by going back to the modified N^2 chart in Figure 12.2. Each of these inputs and outputs needs to have at least one requirement written about it. Those requirements for inputs and outputs associated with fundamental objectives have the performance ranges shown in Figure 12.4 added to the requirements. The functional requirements represent our early definition of the first-level functional decomposition of the Soda Machine's function. The external interface requirements address how the Soda Machine is going to be able to send outputs and receive inputs with the external systems.

Input/Output Requirements

Input Requirements
The Soda Machine shall accept any combination of coins.

The Soda Machine shall have a probability of 0.9 of accepting an old 1-dollar bill.

The Soda Machine shall receive warm sodas.

The Soda Machine shall receive change to return when patrons do not have the exact cost of a soda.

The Soda Machine shall accept a request for up to five types of soda available.

The Soda Machine shall accept test queries that address the following functions: to be determined.

The Soda Machine shall accept repair parts that enable the Soda Machine to return to working condition.

Output Requirements
The Soda Machine shall provide at least five different sodas.

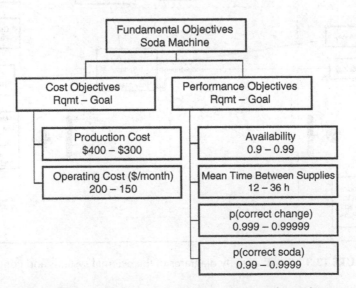

FIGURE 12.4 Fundamental objectives of the soda machine.

The Soda Machine shall have a probability of correct change for any combination of coins of 0.999. the design goal is 0.99999.

The Soda Machine shall have a probability of correct change for a 1-dollar bill of 0.999. The design goal is 0.99999.

The Soda Machine shall provide cold sodas to the Thirsty Patron for the correct change.

The Soda Machine shall provide feedback to the Thirsty Patron that the correct change has been inserted.

The Soda Machine shall provide feedback to the Thirsty Patron that a particular soda selection is not available.

The Soda Machine shall provide feedback to the Thirsty Patron that there is no change available.

The Soda Machine shall provide the money collected for sodas to the Supply Personnel.

The Soda Machine shall provide status reports to Supply Personnel about the availability of change and sodas.

The Soda Machine shall provide test results to the each of the defined test queries.

Functional Requirements

The Soda Machine shall accept change and a soda selection and provide feedback about the status of the Soda Machine and the transaction.

The Soda Machine shall maintain cold sodas.

The Soda Machine shall provide cold sodas.

The Soda Machine shall provide change, if necessary.

The Soda Machine shall provide status reports on the quantity of change and soda for transactions, and responses to diagnostic tests.

External Interface Requirements

The Soda Machine shall use the electric power of the building in which it resides.

The Soda Machine shall use a phone line in the building in which it resides to send status messages.

The system-wide and technology requirements are the non-functional requirements. Here are few candidates for the Soda Machine. All the fundamental objectives that address cost and other system-wide issues should be included in this list with their performance ranges.

System-wide and Technology Requirements

The Soda Machine shall have an operational availability of 0.9. The goal is 0.99.

The Soda Machine shall have a mean time between supplies of 12 h for a scenario in which there are a mean number of customers per hour of 12. The design goal is 36 h.

The Soda Machine shall have an operating cost of $200 or less. The design goal is $150.

(The production cost of the Soda Machine shall be $400 or less. The design goal is $300.)

Next are the trade-off requirements. These requirements address how the fundamental objectives will be used to select the best system architecture. In this case, the swing weights for weighting the relative importance of objectives are given.

Tradeoff Requirements

The relative weights of the cost and performance requirements are 0.6 and 0.4, respectively.

Performance Tradeoff Requirement

The relative weights of the performance requirements are shown on the right side of the hierarchy in Figure 12.5.

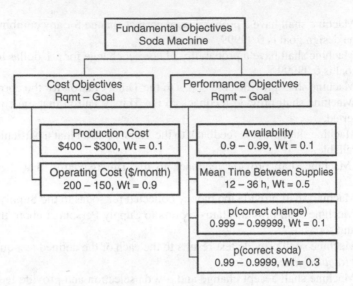

FIGURE 12.5 Fundamental objectives with swing weights.

The value curves for the performance requirements are to be determined.

Cost Tradeoff Requirement

(Note that the production cost is included here for completeness, even though the manufacturing phase has not been addressed.) The relative weights of the cost requirements are shown on the left of the hierarchy above.

Finally, here are some representative test requirements. We are not showing a complete set of test requirements for the above input/output and system-wide and technology requirements, but the reader should understand that this complete set would be needed.

Test Requirement

For each of 5 soda prices every combination of change shall be inserted into the prototype Soda Machine and the response of the Soda Machine shall be determined to be correct in each case.

100 old $1 bills shall be inserted into the Soda Machine for each of 5 soda prices to determine how many times the bill is accepted.

… Additional input test requirements.

Fifty soda selections will be made, and the correct soda output shall be verified by inspection.

For each of 5 soda prices every combination of excess change shall be input 10 times and the change that is output by the Soda Machine shall be verified to be correct by inspection.

… Additional output test requirements.

The functional requirements shall be verified by inspection.

The preproduction Soda Machine prototype shall receive power and be activated. While activated, the test engineers shall raise and lower the frequency of the power by 10% around 60 Hz. In a separate test, the test engineers shall feed the Soda Machine 115 and 125 V and attempt to activate the Soda Machine without sustaining damage.

… Additional external interface test requirements.

The preproduction Soda Machine prototype shall be subjected to life tests that gather reliability and maintainability data that are then entered into simulation models to predict the operational availability of the Soda Machine over a five-year time period.

… Additional system-wide and technology test requirements.

12.6   FUNCTIONAL ARCHITECTURE

The functional architecture (Chapter 7) represents how the system meets its input/output requirements. We return to the context diagram of the IDEF0 model (see Figure 12.2) to determine what the external inputs and outputs are that must be addressed by this functional architecture for the Soda Machine. There are many possible functional architectures for such a system, many of which would work just fine. Here are the functions chosen for this solution; note these come from the functional requirements above:

- Accept change and soda selection and provide feedback
- Maintain cold sodas
- Provide cold sodas
- Provide change
- Provide status reports and responses to diagnostic tests.

The SysML activity diagram for these functions is shown in Figure 12.6. Note that the external inputs, controls, and outputs from the context diagram in Figure 12.2 have been connected to one or more of the five functions. Next, a number of internal items had to be created in order to make this functional architecture "work." The last function collects status reports from two of the other functions and feeds status information to the first function while providing a feedback control loop to maintain the temperature of the refrigeration unit for the cold sodas so that the cold sodas are not too warm or too cold. The first function directs the provision of sodas and changes to the third and fourth functions. In addition, the fourth function triggers the provision of the soda when sufficient change has been received; this extra control is needed to meet the requirement for not giving too much or too little change to the thirsty patron.

The next step in defining the functional architecture is to allocate the input/output requirements to the appropriate items and functions in the SysML activity diagram. Allocating the input and output requirements includes associating each of these requirements with one item and one function. These associations should be obvious. For example, the output requirement "The Soda Machine shall provide at least five different sodas." Should be associated with the output "cold sodas" and its function (Provide Sodas). Similarly, there are five functional requirements in the input/output requirements, each of which is associated with the obvious one of the five functions in Figure 12.6.

Finally, the two external interface requirements need to be associated with the appropriate items and functions. The electric power interface requirement should be associated with the A0 system-level function. However, all the components performing these functions will need electrical power, so all of the functions also need to be associated with this interface requirement. The interface requirement for a phone line is associated with the A5 function since this is how the status reports are transmitted to maintenance and supply personnel.

12.7   PHYSICAL AND ALLOCATED ARCHITECTURES

The physical architecture for this simple problem will just involve naming the five components associated with the five functions in Figure 12.6:

- Patron Selection and Feedback Component
- Soda Chilling Component
- Soda Extraction Component

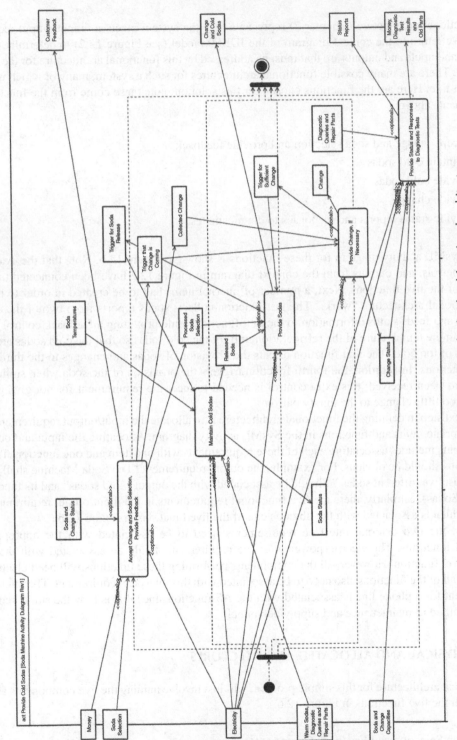

FIGURE 12.6 SysML activity diagram showing the first level functional architecture.

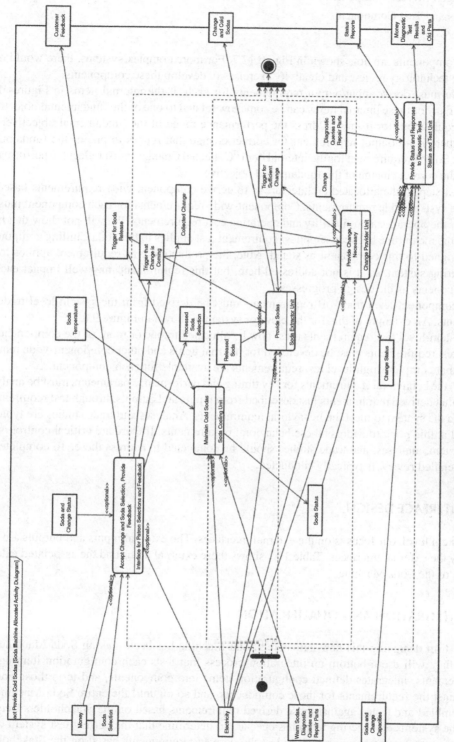

FIGURE 12.7 Allocated architecture for the soda machine.

- Change Provider Component
- Status and Test Component

These components are now shown in Figure 12.7. For more complex systems, there would need to be some technology review and creativity exercises to develop these components.

Now, we must derive input/output requirements for each of the internal items in Figures 12.6 and 12.7. If any of these internal items can be somehow related to one of the fundamental objectives, then the requirement for it must address the performance range of the fundamental objective. For example, there is an output requirement for correct change that relates to one of the fundamental objectives. So, the requirement for the internal item "Collected Change" must reflect the performance range for the correct change of the fundamental objective.

The next step in the allocated architecture is to derive component-wide requirements based on each of the system-wide requirements. Component-wide requirements for each component must be derived for the operational availability and operating cost requirements. We will not show that here. However, the mean time between supplies requirement just applies to the Soda Chilling component, so it is assigned to this component as stated. Note, the production cost requirement applies to the manufacturing system, so it is not addressed here, but this requirement may well impact each of these components as the design progresses.

Next, component-level trade-off requirements must be derived from the system-level trade-off requirements. An example of this for the elevator system is shown in Figure 9.8.

The system-level test requirements must now be allocated to one or more of the components. In addition, test requirements must be developed for internal items and other component-wide requirements so that complete high-level test requirements are available for each component.

Finally, the behavioral requirements for key timing and performance parameters must be analyzed to ensure that derived requirements just described from inputs and outputs through test requirements will enable the system to meet the behavioral requirements. Analyses and trade studies are typically conducted at this point to address these behavioral requirements. If there are critical control issues for the system, analyses, and trade studies should be conducted to address these. To complete this effort, a detailed review is typically conducted.

12.8   INTERFACE DESIGN

At the system level, our focus is on the external interfaces. The external inputs and outputs are used to identify the external interfaces. Table 12.1 shows these external items and the associated external interfaces of the Soda Machine.

12.9   INTEGRATION AND QUALIFICATION

Since the firm designing this new automated Soda Machine has been making Soda Machines for decades, they will use a bottom-up integration process that tests each configuration item against its requirements, integrates defined configuration items into components, and tests those components against the requirements for those components, and so on until the entire Soda Machine has been assembled and tested against those derived requirements based on the Stakeholders Requirements. The systems engineering team has checked to determine that the current test system of the firm has the necessary test equipment to test the derived requirements based on the Stakeholder's Requirements.

TABLE 12.1 External Items and Associated Interfaces

External Items	External Interfaces
Soda selection	Selection device
Money (dollar bills and coins)	Dollar input slot, coin input slot
Electricity	IEEE Std Electric 3 prong plug
Warm sodas	Soda rack inside front cover
Diagnostic queries	IEEE Std USB port
Spare parts	Access to parts inside the front cover
Customer feedback	LCD panels
Cold sodas	Soda delivery bin
Change (coins)	Change delivery bin
Diagnostic test results	IEEE Std USB port and IEEE Std signal
Old parts	Access to parts inside front cover

12.10 BEGINNING THE SUBSYSTEM LAYER

At this point, the systems effort has reached a conclusion and provides a basis for creating five sub-system teams to begin addressing the five subsystems that were identified as the physical architecture and later integrated into the allocated architecture:

- Patron selection and feedback component
- Soda chilling component
- Soda extraction component
- Change provider component
- Status and test component

Each team will be given a subsystem, and all the work will be created at the system level. Each team will then repeat the process described above. In particular, each team will:

1. Operational Concept for the Subsystem (Extended from the System's Operational Concept)
2. External Subsystem Diagram (Derived from External System Diagram)
3. Fundamental Subsystem Objectives (Derived for System's Fundamental Objectives)
4. External Subsystem Requirements (most, if not all, should already exist)
5. Subsystem's Functional Architecture
6. Subsystem's Physical and Allocated Architectures
7. Subsystem's Interface Architecture
8. Integration and Qualification for the Subsystem

The above outline of tasks for each subsystem team should make it clear that the subsystem teams do not begin from scratch but extend the work done at the subsystem level. Success for the system team means that none of the subsystem teams had to submit changes for the work done at the system level, meaning all the subsystem efforts were consistent with the work done at the system level. If one or more of the subsystem teams do submit change requests for the system level, the system's team will need to evaluate and make a decision. If one or more of the change requests are accepted, the system's level documentation will be changed and sent out to all the subsystem teams.

Please pay special attention to how much easier this next layer is when the system's allocated architecture matches one subsystem from the physical architecture with one major function from the functional architecture. If each subsystem team had to use parts of two or more functions from the functional architecture, the subsystem teams would require substantial coordination among themselves about which team was addressing which inputs and outputs for those shared functions. So, if the system's level functional architecture produces functions that are shared with multiple subsystems from the physical architecture, it is imperative that the allocated architecture resolve these issues and produce a product within which each function is uniquely handled by one subsystem. A subsystem may have more than one function, but the function or functions are not shared by other subsystems. In this way, the subsystem engineering teams can work relatively independently of each other.

Part **3**

Supplemental Topics

Chapter 13

The Value of Systems Engineering

13.1 INTRODUCTION

Prior to Honour's work [Honour, 2013], there is very little empirical data about the value that systems engineering adds to the development of a system. Cook [2000] paints a very bleak picture (Table 13.1). True success is achieved only 20% of the time; complete failure occurs 30–40% of the time. There is little evidence to suggest that there have been improvements since 2000.

Honour [2006] provides a very tentative summary of a few, macro-level metrics:

- Better technical leadership correlates to program success
- Better/more systems engineering correlates to shorter schedules by 40% or more, even in the face of greater complexity
- Better/more systems engineering correlates to lower development costs by 30% or more
- Optimum level of systems engineering is about 15% of a total development program cost
- Programs typically allocate about 6% of total development program cost to systems engineering

Honour [2013] states the following research results:

- Positive statistical relationships exist between systems engineering activities and three measures of systems engineering success – cost compliance, schedule compliance, and overall stakeholder success. But no statistical relationship was found between systems engineering activities and technical quality, which is a troubling statement about how systems engineering is practiced today.
- The "optimum" systems engineering effort is between 14% and 15% of total program cost. This "optimum" amount varies between 8% and 19% for different types of programs. Most real programs use significantly less systems engineering than these "optimums" suggest.
- The characteristics about a program that are most relevant in determining the amount of systems engineering to consume are level of problem definition at the start of the program, the autonomy

The Engineering Design of Systems: Models and Methods, Fourth Edition. Dennis M. Buede and William D. Miller
© 2024 John Wiley & Sons, Inc. Published 2024 by John Wiley & Sons, Inc.
Companion website: www.wiley.com/go/engineeringdesignofsystems4e

TABLE 13.1   Summary of Systems and Software Engineering Success and Failure

UK Ministry of Defence	UK Civil Information Technology	US Civil Information Technology
Top 25 programs slipped 35–40 mos. on average	10–20% met success criteria	16% project success
10% of projects missed key technical requirements	40–50% late, over-budget, or did not meet technical goals	53% project challenged
	40% failed or were abandoned	31% project cancelled

of the development team, the level of integration needed at the system level, the size of the system, the difficulty associated with proving the system is the appropriate system, and the risk associated with technologies being integrated into the system.

- There is significant positive return on investment for systems engineering activities, estimated to be between 3.5 and 7 to 1.

The rest of this chapter is going to address the question – How can systems engineering add value to the development, fielding, and upgrade of a system? First, a major contribution is seeking the definition of the right problem to solve and then seeking the right solution to that problem, both of which require a systematic, holistic, and creative process. The type of process that has proven successful at optimizing profits in manufacturing would be the opposite end of the spectrum from that needed for systems engineering. In manufacturing, the product is well-defined, and the process is focused on refining this product to improve its quality and reduce its cost, both of which are very measurable. In systems engineering, we are searching for a definition of the problem and an appropriate solution, little of which is easily measured. Adding to the complexities of this environment are multiple segments of the stakeholders (several types of users, maintainers, suppliers, and bill payers, as well as potential victims) having very different ideas of the problem and the preferred solution. In addition, there are various disciplines of engineers vying to provide the solution; the best solution often involves some integration across the solution space defined by the engineering disciplines. Systems engineers typically resolve the issues of conflicting desires among stakeholders and competing designs among engineering disciplines through trade studies. Central to these trade studies are the objectives (conflicting though they may be) of the stakeholders and alternate design solutions nominated by the members of the systems engineering team.

A second contribution made by systems engineers is to serve as a communication interface among the various segments of stakeholders, the various disciplines of engineers, and between the stakeholders and engineers. This communications interface task includes defining an appropriate language for all to use (at least within the confines of the project) so that understanding about requirements and designs can happen, and agreement can be achieved. While this contribution seems almost trivial, the value added here can be huge.

Showstoppers are those issues in the definitions of the problem and solution, which if not resolved could dramatically end the project or cause an inopportune failure in the future. Examples are the incorrect shape of the primary mirror of the Hubble telescope (case study at the end of Chapter 1), insufficient "error checking" software in the Ariane 5 launch system (case study at the end of Chapter 11), and the insufficient requirements for the Air Bag Safety Restraint system (case study at the end of Chapter 6) that would have saved many lives. Part of the value of systems engineering is finding and fixing these showstopper issues before they cause problems. Almost any testing of the primary mirror of the Hubble telescope would have found a problem. Determining the potential for buffer overflows of unprotected (error checked) data fields on Ariane 5 could have been easily determined by examining the new trajectories planned for Ariane 5. Imagining the devastation of air bags on small children during slow speed accidents that would not otherwise be life threatening is part of what systems engineering is all about.

It is well known in the systems and software engineering literature [Boehm, 1981; Haskins et al., 2004] that the earlier one finds a flaw in the design, the less money it takes to fix the error; see Table 1.7). Finding and fixing errors in the design when the design is quite abstract, incomplete and in flux is the fourth element of value adding associated with systems engineering. A key part of this value-adding element is setting up measurement processes and feedback-control loops so that progress can be checked and determined to be in or out of bounds so that changes can be made quickly and cost-effectively.

Risk reduction is the fifth way in which systems engineers add value. Risk issues are those characteristics of design alternatives that produce great uncertainty about whether a specific solution can meet defined levels of performance within the objectives' hierarchy of the stakeholders. High-risk issues involve substantial uncertainty that a minimally acceptable level of performance will not be achieved. Moderate-risk issues may have less uncertainty of meeting the minimally acceptable performance level or greater uncertainty of attaining a desired level of performance above the minimum acceptable. Systems engineers are supposed to identify these high and moderate risk issues and develop risk mitigation design alternatives. For example, the Hubble telescope team decided to have a second primary mirror built by a subcontractor in case the prime contractor's primary mirror was flawed. Unfortunately, they did not require even a minimum amount of testing of the prime contractor's primary mirror.

Continuous improvement of the systems engineering process is the final contribution. The Capability Maturity Model (CMM) movement in the past 20 years has emphasized this aspect of engineering and project management.

13.2 VALUE PROPOSITIONS FOR SYSTEMS ENGINEERING

13.2.1 Systems Engineering as a Goal-seeking System

Goal-seeking systems generally have defined objectives for success and a feedback-control process (see Chapter 7) for moving toward those objectives. A simple example of such a system is a military fire-and-forget missile that is locked onto its target and attempts to follow it no matter what maneuvers the target performs. For systems engineering, a set of objectives relating to performance, cost, and schedule should be defined so that the systems engineering system can endeavor to meet these objectives with appropriate adjustments, no matter what issues arise.

There is a spectrum of system types for which we could design a goal-seeking system for creation. At one end of the spectrum is a defined product (e.g., manufacturing system) for which we desire to increase the quality. Six Sigma is an example of goal-seeking solution to this problem. At the other end of the spectrum is a largely undefined product that needs to be developed within a given period of time for specified amount of money. Here the largely undefined product might have some use cases, some of which are incorrect, and others are missing. This is the problem that systems engineers are tasked to address.

For the systems engineering problem, there must be at least two discovery portions to the solution as well as the targeting solution associated with the fire-and-forget missile once the discovery phases are complete. The first discovery phase deals with identifying the right problem, that is the right set of use cases together with the right rough-order-of-magnitude estimates for cost, schedule, and performance parameters. The second discovery phase deals with the assessment of a wide range of alternate solutions to the definition of the right problem. For this discovery of the right solution the rough-order-of-magnitude estimates for cost, schedule, and performance parameters should be used, but augmented with a set of parameters associated with negative unexpected consequences for cost, schedule, and performance parameters that have not been defined. Finding the right solution in this second discovery phase has to deal with additional issues that address compromises among groups of stakeholders that have differing opinions about the what the solution is and how it should be tailored, as well as compromises among the engineering disciplines involved in producing the solution about which technologies should be employed in the solution. The third and final phase of this goal-seeking solution for systems engineering deals with finding and solving problems in the definition of the right solution as the solution is being matured and tested.

Some successful examples this goal-seeking system for systems engineering include the redesign of power tools by Black and Decker in the early 1970s, the design and launch of Pioneer 10 by NASA in 1972, and the design of the Sidewinder missile by the Navy around 1950. More recent examples include the Mars Exploration Rover system and search engines for the Internet.

The Black and Decker story has been told in detail by Meyer and Lehnerd [1997]. In 1970, Black and Decker had about 20% of the U.S. market share for power tools, earning roughly $200 million per year. However, their product designs, materials, and technology were highly nonstandard even though they had a good reputation in the industry among consumers. They faced several serious threats at this time. First offshore manufacturers were entering the U.S. market with better products at lower prices, thus threatening the profit margins of Black and Decker. But of even greater concern to management was a movement in the U.S. Congress to require higher safety standards, namely double insulation between the user and raw electricity inside the power tool. There were some great product improvement opportunities available at this time as well: semi-automated manufacturing is gaining market share and new materials (e.g., plastics) are being used. So Black and Decker management selected a broad definition of the problem and decided to bet the company by redesigning both the manufacturing system and the power tool family of products at the same time. Black and Decker had to borrow about $6 million in capital. They also reorganized their corporate structure by abolishing the stovepipes of engineering and manufacturing (each with their own vice president) and creating a single vice president of operations. A war room for the redesign of the manufacturing and the power tool suite was set up at headquarters with everyone working on the project reporting there. In a sense Black and Decker redesigned the "engineering system" before it engineered the manufacturing and power tool suite. The result was a resounding success. The resulting profits paid back the investment in four years (rather than the planned seven years) and enabled Black and Decker to maintain its profit margins in the face of overseas competition in the U.S. while most other U.S. manufacturers left the product market for some number of years.

The Pioneer 10 story has been told by Wolverton [2001]. In this case, NASA engineers realized in 1969 that there was an infrequent launch window in 1972 that would enable a spacecraft to get to Jupiter in a year. The Pioneer 10 program started quickly with Congressional approval. The NASA program manager put a major emphasis on schedule since the launch window in March–April of 1972 was so close. NASA engineers selected a high-level design that included reusing a great deal of previous spacecraft. TRW was given a sole source contract since it had designed most of the reused parts. In this case defining the problem was quite straightforward, but the NASA program manager excelled because he let the NASA and TRW engineers know that any design changes were going to have to improve the schedule or solve schedule problems; engineers bringing other design changes to him would be summarily dismissed. Even so there were a number of difficult design issues associated with balancing the scientific instruments and the radioisotope thermoelectric generators (small nuclear power sources) around the back of the communications antenna. This balancing problem required a solution so that the spacecraft could achieve a stable spin rate of five revolutions per minute. The nuclear devices and some had to be separated with sufficient distance to minimize any negative impacts on the performance of sensors. Many trade studies were performed in a short period of time with limited computing resources to achieve success. Pioneer 10 continued to relay scientific data back to the Deep Space Network until 23 January 2003, about 30 years from its successful flight by Jupiter.

Perhaps the most compelling "goal-seeking system for systems engineering" occurred about the time that the phrase "systems engineering" was coined at Bell Labs (see Section 1.2). The Sidewinder missile was developed by a team at the Navy's base at China Lake, CA, starting in 1947 and completing nine years later in 1956. This team was headed by Dr. William Burdette McLean, a gifted technologist and exceedingly gifted systems engineer. McLean served as the project engineer and system engineer throughout Sidewinder development, managing all aspects of system design and development. He and his team were not initially commissioned to design Sidewinder and therefore had no externally imposed specifications during the first couple years [Kopp, 1994]. Furthermore, McLean believed that a more creative approach to the design of a system could be utilized in the absence of a set of definitive specifications. Specifications, in his view, tended to channelize thinking along one approach [McLean, 1960, 1962]. This philosophy realized the need for a *creative system development approach* – the use of goals in the design and development of the Sidewinder missile.

As stated, a *goal* may be defined as "the purpose toward which an endeavor is directed; an objective that is not fixed in time; an end toward which effort is directed and *under no obligation* (italics ours)." In contrast to requirements, the use of goals delays the setting of performance targets and thresholds (or requirements) early in the design process. Requirements, however, are often defined early before many unexpected issues arise and are routinely tied to a development schedule.

McLean's process of beginning with goals has proved in the development of the Sidewinder missile system. This development process is referred to as the Goal-based Definition and Design Process (see Figure 13.1). First, users are interviewed to determine user needs. A set of system goals is then derived from the defined user needs. Interactions with users continue until the concept is proven through testing that the system design meets the critical user needs. *This process requires that the designer have direct contact with the ultimate user throughout development*. Note, this approach

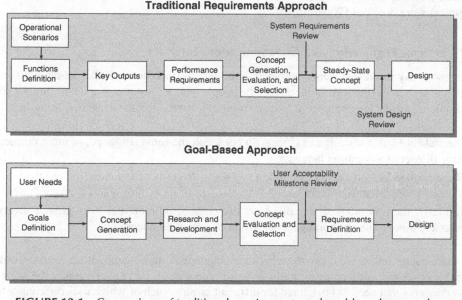

FIGURE 13.1 Comparison of traditional requirements and goal-based approaches.

(and all others) is more difficult to use successfully when there are multiple, conflicting groups of stakeholders. The need becomes clearer as a result of the continual contact between the designer and the ultimate user. Once the concept is proven, the development process in meeting user need is complete. Requirements are then formulated, and the system is now ready for mass production [Powell, 2002].

The Sidewinder was a success, and as a result, its creative system development approach continued in the design of two additional missile systems – the Polaris and Poseidon missile systems. In contrast to reasons why beginning with requirements may not be prudent, the following points clarify the significance of beginning with goals.

- The use of goals precludes changes in requirements.
- The use of goals coupled with continual customer interaction works to prevent customer dissatisfaction.
- The use of goals results in a system that meets user needs.
- The use of goals minimizes the constraints on creativity in system design.
- The use of goals precludes adherence to formal specifications thus allowing for early and rapid testing.
- The use of goals supports experimentation and encourages the best solutions to technical problems to surface
- The use of goals does not constrain system design

Requirements were defined only when system goals were met or when a system had been designed, developed, and proven. The absence of requirements, McLean thought, allowed the best solution for the users in terms of what could be accomplished. **A requirement implied a guarantee** when in fact no one knew what was possible until research and experimentation had explored the possibilities. McLean felt, beginning with requirements instead of goals, did not allow for such an approach and did not guarantee long-standing success. He strongly believed in trying things out rather than figuring them out, which was one of the reasons formal requirements in advance of development were not used [Westrum, 1999].

13.2.2 Systems Engineering as a Communications Interface

One common way to view systems engineers is as the interface between the people who have the problem and resources (the stakeholders) and the people who are designing and building the solution to the problem (the design engineers). Such a communication interface is needed because:

- The stakeholders and design engineers do not speak the same language, or more correctly, the same dialect of a common language
- There is often great diversity of opinion and some diversity of language among the nonhomogeneous group of stakeholders
- Similarly, there is often an equally great diversity of opinion and even greater diversity of language among the nonhomogeneous group of design engineers

These language issues are a problem because language contains ambiguous terms, words, and phrases that have multiple meanings; multiple words and phrases that mean roughly the same thing; and new concepts that need to be defined for different groups, each of which uses different terms for the same concepts and the same terms for different concepts.

Determining Needs Assessments	Gaining Agreements on a Path Forward

• Need individual sessions with homogeneous stakeholder groups
  – Structure their needs
  – Convince them they have been heard
• Analyze the divergence of needs across homogeneous groups
  – Find agreements
  – Find divergence (nonconflicting differences)
  – Find the conflict areas
	• Need group sessions that span homogeneous stakeholder groups
	  – Explain the zones of agreements, divergence, and conflicts
	  – Seek their agreement to resolve these issues
	• Convince them that incremental development is part of the solution
	  – Get their agreement

FIGURE 13.2 Gaining closure with stakeholders.

Figure 13.2 describes a process for interacting with diverse groups of stakeholders so as to gain some degree of closure with them. These sessions begin with homogeneous subsets of stakeholders and then move to the broader set of stakeholders. As part of this process the stakeholders need to understand each other's perspectives, their differences in priorities for the design trade-offs that must be resolved, and the value of incremental development in fielding a basic solution early followed by numerous enhancements.

Figure 13.3 extends the process to address design engineers. The design engineers need to understand and accept the trade-offs needed to meet the needs of the stakeholders. These design engineers also need to be motivated to user their creativity to achieve the stakeholders' objectives.

The use of Design-Build Teams during the design of the Boeing 777 provides some good examples of these processes. One such issue, which was uncovered in discussions with the stakeholders (airline companies), was that the trailing edge flap was designed to be 43 feet long. However, the airline companies routinely use autoclaves (ovens) to heat up and "smooth out" the trailing edge flaps of aircraft after they had been in service for years. However, the autoclaves owned by airline companies were only 25 feet long. Boeing did not like the suggestion from the airline companies that the trailing edge flap should be cut in half, so Boeing suggested redesigning the trailing edge flap so that it could be cut in half by the airlines when they desired to do the first maintenance. The airlines were very happy with this solution.

Include Representatives from the Beginning	Document the Decisions for Derived Requirements

• Design Engineers need to understand the "trades" leading to the design concept and high-level system architecture
  – Meta-system measures of effectiveness (MOEs)
  – System-level key performance parameters (KPPs)
• Those design engineers included at the start will become the evangelists for the system's success being defined by KPPs
	• Document derived requirements as they are defined at lower and lower levels of components
	  – Provides rationale for all engineers to see why their design problem is what it is
	  – Provides continuity for those that follow

FIGURE 13.3 Gaining closure with design engineers.

13.2.3   Systems Engineering to Avert Showstoppers

We separate a "showstopper" from errors or regular risk issues in this discussion. A showstopper is an event related to the systems engineering efforts that "stops or could stop the program or major processes that are part of the program, thus undermining the entire system development."

We have not found any examples of the program being stopped early due to a showstopper surfacing, but there are several examples where the program should have been stopped early because the showstopper should have surfaced. The first example is Iridium – commercial competitors were developing mobile phones that would dominate Iridium's technology and pricing. Then we examine design incompatibilities for an air defense gun system, issues of requirements for air bags safety restraints, and design decisions for a communications satellite and a personal computer.

Iridium is a satellite-based communications company that is doing well. However, the current owners purchased the company for a little more than $25 million in 2000 after the original owners had spent over $5 billion from the late 1980s to 2000. The concept began when the wife of a Motorola employee complained that she could not complete a mobile phone call from the Caribbean while on vacation from her real estate business. Engineers and executives from Motorola put the system concept of a satellite-based system with earth-based communication gateways that would enable phone calls from anywhere in the world to anywhere else in the world at any time of the day. Real funding and development started in 1990. In addition to the satellites and earth-based communication gateways, this complex system used two segments of code division multiple access as well as time and frequency division multiple access and required agencies of many countries to provide spectrum allocations for each of these two segments. Launches of the required 66 satellites began in 1997 and reached 95 satellites in orbit when service began in 1998. (Note the original design required 77 satellites, which is why the name Iridium [chemical element with 77 electrons] was chosen). Unfortunately, Iridium's competitors included the rapidly expanding general mobile phone industry. Most potential customers did not need to make calls from or to unusual places and did not want an outsized mobile phone with a data rate of 2–4 kilobits per second at a price of $6–30 per minute. Sales of the Iridium phone were sufficiently low that the company went bankrupt in 1999. The showstopper was that the Iridium engineering team did not track and understand the customers' preferences and the capabilities of their competitors.

A somewhat similar story is the Army's Sergeant York Division Air Defense Gun (DIVAD Gun), about which a movie was made. Focusing earlier than the testing problems covered in some detail in the movie were the threat and the design. The primary threat that the DIVAD Gun was to counter was pop-up helicopter. The Army sought many concept designs from industry with cautions that money was tight, and the reuse of existing technologies was critical. Unfortunately, there were several critical technologies and no one contractor had the right mix: a vehicle chassis that the government would supply to keep up with the new tank and troop carrier; a turret system for the government supplied chassis; a radar system that could search for and track pop-up helicopters quickly in substantial ground clutter; all weather capability was needed, which meant a FLIR and laser range finder, and a gun system that could begin firing quickly with accuracy. The winning design from Ford Aerospace used a tracking radar that had been used for the F-16 aircraft. The radar could not be modified to address the clutter problem. In addition, the ground vehicle environment, including dust in the air and vibrations, reduced the reliability of the track radar. The new turret designed by Ford Aerospace was heavier than the turret it replaced, which also reduced reliability for the ground vehicle furnished by the government. The incompatibilities of reused parts proved too great an obstacle. Finally, the threat requirements were changing, which made the new system design less valuable even if everything had worked.

The air bag safety restraint system is described at the end of Chapter 6. The main point related to showstoppers was that the requirements focused on a single safety scenario – the median male traveling at 30 miles per hour into a frontal collision. Test data existed that demonstrated the ultimate design would be fatal to children. Simple logic was compelling that short drivers too close to the steering wheel (and air bag) would also be hurt or killed with high frequency. Yet, no one in industry or the government raised objections to these issues. Industry focused on the increased cost of autos and government focused on getting a solution fielded.

Hooks and Farry [2001] describe the design of a communications satellite that is needed to bolster a constellation of such satellites for a limited period. Due to the limited operational period and the fact that battery replacement is not possible in space, the battery pack design used nonrechargeable batteries in a compartment with a hundred or more screws. Unfortunately, testing did not proceed as quickly as scheduled so that the batteries needed to be replaced instead of being recharged and the hundred screws needed to be removed and replaced. This happened many times. Finally, the initial battery packs were redesigned.

The final example is the LISA, a personal computer designed by Apple and released in 1983 for $9995. The LISA program was stopped in 1985 due to insufficient demand. LISA had the first commercial graphical user interface. But Apple changed their practice of giving the software development industry early design kits so that plenty of software was available. Instead, Apple created six software programs and hoped that would satisfy the LISA purchasers. It did not satisfy the LISA purchasers. Shortly after LISA was ended Apple introduced the Macintosh, which made use of lot of the LISA design and was sold for just over half the price. There was an active software industry ready for the release of the Macintosh.

13.2.4 Systems Engineering to Find and Fix Errors

Recall "error" was defined in Chapter 7:

> *Error:* a subset of the system state, which may lead to a failure. The system can monitor its own state, so errors are observable in principle. Failures are inferred when errors are observed. Since a system is usually not able to monitor its entire state continuously, not all errors are observable. As a result, not all failures are going to be detected (inferred).

Now, however, the focus here is on the systems engineering system rather than the product system. The systems engineering system is primarily a system comprised of people, so we are primarily trying to find errors made by people. One of the great values of model-based systems engineering approach is that the models can help monitor the entire state of the systems engineering system continuously, enabling the people find these errors better than the people could do on their own.

The focus of the goal-directed system approach discussed above was to define the "problem" being solved and then define the "solution" to that problem. Here our focus is on identifying errors in the definitions of either or both of the "problem" and "solution." The value of this exercise can be traced directly to Figure 1.2 and Table 1.7 from Chapter 1. Table 1.7 directly demonstrated that the earlier errors are discovered, the cheaper they are to fix. This is primarily a function of the rapid rise of the "cost committed" curve at the beginning of the systems engineering effort, as shown in Figure 1.2.

Some examples of the types of errors that should be discovered as early as possible by the systems engineering system are:

- New requirements (or goals) that were missed earlier in the problem and solution definition activities. Note the systems engineering system should be designed to detect new requirements as early as they become recognized by the stakeholders; the best systems engineering system will stimulate the stakeholders to identify these requirements before they would normally do so on their own.

- Changes in requirements (or goals) as soon as those changes become apparent to the stakeholders.
- Changes in derived requirements that were missed.
- Engineering design changes that were not communicated properly and are being misinterpreted
- Engineering design changes that should have been made but have not been identified
- Engineering design changes that were communicated properly but are being misinterpreted
- Test that are not being performed but should be
- Tests that are being conducted incorrectly
- Tests that being conducted correctly but the results are being recorded incorrectly
- Test results that are being misinterpreted
- Tests that are being performed beyond the point that is necessary, wasting resources
- Not performing regression testing properly or to the extent needed

A more general taxonomy of sources for errors that might guide the interested reader a larger list of the systems engineering errors is

- Not looking for errors that exist
- Looking for errors but not looking for all of the signals associated with the errors
- Ignoring signals of errors that are being seen (very common human problem)
- Looking for signals of errors but missing them (e.g., collecting the wrong measures or collecting the right measures but doing so incorrectly)
- Misreading the signals associated with errors, which may result in confusing one error for another or thinking an error that is not an error is an error

The Apollo 13 case study is described at the end of Chapter 6 and provides several great illustrations of missed errors. First, the Apollo systems engineers missed an operational concept that addressed connecting the Apollo system to the launch pad during the pre-flight system check conducted the day before the launch. During this pre-flight system check, the Apollo electrical system was bypassed since it was not operating; the launch pad electrical system was used instead. The launch pad electrical system was designed to operate at 65 volts direct current rather than the 28 volts direct current chosen as a design decision for the Apollo system. So, the Apollo systems engineers added a new requirement that the Apollo system had to work with 65 volts direct current, which was flowed down to all the subsystems and configuration items (hydrogen and oxygen tanks). The 65-volt direct current requirement was not flowed down to the thermostat in the oxygen tanks, so this was a second error. However, a third error was not realizing that there was a simpler solution than changing the direct voltage requirement from 28 to 65, namely inserting a step-down circuit the between the launch pad and the Apollo system that reduced the 65 volts of the launch pad down to 28 volts.

Another case study to illustrate that system engineering is an error-correcting system is the Hubble Telescope (see the end of Chapter 1). Recall that the primary mirror was ground to a shape that did not meet the requirements. One can argue that adequate tests were done, but the results were misinterpreted, or that inadequate tests were done but the systems engineers thought they were adequate, or both inadequate tests were done and those that were done were interpreted incorrectly. Based on the research the first author has done, he believes the latter argument.

A positive example of systems engineering as error reduction is the major error reduction effort was undertaken during the design of the Boeing 777. A well-known source of engineering design changes for commercial airliners was moving the placement of the external doors that permit passengers to enter and exit the airliner during the design process. On the previous Boeing airliner (the 767) there were 13,341 design changes associated with the external doors. This is a result of the fact that the fuselage shape changes from place to place. The Passenger Door Design-Build Team for the 777 was asked to address this problem by the system engineering team. As a result of a design effort that took about two months this design-build team was able to develop a door design such that the hinges and 98% of the door mechanism would not change as the door placement changed. As a result, a major source of engineering design changes was dramatically reduced, leading to an overall reduction of engineering design changes from the 767 to 777 of 90%.

13.2.5 Systems Engineering as Risk Mitigation

Risk was defined in Section 14.5.4, as were several strategies for dealing with risk: avoidance, transference, management, and analysis. A risk matrix (shown in Figure 13.4) is a common approach for visualizing risks. The horizontal axis captures the likelihood or probability that a bad outcome will occur; the vertical axis captures the consequence of the bad outcome. High-risk issues (shown as dots) are in the upper right corner, and low risk issues are in the lower left corner. Risk management actions are shown as arrows. Risk transfer (e.g., buying insurance) reduces the consequence but does not change the probability of the bad outcome. Implementing redundant resources may reduce the probability but not the consequence of the bad outcome. More complex risk management strategies may reduce both the probability and the consequence of the bad outcome. Risk acceptance is commonly chosen for the lower risk issues. For moderate- to high-risk issues where there is no cost-effective risk management approach, a risk watch effort is undertaken, meaning no effort is taken to reduce the risk of the issue, but appropriate engineers are designated to keep track of the risk issue to make sure it is not getting worse. Note, the Hubble design team had designated the shaping of the primary mirror as a major risk issue and had mitigated this risk issue by having another contractor build a second primary mirror (a redundant resource which decreased the probability of the bad outcome). However, the Hubble design team did not discover the error in the shape of the

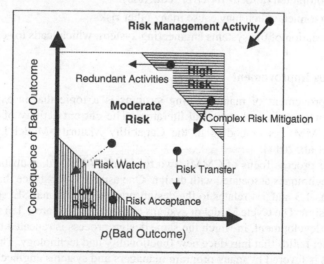

FIGURE 13.4 An exemplary risk matrix.

primary mirror built by Perkin Elmer, so they did not make use of the risk mitigation effort for which they paid.

This discussion highlights the need for the systems engineering system to have a focus on the risk associated with both the systems engineering system itself as well as the product system. In fact, the risks associated with the system throughout the system's life cycle should be addressed, just as the goals, showstoppers, and errors need to be addressed throughout the life cycle.

Returning to the Boeing 777, we find a positive example of risk identification and mitigation. The systems engineers identified weight savings as a major performance goal, as is done during the design of any airplane. For the 777, one way to achieve major weight savings was to use a new aluminum-lithium compound for the non-weight bearing elements of the Boeing 777 frame. The systems engineers estimated that using this new compound rather than the traditional compound would reduce the empty weight of the 777 by 280–400 pounds. The price the systems engineering team was willing to pay reduce the empty weight by a pound was $600. This aluminum-lithium compound only cost $100 more per pound than the traditional compound so this was a great deal. Unfortunately, a risk issue was identified. The materials engineers found that after machining, cutting, or drilling this aluminum-lithium compound, small surface cracks appeared. The materials engineers were convinced that these cracks were surface cracks and posed no structural risk. However, the lead systems engineer decided that the likelihood of these cracks being discovered by maintenance personnel sometime after the 777 began to fly was very high. This discovery would lead to all 777s being grounded while inspections took place. This economic loss to Boeing as well as the bad press was not worth the weight savings in this case.

As a stimulus for the reader, DeMarco and Lister [2003] have identified five common or "core" risks faced by all programs. Some of these risks are more relevant to the project and program managers, but systems engineers would be well advised to remember them as well.

- Wishful thinking about the number of problems that are going to arise, the intrinsic schedule flaw that causes all of us to believe we will finish something earlier than we will
- *Requirements or Specification Failure*: our inability to achieve consensus among stakeholder groups (one of the foci above for communication)
- *Scope Creep*: too many requirements (goals) were missing at the beginning and show up before the system is completed (a focus for error reduction)
- Loss of key personnel, which may cause many other risks
- Productivity variation of the systems engineering system, which leads to schedule problems

13.2.6 Continuous Improvement

The continuous improvement of manufacturing systems is a topic that is explored throughout the engineering, management, and statistical literature. The current activity of note is Capability Maturity Models (CMM), as embodied in the Capability Maturity Model Integrated (CMMI) approach [Chrissis et al., 2011].

In addition to the process focus of CMMI, two topics of note are the substance of design and the methods-tools-techniques associated with design. One aspect of substance that has already been discussed in Sections 1.3 and 1.4 relates to the design process: traditional SE, spiral development, and model-based design. The Cycle Model of systems engineering (Figure 1.18) introduces the concept of incremental development, in which the same design process is repeated to create additional increments or product builds that introduce new functionality and technology. The concept of incremental development is favored by many program managers and systems engineers but not accepted by many stakeholders who do not appreciate the complexity and risks in system development.

Another topic of substance is increasing the skills and capabilities of the engineers and other personnel working as part of the systems engineering system. This includes not only education and training efforts but also "bottling and distributing" systems engineering expertise. This is commonly done by documenting and following best practices. What is important for the future is building "expert systems" based on notable systems engineers to serve as critics, mentors, and replacements for less expert systems engineers. When such "bottling and distributing" activities are undertaken, the major areas of interest are process expertise and content expertise. Design patterns are a good example of harnessing content expertise [Fowler, 1960; Brown et al., 1998; Mowbray and Malveau, 1997; Douglass, 2002]. But one critical topic of expertise (or partial expertise) is working with people, which is seldom addressed. Yet, one of the major reasons for the systems engineering system being comprised of people is that systems engineering is the communications interface between stakeholders and design engineers, all of whom are people.

Another critical element of substance relates to an increasing emphasis on using product platforms [Meyer and Lehnerd, 1997; Gawer and Cusumano, 2002; Simpson et al., 2007; Ericsson and Erixon, 1999] when designing a system. A product platform is a set of products that are built from a set of common components performing common functions. One of the most common examples of a product platform is the set of power tools (both indoor and outdoor) manufactured by companies such as Black and Decker. These power tools are used by homeowners for occasional repairs as well as by professionals who work full time. So there are often several versions of each component that the power tool company can select when going to market with "good," "better," and "best" versions of each tool. Other examples of product platforms are Sony Walkman and HP Deskjet printers. Today, product platforms exist in the auto industry for the chassis, engine, and power train (both hardware and software), the commercial airline industry (e.g., DC-3, Boeing 777), the camera industry, the computer industry (e.g., personal computers), the printer industry, the integrated circuit industry, the software industry (e.g., Microsoft Office), the engine control software (e.g., Bosch) industry, the wireless sensor industry, the pharmaceutical industry (e.g., Tylenol), and the entertainment industry (e.g., Disney World). There is even work in product platforms for augmented cognition systems.

The benefits of a product platform are typically reduced production and operating costs; improved availability, reliability, safety, and security; improved ease of use; and more rapid introduction of new products. There are negatives (as usual): slower introduction of new technologies; increased cost of implementing new design in changes in the short term but not in the long run; and not all products benefit when a component in the platform is upgraded.

McLean [1960, 1962], of Sidewinder fame, provided some suggestions that ought to guide most continuous improvement processes. First, he felt there needed to be a single person (chief systems engineer) who could visualize the design throughout the life cycle of the system in sufficient detail to be right about design questions most of the time. Note the Sidewinder was a very complex design with most parts being used for at least two functions. However, it had relatively few parts. He also felt that trade-offs needed to be studied throughout the design process, high level, abstract trade-offs early in the design, and less abstract trade-offs as the design is being finished. With today's systems, there are many parts and many technologies being integrated so there are very few chief systems engineers who keep a detailed design in their head. But all should be able to use current computer technology to drill down as deeply as needed. The question is, how many of today's chief systems engineers use computer technology this way.

More importantly, McLean believed in a great deal of prototyping at the system and subsystem levels early in the design process as part of doing the high level, abstract trade-offs. He felt the systems engineering team should be small early so that everyone involved in the early prototyping and testing was communicating and learning from each other. This process should produce a working model of the system early for rigorous testing and evaluation. Note, this early model may not contain the subsystem designs that are the final designs, but the system-level issues are being identified so that they can be addressed as the design decisions move to the subsystems and below. (Note this is the top-down testing described in Section 11.4.) As the design decisions move to the subsystems and below, the systems engineering team begins to expand in size. Currently, the systems engineering team balloons to an unmanageable size very early in the development process and then grows bigger. Testing of high-level prototypes has been replaced by simulation modeling of high-level prototypes for many good reasons. But insufficient communication typically occurs among the simulation modelers, the technology experts, and the chief systems engineer. Testers should also be part of this process so that they can design and conduct tests to augment the simulations.

As the design trade-offs move to the subsystem and below, McLean felt that systems engineering system should foster competition among teams working the same problem. McLean used contractors to create this competition. Today, the most common competition occurs at the system level, very early in the design process; two to four contractor teams are given early design contracts to compete very different concepts. In the rush to develop a system as quickly as possible, these teams are only given one to two years typically and cannot possibly explore all the design issues associated with a specific concept. In addition, the organization funding the competition seldom understands the design issues in detail across the all the concepts being funded yet has to make a choice of one concept in one to two years. McLean took three to four years to select the Sidewinder concept. McLean even believed that there should be two system-level teams competing during the development process, one with 10% of the funding that the main team has. There is at least one positive example of this in practice as described in *The Soul of a New Machine* by Kidder [1981].

13.3   SUMMARY

This chapter has explored six value-adding propositions that systems engineering provides to program and project managers and funding agents: goal-seeking system that is driven the stakeholders goals; a communication interface among and between the stakeholders and design engineers so that each group can affect the objectives and constraints of the other; a means to find showstoppers early and avert them; an error discovery and resolution system so that errors are fixed when the fixes are inexpensive; a risk management system that identifies and finds risk mitigation solutions when mitigation is cost-effective but watches risks when mitigation is not cost-effective; and continuous process improvement.

Chapter 14

Decision Analysis for Design Trades

14.1 INTRODUCTION

Decision-making is a process undertaken by an individual or organization. This process intends to improve the future position of the individual or organization in terms of one or more criteria. Most scholars [Howard, 1968] of decision-making define this process as one that culminates in an irrevocable allocation of resources to affect some chosen change or the continuance of the status quo. The most commonly allocated resource is money, but other scarce resources are goods and services and the time and energy of talented people.

Watson and Buede [1987] have identified three primary decision modes: *choosing* one alternative from a list, *allocating* a scarce resource(s) among competing projects, and *negotiating* an agreement with one or more adversaries. Decision analysis is the common analytical approach for the first mode, optimization for the second, and a host of techniques have been applied to negotiation decisions [Jelassi and Foroughi, 1989]. Concepts of decision analysis are relevant to the second and third of these modes.

Section 14.2 provides a philosophical discussion of decision-making and the elements of decision-making: values, alternatives, and facts. Section 14.3 explains the rational basis of decision analysis in terms of a set of axioms that provide a compelling structure for some decision-makers. Section 14.4 provides an analytical basis for modeling stakeholder values in the face of conflicting objectives, a critical element in design decisions when faster, better, and cheaper are all desired but not mutually compatible. Section 14.5 discusses the modeling of uncertainty and risk preference for design decisions; decision trees, relevance diagrams, and influence diagrams are introduced as modeling tools. A sample application focused on the development of trade-off requirements consistent with an objectives hierarchy and performance requirements is presented in Section 14.6; this sample application is based upon a real application of decision analysis to requirements development.

This chapter describes a model of uncertainty (probability theory), a model of value (multiattribute value theory), a model of risk preference (utility theory), and a normative model for

---

The Engineering Design of Systems: Models and Methods, Fourth Edition. Dennis M. Buede and William D. Miller
© 2024 John Wiley & Sons, Inc. Published 2024 by John Wiley & Sons, Inc.
Companion website: www.wiley.com/go/engineeringdesignofsystems4e

incorporating uncertainty, value, risk preference, and complexity for aiding the thought and conversation process needed to make explicit, rational decisions.

14.2   ELEMENTS OF DECISION PROBLEMS

Decision analysis is a normative theory for making a *decision* (an irrevocable allocation of scarce resources). The three major elements of a decision that make its resolution troublesome are the creative generation of alternatives, the identification and quantification of multiple conflicting criteria, and the assessment and analysis of uncertainty associated with what is known and not known about the decision situation. Howard [1993] has drawn an analogy between the model-building and analysis processes inherent in decision analysis and a conversation with a decision maker. The conversation (or modeling) needs to address what the decision maker (stakeholders in systems engineering) cares about (values), what the decision maker can do (alternatives), and what the decision maker knows (facts or absence thereof).

Many stakeholders and systems engineers claim to be troubled by the feeling that there is an, as yet unidentified, alternative that must surely be better than those so far considered. The development of techniques for identifying such alternatives is receiving considerable attention [Elam and Mead, 1990; Friend and Hickling, 1987; Keller and Ho, 1988; Keeney, 1992; West, 2007].

Ample research [von Winterfeldt and Edwards, 1986] has been undertaken to identify the pitfalls in assessing probability distributions that represent the uncertainty of a stakeholder. Research has also focused on the identification of the most appropriate assessment techniques. Similar research [von Winterfeldt and Edwards, 1986] has focused on assessing value and utility functions. Keeney [1992] has recently advanced concepts for the development and structuring of a value hierarchy for key decisions. While it will never be possible to turn decision support via decision analysis over to a computer, the vast number of real-world applications of decision analysis [Kirkwood and Corner, 1993] demonstrate that this analytic modeling support is well worth the time and effort.

14.3   AXIOMS OF DECISION ANALYSIS

There are five basic rules of thought [von Neumann and Morgenstern, 1947; Howard, 1992] that establish decision analysis: probability, order, equivalence, substitution, and choice. Probability is adopted as the representation of uncertainty. This is a well-founded discipline for addressing uncertainty and is the common approach within engineering.

The order rule states that our preferences are sufficiently well defined that any possible list of outcomes associated with the design alternatives can be ordered from least preferred to most preferred on each objective in the fundamental objectives' hierarchy. In addition, once our preferences are aggregated across all objectives, there is a single list of outcomes ordered by our preferences. Naturally, it is possible to be indifferent between two outcomes on a specific objective or on the aggregate. Our preference order does not need to be the same from one objective to the next; in fact, there would be no need to have multiple objectives if this were the case. The ordered list must be transitive, which is to say that any outcome can only appear once on any ordered list. If this is not the case, we become subject to the "money pump" argument; a disinterested party could entice us to put up an infinite amount of money by offering us a sequence of trades among three alternatives. For example, I would be intransitive if I stated that I preferred a Lexus to a Cadillac, a Cadillac to a BMW, and a BMW to a Lexus. With these preferences and ownership of a Lexus, I would pay to swap for your BMW, pay again to swap the BMW for your Cadillac, and then pay a third time to swap the

Cadillac for the Lexus I originally owned. By this time, I should realize there was something wrong with my preference structure.

The equivalence rule sets up a situation with three outcomes, A, B, and C, where A is preferred to (>) B, and $B > C$. This rule states that there is some lottery containing a probability, p, of obtaining outcome A and a probability of $(1 - p)$ of obtaining C that will make us indifferent to obtaining outcome B for sure.

The substitution rule states that we are willing to substitute any combination of outcomes in a decision-making situation if we are indifferent between them. This is just the operational definition of equivalence.

Finally, suppose we have two alternatives, each with exactly the same outcomes, and the probabilities of the outcomes are the same for all but two. If one of the alternatives has a higher probability associated with the outcome that is most preferred, then we should be happy to choose this alternative. This is the choice rule.

Given these four rules plus the axioms of probability theory, a normative theory of decision-making results that dictates the maximization of expected utility. Utility in this case needs to be measured on an interval scale; an interval scale preserves equal intervals of measure and can be multiplied or divided by a constant and can have a constant added or subtracted from it. A ratio scale of measurement for utility could be used but is not necessary. Note that probabilities are constructed on a ratio scale.

14.4 MULTIATTRIBUTE VALUE ANALYSIS

Multiattribute value analysis is a quantitative method for aggregating a stakeholder's preferences over conflicting objectives to find the alternative with the highest value when all objectives are considered. (Note that the phrases "multiattribute utility analysis" and "multiple objectives decision analysis" are also often used. In this book, the word utility is reserved for situations in which uncertainty has been explicitly modeled and the stakeholder's risk preference is being included in the analysis.) Multiattribute value analysis can be addressed simply as is done in this chapter or with a great deal more sophistication [see French, 1986; Keeney and Raiffa, 1976; Parnell et al., 2013]. Additional insights can be found in Kwinn and Parnell [2007]. Other approaches to value computations are also available: analytical hierarchy process (AHP) [Saaty, 1980, 1986], percentaging [Nagel, 1989], the technique for order preference by similarity to ideal solution (TOPSIS) [Yoon, 1980], a fuzzy algorithm [Yager, 1978], quality function deployment (QFD) [Akao, 1990], Pugh matrix [Pugh, 1991]. None of these other approaches are based on an underlying set of axioms that provide a foundation for justifying an analytical process except the AHP. However, there are several analytical concerns that have been raised about AHP, percentaging, TOPSIS, and similar approaches [Buede and Maxwell, 1995; Dyer, 1990; Harker and Vargas, 1990].

The process for defining the objectives of interest for a system has been defined in Chapter 6. For the systems engineering application addressed in this book, the objectives are the performance requirements that have been defined as described in Chapter 6, as well as derived performance requirements that have been defined as part of the development of the allocated architecture.

Following the definition of the objectives, a value scale must be defined for each objective at the bottom of the objectives' hierarchy. This value scale definition begins by defining the minimum acceptable value of performance for a given objective (constraining requirement) and the most desired value of performance for the objective (the design goal). Then the relative value of improving from the minimum acceptable threshold to the design goal is quantified in the form of a value curve. Objectives that are a combination of bottom-level objectives are in the hierarchy for ease of aggregation and communication, as a result these intermediate and the top-level (or fundamental) objectives are computed from lower-level objectives.

After value scales are defined for each bottom-level objective. value weights that address the relative value associated with improving from the bottom (minimum acceptable threshold) of the value scale to the top (design goal) must be assessed from the stakeholder for all bottom-level objectives as well as the intermediate objectives. The discussion in this chapter is going to address the common, but not universal, case in which the values can be aggregated across objectives by using a weighted-average formula. The books by French [1986] and Keeney and Raiffa [1976] address the general aggregation process and the assumptions required for various aggregation formulas.

The assumption that the general value function over the vector x of n bottom-level objectives can be written as a weighted additive function of value functions on the individual objectives:

$$v(x) = \sum_{i=1}^{n} w_i v_i(x_i) \tag{14.1}$$

will be adopted from here on out. Here

x_i represents how well an alternative does on each bottom-level objective i,

$v_i(x_i)$ represents the value associated with x_i for an alternative on objective i, and

w_i represents the "swing" weight that differentially values each objective i.

Note the weights are commonly normalized to sum to 1.0, and the value functions are normalized to range from either 0 to 1, or 0 to 10, or 0 to 100.

14.4.1 Eliciting Value Functions

The axioms of decision analysis produce the result that the value function over the vector x of bottom-level objectives must only be an interval function when the decision maker is risk-neutral (the assumption made here). As a result, the individual value function v_i over bottom-level objective x_i must also be an interval-scaled function of x_i. This interval property is the key to eliciting value functions from stakeholders about the relative value they assign to improving from the threshold of acceptable performance of x_i, x_i^0, to the most desired value of x_i^*. Watson and Buede [1987] present the bisection and the equal differences methods for eliciting these functions.

These value functions take four general forms (see Fig. 14.1): decreasing returns to scale (RTS), linear RTS, increasing RTS, and an S-curve. The decreasing RTS signifies a satiation of preference near the most desired value. Decreasing RTS is commonly encountered when the threshold of acceptable performance is within the key performance range of interest to the stakeholders and the most desired value is outside this key performance range where satiation takes over. The linear RTS is commonly found when both the threshold of acceptable performance and the most desired value are within the key performance range of interest, or when there is no possible satiation of preference. The increasing RTS occurs when (1) the threshold of acceptable performance has been pushed below (in a value sense) the key performance range and (2) there is a technological or other cap on the most desired value, so satiation of preference has not begun. Pushing the threshold of acceptable performance below the key performance range in a value sense means limited value is obtained by small increases in the performance parameter until some significant change is achieved. The S-curve reflects a joining of decreasing and increasing RTS and reflects the case in which the key performance range lies between the threshold of acceptable performance and the goal. The S-curve indicates that the range of possible designs has been maximized.

Note, no value curves that increase and then decrease, or decrease and then increase, have been shown. When value functions that are not monotonic (always increasing or always decreasing)

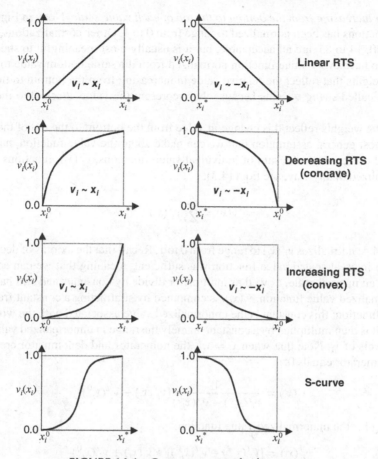

FIGURE 14.1 Common types of value curves.

are elicited, it is highly likely that there are two underlying objectives that have been combined. These two objectives should be separated so that the stakeholders are only considering one objective at a time when being asked to specify their preferences.

Exponential functions are commonly used to approximate the value functions of stakeholders [Kirkwood, 1997]. Equation (14.2) shows a standard form for variables on which more is better, and that is normalized to be 0 when the minimum acceptable threshold is met and 1.0 when the design goal is met. When a is greater than 1.0, this equation demonstrates decreasing RTS. When a is equal to 1.0, this equation becomes a straight line. When a is less than 1.0, this equation demonstrates increasing RTS.

$$v_i(x_i) = \frac{1 - e^{-\alpha(x_i - x_i^0)}}{1 - e^{-\alpha(x_i* - x_i^0)}} \tag{14.2}$$

Wymore [1993] has suggested a value function (or figure of merit) family that can accommodate all of the above value curves to some degree.

14.4.2 Eliciting Value Weights

Before discussing how to elicit the weights that are used in the additive value function of Eq. (14.1), the meaning of these weights must be made clear. In words, *the weights must reflect the relative value associated with increasing from the bottom to the top of each value scale*. Note, in Figure 14.1 each of the value functions has been normalized to range from 0 to 1. Other normalizations, for example, 0 to 10, 0 to 100, 14 to 85, are all acceptable, but it is usually most meaningful to stakeholders and everyone else to have every value function normalized from the same bottom value to the same top value. Value weights that reflect the relative value in increasing from the bottom to the top of each value scale are called *swing* weights because they represent the value attached to the swing from bottom to top.

Why must the weights reflect this change in value from the bottom to the top of the value scale? Consider the most general assumption that we can make about the value function, namely that the value across all objectives is the sum of individual value functions, $v_i'(x_i)$, functions that have not yet been normalized *in* any way; see Eq. (14.3):

$$v(x) = \sum_{i=1}^{n} v_i'(x_i) \tag{14.3}$$

Equation (14.4) normalizes $v_i'(x_i)$ to range from 0 to 1. Recall that the axioms of decision analysis implied that an interval-scaled value function was sufficient, meaning that we can add or subtract constants from an interval scale, as well as multiply or divide by constants and still have an interval scale. The normalized value function, $v_i(x_i)$, is computed by subtracting a constant from the unnormalized value function; this constant is the unnormalized value associated with the worst value (x_i^0) of x_i. This result is then multiplied by a constant, namely the range in unnormalized value from worst to best (x_i^*) levels of x_i. Note that when $x_i = x_i^*$, the numerator and denominator are equal. When $x_i = x_i^0$, the numerator equals 0.

$$v_i(x_i) = \frac{1}{v_i'(x_i^*) - v_i'(x_i^0)}[v_i'(x_i) - v_i'(x_i^0)] \tag{14.4}$$

Now solving for the unnormalized value function:

$$v_i'(x_i) = [v_i'(x_i^*) - v_i'(x_i^0)] * v_i(x_i) + v_i'(x_i^0) \tag{14.5}$$

Substituting (14.5) into (14.3) we get

$$v(x) = \sum_{i=1}^{n}(v_i'(x_i^*) - v_i'(x_i^0)) * v_i(x_i) + v_i'(x_i^0)$$

$$= \sum_{i=1}^{n}(v_i'(x_i^*) - v_i'(x_i^0)) * v_i(x_i) + \sum_{i=1}^{n} v_i'(x_i^0) \tag{14.6}$$

The last summation is a constant that has no relevance to distinguishing among alternatives, so it can be subtracted from both sides of the equation.

Now divide both sides by the constant

$$\sum_{i=1}^{n} [(v_i'(x_i^*) - v_i'(x_i^0))]$$

and distribute this term throughout the summation on the right side of the equals sign. The weights for each objective are defined to be

$$w_i = \frac{v_i'(x_i^*) - v_i'(x_i^0)}{\sum_{i=1}^{n}(v_i'(x_i^*) - v_i'(x_i^0))} \tag{14.7}$$

Substituting Eq. (14.7) into (14.6),

$$\frac{v(\boldsymbol{x}) - \sum\limits_{i=1}^{n} v_i'(x_i^0)}{\sum\limits_{i=1}^{n} [v_i'(x_i^*) - v_i'(x_i^0)]} = \sum\limits_{i=1}^{n} w_i v_i(x_i) \tag{14.8}$$

which is a linear transformation of the original value function and therefore equivalent to Eq. (14.1). So, the value weights in Eq. (14.1) must be defined to be the relative swing in value from the worst point x_i^0 to the best point x_i^* across all objectives.

Any mathematical approach employing interval scales that uses Eq. (14.1) to compute value but does not explicitly call for the use of swing weights is doing the equivalent of changing money from one currency to another by picking a random set of exchange rates rather than using the current market-derived *exchange rates*. The use of weights that are not swing weights may well suggest an alternative as best that is not consistent with the stakeholders' preferences. Some methods such as the Pugh methodology [Pugh, 1991] hope the objectives can be developed so that they are nearly equal in relative weight, without even defining what the weight means. No application of the authors (out of over a hundred) has generated a set of objectives that were nearly equal in importance.

While value functions only need to be interval scales, weights must be defined on a ratio scale. A *ratio* scale is one on which zero means zero value. In this case the value at the design goal must be equal to the value at the minimum threshold: $v_i(x_i^*) = v_i(x_i^0)$. A weight of zero means that the objective can be ignored.

Weight elicitation techniques can be divided into two categories: those that ask directly for numbers and *those* that ask for indirect ordinal or interval judgments that are used to derive a ratio scale.

14.4.2.1 Direct Weight Elicitation Techniques.

The most common direct elicitation technique for ratio scale numbers is to ask people to *spread 100 points* among the objectives at any given level of the objectives' hierarchy. This is a typical technique for eliciting weights in any multiattribute value application. The research literature [Watson and Buede, 1987] is not kind to this technique, and our experience confirms the literature findings. While it is relatively easy to do, people assign numbers that are far too close together to meet any ratio scale requirements; this is true no matter how many caveats the assessor presents to the participants to remember the ratio scale requirements [Stillwell et al., 1981].

Two other common direct assessment techniques involve *anchoring* on either the most important or least important objective. The stakeholder is then asked to *assign the most (least) important a score of 100 (1) and scale the remaining down (up)* based upon ratio scale requirements. The research literature has not really examined this method. In practice, it has not worked well for making the initial assessment queries but has worked reasonably well when it is introduced later in the assessment process. By this point, the stakeholders have become accustomed to thinking about ratio scale properties based upon a more detailed assessment process. The advantage of starting with the most important objective is that the stakeholders are probably most familiar with it and therefore, it is a useful anchor. The least important objective may not be that familiar to the stakeholders. In either case, the weights are normalized to sum to 1.0 at the end.

Edwards [1977] introduced a multiattribute utility technique called SMART that was based upon importance weights. (Edwards describes this as a self-recognized intellectual error [Edwards and Barron, 1994].) Edwards and Barron [1994] introduced SMARTS and SMARTER. SMARTS is simply SMART recast with the intellectually proper swing weights. SMARTS employs anchoring on the best objective at 100 points and scaling the rest down, then normalizing the weights to sum to 1.0.

SMARTER involves using the *rank-order centroid* technique of transforming the swing ranks of criteria into swing weights. Stillwell et al. [1981] offered several ad hoc ways to translate rank orders into weights. In the following equations, r_i is the rank of the ith objective, K is the total number of objectives, and w_i is the normalized approximate ratio scale weight of the ith objective.

Rank sum:

$$w_i = \frac{K - r_i + 1}{\sum\limits_{j=1}^{K} K - r_j + 1}$$

Rank exponent:

$$w_i = \frac{(K - r_i + 1)^z}{\sum\limits_{j=1}^{K} (K - r_j + 1)^z}$$

where z is an undefined measure of the dispersion in the weights. The larger z is, the larger is the ratio of the most important objective to the least important objective.

Rank reciprocal:

$$w_i = \frac{1/r_i}{\sum\limits_{j=1}^{K} (1/r_j)}$$

Rank-order centroid (ROC):

$$w_i = (1/K) \sum_{j=i}^{K} (1/r_j)$$

$$w_1 = (1 + \tfrac{1}{2} + 1/3 + \cdots + 1/K)/K$$

$$w_2 = (0 + \tfrac{1}{2} + 1/3 + \cdots + 1/K)/K$$

$$w_3 = (0 + 0 + 1/3 + \cdots + 1/K)/K$$

$$\vdots$$

$$w_K = (0 + 0 + 0 + \cdots + 1/K)/K$$

Barron and Barrett [1996] show that ROC weights accurately define the best alternative 75–90% of the time based upon a set of true swing weights elicited some other way. When the incorrect alternative was identified, the loss of utility averaged 3–7%. The ROC results were at the worst ends of these ranges when the attribute values of the alternatives were negatively correlated, which unfortunately is the most common situation in practice. Barron and Barrett [1996] show that the rank-reciprocal and rank-sum weights were nearly always worse than the ROC weights. Kirkwood and Corner [1993] use an actual application by Ulvila and Snider [1980] on oil tanker standards to provide some results that contradict claims concerning the effectiveness of rank-sum, rank-reciprocal, and rank exponent weights.

SIDEBAR 14.1: ILLUSTRATION OF WEIGHTING TECHNIQUES

To illustrate the weight elicitation techniques, consider the following engineering design sample problem. Suppose a communication system to be deployed as part of a data collection system is being designed. As part of our requirements analysis the following five major performance parameters that determine successful and profitable data collection operations (our measure of effectiveness) have been identified and ranked based upon the importance of the swing from minimum acceptable to ideal performance:

Performance Parameter	Minimum Acceptable Performance	Design Goal	Rank Order
Throughput, mbits/s	100	120	1
Availability	0.85	0.95	2
Operating life, years	5	7	3
Procurement cost, $	100	85	4
Operating cost, $/months	1.00	0.70	5

For the rank-based techniques the results in the table below are obtained. (Note that a 0.4 was used for the parameter in the rank exponent method.)

Rank Method	Throughput	Availability	Operating Life	Procurement Cost	Operating Cost
Rank sum	0.33	0.27	0.20	0.18	0.07
Rank exponent	0.25	0.23	0.21	0.18	0.13
Rank reciprocal	0.44	0.22	0.14	0.11	0.09
ROC	0.45	0.26	0.16	0.09	0.04

14.4.2.2 Indirect Weight Elicitation Techniques. Indirect assessment of weights can be obtained via one of several paired comparison techniques and the use of graphical adjustments on a computer. These techniques are generally far superior to any of the direct techniques in their ability to capture the decision-maker's tradeoffs across objectives.

The *paired* comparison *techniques* are the most common and include the AHP [Saaty, 1980], tradeoffs [Watson and Buede, 1987], balance beam [Watson and Buede, 1987] judgments, and lottery questions [Keeney and Raiffa, 1976].

AHP (see Sidebar 14.2) can be used to assess the weights of the objectives. In the full implementation of AHP, it is not easy to elicit swing weights because the AHP does not use the full value scale from 0 to 1. In AHP the stakeholders are asked to compare each objective with every other objective; note, it is possible to skip some comparisons, but the accuracy of the results decreases rapidly as the number of skipped comparisons grows. The AHP commonly does not ask the stakeholders to rank order the objectives in terms of overall benefit but begins by asking the stakeholders to compare objectives two at a time in whatever order they appear. The stakeholders are given the option of using a verbal scale. A numerical scale or adjustable bar graphs. The numerical scale ranges from nine times more valuable to one-ninth as valuable. The verbal choices have numerical equivalents that also vary from nine to one ninth. If there are K objectives, AHP would pose $K(K-1)/2$ questions of this sort. These responses are used as input to form a matrix upon which an eigenvector calculation is performed; these mathematical operations are justified by a set of axioms that Saaty [1980, 1986] has developed. It is possible that the stakeholders' judgments have inconsistencies embedded in them. Saaty [1980] has developed an inconsistency index based upon the mathematical operations he developed. Typically, the stakeholders are asked to rethink selected judgments if the inconsistency index is greater than 0.1. This approach seems to work well when the number of objectives is greater than 3 and less than 7 or 8. Naturally, it is possible to break a large number of objectives into subsets too – this approach is more efficient.

SIDEBAR 14.2: AHP EXAMPLE

Returning to the example of design trade-offs for a communication system, suppose the stakeholders provide the judgments shown in the following table in the AHP verbal mode.

	Throughput	Availability	Operating Life	Procurement Cost	Operating Cost
Throughput	(Equal) 1	2	4	6	(Absolutely) 9
Availability	½	(Equal) 1	4	(Strongly) 5	(Absolutely) 9
Operating life	¼	¼	(Equal) 1	(Weakly) 3	(Strongly) 5
Procurement cost	1/6	1/5	1/3	(Equal) 1	2
Operating cost	1/9	1/9	1/5	½	(Equal) 1

The normalized eigenvector of the largest eigenvalue for the numerical version of the above matrix is 0.45, 0.33, 0.13, 0.06, and 0.03. (Note that the AHP process associates a 9 with absolutely, 7 with very strongly, 5 with strongly, 3 with weakly, and 1 with equal.)

Trade-offs are used for swing weights and involve using the scores to help elicit the weights of the objectives. First, the objectives are ranked in order of their overall swing in value. Next, the stakeholders are asked if the overall swing weight of the second objective is as great as the swing from the lowest to some intermediate point of the value scale of the first ranked objective. For example, the stakeholders are asked whether the overall swing in value of the second-ranked objective was closer to 80% or 60% of the swing in value of the first-ranked objective. Suppose after some discussion the stakeholders agreed that the swing in value on the second objective was roughly equivalent to a swing from 0 to 0.7 on the value scale (normalized to a high of 1.0) of the first objective. This establishes that the weight of the second objective is 70% that of the first objective. The third-ranked objective could now be compared to intermediate points on either the first or second-ranked objectives. This method works very well when the value curves are firmly established and the value curves are continuous. If the value curves change significantly after trade-offs have been used, the weights have to be reassessed.

The *balance beam approach* is another approach for assessing the weights of the objectives (see Sidebar 14.3). The stakeholders are initially asked to establish a rank order of the overall swing weights of the objectives. Next, a series of questions is posed to the stakeholders that begins with "Is the overall swing in value of the first objective (a) greater, (b) less than, or (c) equal to the combined overall swing in values of the second and third most important objectives." To illustrate this question, a balance beam analogy (see Fig. 14.2) is used. If the stakeholders respond that the first-ranked objective has the highest overall swing weight, the attractiveness of the other choice is increased by adding the fourth-ranked objective to the package of the second- and third-ranked objectives. If the stakeholders say the package of second- and third-ranked objectives has a higher swing value than the first-ranked objective, the attractiveness of the combination package is decreased

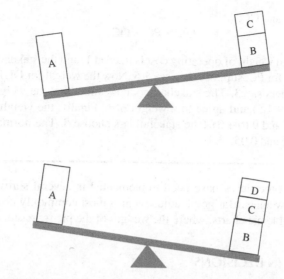

FIGURE 14.2 Balance beam analogy for paired comparisons.

by dropping the third-ranked objective and adding the fourth-ranked objective. This process is continued until the stakeholders have found a package of objectives with an overall swing in value that is comparable to the first-ranked objective. Next, the second-ranked objective is compared with the third and fourth-ranked objectives. This continues until only the last two objectives remain. The process creates a set of inequality and equality equations that relate the swing weights of the objectives. Typically, a weight of 1 is assigned to the least weighted objective, the stakeholders are asked to assign a swing weight to the second least weighted objective, and then the equations are used to bound the swing weights of the remaining objectives. It is possible that there will be an inconsistency in a subset of the equations. If such an inconsistency exists, the balance beam questions posed by this subset of equations are reexamined until the stakeholders identify their inconsistency and make an adjustment. This approach generally produces a widespread in the swing weights for the objectives.

SIDEBAR 14.3: BALANCE BEAM EXAMPLE

Using the balance beam approach for the communication system design, the stakeholders are asked to compare the swing in benefit of throughput (T) to that of the combined swings of availability (A) and operating life (OL). The stakeholders respond the combination is greater than that of throughput, or

$$T < A + OL$$

However, throughput (T) is preferred to availability (A) and procurement cost (PC):

$$T > A + PC$$

Availability is preferred to OL, PC, and operating cost (OC):

$$A > OL + PC + OC$$

OL is preferred to PC and OC:

$$OL > PC + OC$$

Next, the unnormalized weight of operating cost is fixed at 1, and the stakeholders are asked to provide a ratio weight for PC; suppose they say 1.5. Now the weight for OL is greater than 2.5, suppose the stakeholders say 3. The stakeholders now know that the weight for availability is greater than 4.5 (3 + 1.5) and agree to a weight of 6. Finally, the weight of throughput is between 7.5 (6 + 1.5) and 9 (6 + 3). The stakeholders choose 8. The normalized weights are 0.41, 0.31, 0.15, 0.08, and 0.05.

Graphical elicitation procedures have been implemented in several software packages for the elicitation of scores and weights. Bar graph adjustment is most commonly used, but some software packages contain adjustable pie charts, where the wedges of the pie represent different objectives.

14.5   UNCERTAINTY IN DECISIONS

This section addresses the analysis of decisions when there is substantial uncertainty associated with outcomes impacting the relative value of the decision's alternatives. In systems engineering, this uncertainty could be associated with the state of technology at some time in the future; the stakeholders' needs now and in the future; the ability to achieve cost, schedule, or performance goals; and environmental variables associated with the use or testing of the system.

Probability theory is discussed in Section 14.5.1 to refresh the reader's knowledge of this subject. Section 14.5.2 discusses the use of relevance diagrams to represent joint probability distributions. Influence diagrams are introduced in Section 14.5.3 as a way of representing a decision. The calculations of expected utility are described in terms of decision trees. Section 14.5.4 addresses risk preference.

14.5.1   Probability Theory

This section is not meant to be a detailed introduction to probability theory; for such an introduction, see Roberts [1992] and Ghahramani [1996]. The reader is assumed to be familiar with the concepts of probability density functions for continuous random variables and probability mass functions for discrete random variables, the difference between marginal and conditional probability distributions, the notion of cumulative probability distributions, and joint probability distributions of two or more random variables. First, the concepts of probabilistic independence and dependence are discussed. Then, two important equations, the law of total probability and Bayes rule, are provided. Finally, relevance diagrams are introduced to describe the probabilistic dependencies among a set of random variables. This entire discussion will be conducted in terms of discrete random variables because the mathematics is easier to convey, and discrete random variables are more commonly encountered is systems engineering problems. In addition, decision analysis commonly discretizes continuous random variables for computational ease.

The *probabilistic independence* of two random variables, X and Y, is defined to occur when the conditional probability distribution on X given Y equals the marginal probability distribution on X. It can be shown that when the preceding is true for X, then the probability distribution on Y given X must also equal the probability distribution on Y. As a result, the joint probability distribution of instances of X, x_i, and Y, y_i, can be written as

$$p(x_i, y_j) = p(x_i \mid y_j)\,p(y_j) = p(y_j \mid x_i)\,p(x_i) = p(x_i)\,p(y_j) \qquad (14.9)$$

when X and Y are probabilistically independent, intuitively, probabilistic independence means that learning the value of X does not cause us to change our probability distribution about Y.

The *law of total probability* allows the computation of a marginal probability distribution of one random variable by summing over all possible values of a second random variable that is probabilistically dependent on the first. This law is used to compute $p(x_i)$ when the probabilities on the right-hand side of Eq. (14.10) are known better than $p(x_i)$ (shown in Fig. 14.3):

$$p(x_i) = \sum_{j=1}^{m} p(x_i \mid y_j)p(y_j) \tag{14.10}$$

Bayes rule is used to update our uncertainty on one random variable when information about another random variable becomes available, assuming the two random variables are probabilistically dependent on each other.

$$p(y_j \mid x_i) = \frac{p(x_i \mid y_j)p(y_j)}{\sum_{i=1}^{n} p(x_i \mid y_j)p(y_j)} = \frac{p(x_i \mid y_j)p(y_j)}{p(x_i)} \tag{14.11}$$

In the case of Eq. (14.11), information about the value of random variable X is obtained and is used to update our uncertainty about Y. The left-hand side of Eq. (14.11) is called the posterior probability distribution of Y when all values of $j = 1, 2, \ldots, m$ are considered. The $p(y_j)$ in the numerator on the right-hand side of (14.10) is called the prior probability, the probability of Y before information on X became available. The values of $p(x_i|y_j)$ in the numerator and denominator are called the likelihood values of getting information on X given values of Y. Finally, the denominator of Eq. (14.10) is called the preposterior and is in fact equal to $p(x_i)$, as computed by the law of total probability (Eq. (14.10)). The contrast between the law of total probability and Bayes rule can be seen by revisiting Figure 14.3. With the law of total probability, the task is to compute the probability of a subset of the universal event using conditional probabilities that partition the universal event. With Bayes rule the universal event has been redefined based upon a new state of information, namely, x_i is known to be true. Bayes rule provides the process for updating the probability of any variable based upon this new information.

Adoption of Bayes rule in practice requires a philosophical shift in the meaning of probabilities for most people. The most common philosophical interpretation of probability among engineers and statisticians is that of a long-run frequency associated with a set of events that have been or could be repeated many times, for example, flipping coins and removing production samples from a production line. However, in systems engineering, the engineer of a system is typically involved in very early design decisions regarding the operational system, the test system for the operational system, the manufacturing system of the operational system, the test system for the manufacturing

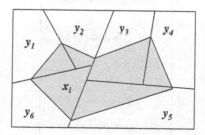

FIGURE 14.3 x_i as a subset of the universal event, which is partitioned by Y.

system of the operational system, and so forth. In these early design decisions, there is typically a great deal of uncertainty about specific outcomes related to these decisions and very little data. In fact, it is often not possible to contemplate repeating experiments to develop long-run frequencies within a reasonable amount of time and money. Bayesian, or subjective, probability interprets a probability as a state of information about the uncertainty regarding a variable. Powerful mathematical and logical arguments have been put forward by Savage [1954], De Finetti [1974], Lindley [1994], and others for this interpretation of probability. Now that the computational power that we have on our desks is quite sizable, many theoreticians are becoming Bayesians due to the theoretical justification of the Bayesian argument. Yet many of these Bayesian converts still prefer to put uniform priors on the random variables and let the data shape the posterior distributions. This is fine when there is a lot of data, as there is late in the systems engineering development process. Early in the development process, there is precious little data, and uniform priors are not consistent with engineering judgment and likely to lead to poor design decisions. There is a vast amount of research available on the ability of humans to provide probability judgments [Hogarth, 1980; Kleindorfer et al., 1993; Wright and Ayton, 1994]. Serious probability elicitation processes have been developed and used extensively with successful results [Spetzler and von Holstein, 1975; Merkhofer, 1987].

Bayes rule is useful during the design phase in systems engineering when there is little hard data available. During this phase, there are often significant results available from analyses and simulations; these results are appropriately considered as data, making Bayes rule an appropriate tool.

Bayes rule has wide applicability in the world of testing. Before the test we have some uncertainty about the ultimate value of certain performance, cost, or schedule parameters. Data is collected during the test regarding the values of certain system or project characteristics that relate to the parameters of interest. These data should then be used to update our uncertainty about the parameters of interest. Test data should always be viewed as likelihood measures. All too often, the test result is viewed to be the answer, and only the data parameter associated with the largest likelihood value is reported.

14.5.2 Relevance Diagrams

A *relevance diagram* is a directed graph, or digraph, that is a statement of the joint probability distribution among a set of random variables as a factorization of conditional and marginal probability distributions. For example, the three possible factorizations of two random variables, X and Y, are shown in Figure 14.4. Each random variable is shown as a node with an oval encapsulation. The top case shows two probabilistically independent random variables; the absence of an arc indicates this

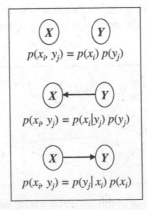

FIGURE 14.4   Relevance diagrams for two variables.

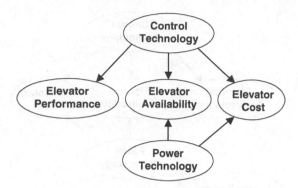

FIGURE 14.5   Notional relevance diagram for elevator design.

independence. The next two cases show dependence or relevance in a Bayesian sense of probabilistic updating; the arc can go in either direction, with the direction reflecting a different conditional and marginal distribution that defines the joint distribution. It is obvious from this simple graph that the arc in the bottom two graphs can be flipped (have its direction changed) without any repercussions. However, this is not true in general. A relevance diagram cannot have a cycle (see Chapter 5 for a definition), so flipping an arc that causes a cycle to form is never possible. In addition, when flipping an arc does not cause a cycle to be formed, it is possible that arcs will have to be added to the digraph [see Shachter, 1986].

As an example of relevance diagrams for systems engineering, consider an elevator design in which the state of technology related to control systems and power systems is highly uncertain in the time frame of the development effort (Fig. 14.5). The key performance requirements (design objectives) are elevator performance in terms of mean wait times, the operational cost of the system, and the availability of the elevator system. A relevance diagram depicting the probabilistic dependencies is shown in Figure 14.5. Note that there is no dependence between the three key performance requirements; these three variables are probabilistically independent of each other given the states of control technology and power technology. This is called conditional independence; if the variables for the control and power technologies were not present, there would be edges between the three requirements nodes (performance, availability, and cost). As discussed in previous chapters, there is great power to be gained in communicating the structure of reasoning (modeling) about design issues by using a graphical representation such as relevance diagrams.

As mentioned above, test results always provide likelihood information for Bayes rule. As a result, a relevance diagram that includes test results will have arcs going to the test result from the variable relevant to the test. A survey of power technology to assess the possible state of power technology in two years is an example of test data for the elevator design problem. This test data would be shown as a node with an arc coming to it from the Power Technology node in Figure 14.6. Bayes rule would then be used to flip this arc so that the survey results could be incorporated in the decision being made.

14.5.3   Influence Diagrams, Decision Trees, and Value of Perfect Information

Consider a standard design decision faced by systems engineers: Should a component for the system be bought from an existing supply source or be developed from more basic components? The uncertainty that may be most troublesome in this decision is how long it will take to develop the major component and how much it will cost. The schedule and cost results could be better than, equal to, or worse than the result associated with purchasing the component. For this simple example, assume the performance of both alternatives is equal. A *decision tree* depicting this decision is shown in

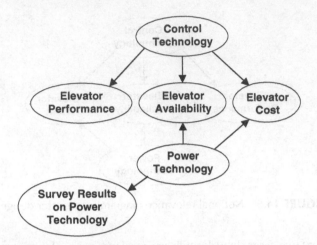

FIGURE 14.6   Relevance diagram with survey data on power technology.

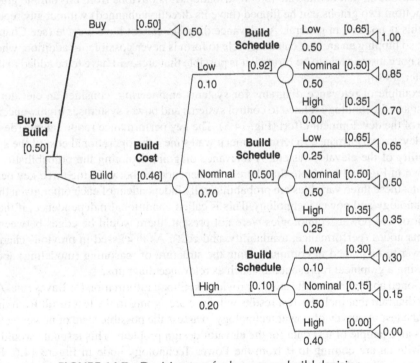

FIGURE 14.7   Decision tree for buy versus build decision.

Figure 14.7. The value computation is at the end of each branch of the tree and addresses the cost and schedule issues via a multiattribute value formulation. The decision node at the beginning of the tree depicts the two alternatives as branches emanating from a small square. After the Build alternative there are chance nodes (little circles) that represent the uncertainties concerning cost and schedule. The name of each branch is above the line while the probability for each branch is under the line. The tree Is "rolled bac" by multiplying the value at the end of each branch times the probability value on the branch just before it. These probability-weighted values are summed at each chance node to get an expected value at that node, which are shown in brackets, e.g., [0.5]. These expected values are then multiplied by the probabilities on the branches before them and summed again. This process continues until the expected value of each alternative is available at the decision node. The preferred alternative should be the one with the highest expected value.

Decision analysis offers insight into what the value of spending time and money would be if an uncertainty about an uncertain node if that uncertainty could be resolved. This section demonstrates how to calculate the *Value of Perfect Information*. The interested reader can consult [Clemen and Reilly, 2001] to find out about calculations for imperfect information.

The definition of the Value of Perfect Information (VoPI) is: Value with Perfect Information (VwPI) – Value without Perfect Information (Vw/outPI). The Vw/outPI is simply the value of the best decision strategy that has been calculated for the base decision, which is 0.5 for the Buy versus Build decision in Figure 14.7. To calculate the VwPI we take the following steps:

1. Identify the chance node that we would consider collecting information (we will look at collecting information on the Build Cost chance node in the Buy versus Build decision)
2. Modify the decision tree or influence diagram by moving the selected chance node before the first decision node (see the revised decision tree in Fig. 14.8)
3. Use Bayes rule to update the probabilities of the chance nodes if needed (this is not needed for the Buy versus Build decision because the two chance nodes are probabilistically independent)
4. Roll the expected value for the revised decision tree back to the base node (Build Cost in Fig. 14.8). We get an expected value of 0.54, which is higher than the 0.5 in Figure 14.7.

So, the VoPI is $0.54 - 0.5 = 0.04$.

Now, let us discuss some important concepts behind the VoPI. First, there is **no** source of perfect information. The practical value of knowing the VoPI is that the decision-maker should not pay more than this value. If this value is zero or near zero, then the decision maker can ignore suggestions to seek data or experts that can provide partial information. Note, 0.04 in the above example may seem small, but recall that the value scale goes from 0 to 1 and represents the total profit that could be made for this product. So, in fact, 0.4 could be a very large number.

Second, the value of VoPI cannot be negative. The decision maker always has the choice of not paying for any information if the information does not add value. In Figure 14.8, there are two branches for Build Cost where the best-expected value is equal to the Vw/outPI, which means there is no value to perfect information for these two branches. The third branch, when Build Cost is Low, has positive VwPI.

Finally, for complex decision problems, these calculations can be complicated. Fortunately, there are software tools available that perform these calculations with a few keystrokes from the user.

Influence diagrams are a graph-theoretic representation of a decision. Shachter [1986, 1990] presented the requirements and algorithms needed to transform an influence diagram from solely a communication tool into a computation and analysis tool capable of replacing the standard decision analysis tree. Significant additional research continues into influence diagrams for structuring decision problems, defining the underlying mathematics and graph theory of influence diagrams, and analyzing decision problems. When properly implemented, decision trees and influence diagrams provide identical solutions to the same problem. They are referred to as isomorphic since the decision tree can be converted to an influence diagram, and vice versa.

An Influence diagram may have four types of nodes (decision, chance, value, and deterministic), directed arcs between the nodes, a marginal or conditional probability distribution defined at each chance node, and a mathematical function associated with each decision. value, and deterministic node. Each decision node, represented by a box, has a discrete number of states (or decision alternatives) associated with it; chance nodes, represented by an oval, must be discrete random

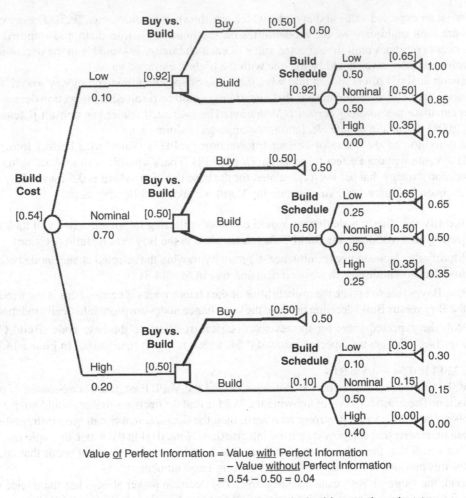

Value <u>of</u> Perfect Information = Value <u>with</u> Perfect Information
− Value <u>without</u> Perfect Information
= 0.54 − 0.50 = 0.04

FIGURE 14.8   Decision tree for calculating VwPI for build versus buy decision.

variables. Deterministic nodes are represented by a double oval. A value node may be represented by a roundtangle, diamond, hexagon, or octagon.

An arc between two nodes (shown by an arrow) identifies a dependency between the two nodes. An arc between two chance nodes expresses relevance and indicates the need for a conditional probability distribution. An arc from a decision node into a chance or deterministic node expresses influence and indicates probabilistic or functional dependence, respectively. An arc from a chance node into a deterministic or value node expresses relevance; that is to say, the function in either the deterministic or value node must include the variables on the other ends of the arcs. An arc from any node into a decision node indicates information availability; that is. the states of these nodes are known with certainty when the decision is to be made.

Figure 14.9 shows an influence diagram for the buy versus develop decision described in the decision tree of Figure 14.7. The decision is represented in the box, the value node in the box with rounded corners, and the two chance nodes in ovals. Note that the alternatives and chance outcomes that were shown in the decision tree are not visible in the influence diagram. However, the edges in the influence diagram provide new information that was not readily available in the decision tree,

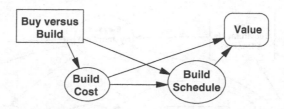

FIGURE 14.9 Influence diagram for build–buy decision.

namely the probabilistic and value dependencies inherent in the decision. Both cost and schedule are dependent on which alternative is selected. Cost and schedule are also probabilistically dependent on each other, with the influence diagram showing an arc from Build Cost to Build Schedule. Value only depends on cost and schedule.

The decision node represents a logical maximum (minimum) operation; that is, choose the alternative with the maximum (minimum) expected value or utility (cost). A deterministic node can contain any relevant mathematical function of the variables associated with nodes having arcs into the deterministic node. A value node can also contain any mathematical function of the variables with arcs entering the value node. In addition, the mathematical function in the value node defines the risk preference of the stakeholder.

A well-formed influence diagram meets the following conditions: (1) the influence diagram is an acyclic-directed graph, that is, it is not possible to start at any node and travel in the direction of the arcs in such a way that one returns to the initial node; (2) each decision or chance node is defined in terms of mutually exclusive and collectively exhaustive states; (3) there is a joint probability distribution that is defined over the chance nodes in the diagram that is consistent with the probabilistic dependence defined by the arcs; (4) there is at least one directed path that begins at the originating or initial decision node, passes through all the other decision nodes, and ends at the value node; (5) there is a proper value function defined at the value node (i.e., one that is defined over all the nodes with arcs into the value node); and (6) there are proper functions defined for each deterministic node. An influence diagram that is well-formed can be evaluated analytically to determine the optimal decision strategy implied by the structural, functional, and numerical definition of the influence diagram. The analytic operations needed to evaluate an influence diagram numerically are evidence absorption, deterministic absorption, null reversal, arc reversal, and deterministic propagation [Shachter, 1986].

The influence diagram in Figure 14.10 shows an example of an influence diagram for a requirements allocation decision for the design of a new elevator system. The systems engineer is considering the use of one of two new technologies (power or controller); the large decision node (center left of Fig. 14.10) defines the three alternatives. The requirements allocation (shown as three separate decision nodes) of costs, performance, and availability will be different if one or neither of these technologies is included in the design. Since this initial decision will be known when the three requirements allocation decisions are made, there are arcs from the initial decision node to the three requirements allocation decision nodes. The other arcs between the three requirements allocation decision nodes indicate the order in which the decisions will be made: performance, availability, and cost. (The decision maker is free to select any order among these three nodes.) These allocations and the prior uncertainty of the systems engineering team about the power and controller technologies will affect the uncertainty about the elevator's cost, performance, and availability. The arcs between the chance nodes are identical to those shown in Figure 14.5. Note, this diagram shows the uncertainty of elevator performance to be independent of power

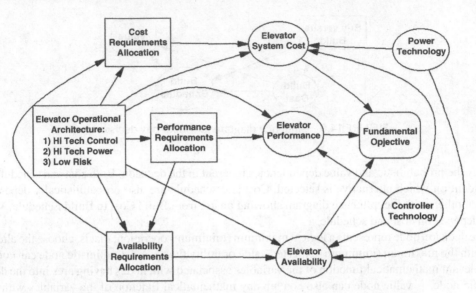

FIGURE 14.10 Sample influence diagram for requirements allocation.

technology. In this simplified example, the fundamental objective is comprised of three elements: cost, performance, and availability.

The results of a case study analysis of the above elevator architecture and requirements allocation decision are shown in Figure 14.11. First, the value functions for elevator performance (an index of various passenger waiting times), life cycle cost, and availability and their weights are shown. Note that marginal decreasing RTS is shown in each curve as capability moves from the minimum acceptable threshold to the technological maximum. Next, the uncertainties associated with the two technologies in question are shown. The other uncertainties encoded as part of the analysis are not shown here. The analytical results show that the allocated architecture and the requirements allocation associated with the advanced power technology should be chosen to be consistent with the requirements (the value structure captured by the trade-off requirements) and the uncertainty about the technologies. The alternative associated with the control technology is very close; in fact, too close to be confident that the power technology is preferred given the limitations of value and probabilistic assessments. The low-risk alternative is clearly inferior; the design team could feel comfortable choosing either of the new technologies. The choice of technology would significantly change the requirements allocation decisions made in the three subsequent decision nodes.

14.5.4 Risk Preference and Expected Utility

Webster's dictionary defines risk simply as the "exposure to the chance of loss," and most people have at least an intuitive sense of what risk means to them. But, from a decision-making perspective, it is essential to provide a more formal definition. The Defense Systems Management College (DSMC) [1989] in their *Risk Management Handbook,* defines *risk* as "the combination of the probability of an event occurring and the significance of the consequence of the event occurring" and defines risk management as "the various processes used to manage risk."

There are several strategies used for dealing with risk: avoidance, transference, management, and analysis. *Risk avoidance* is the selection of the low-risk alternative; unfortunately, what seems to be low-risk intuitively is high-risk in some cases. For example, consider a situation in which you have a sizable portfolio of U.S.-based stocks and are considering purchasing either another

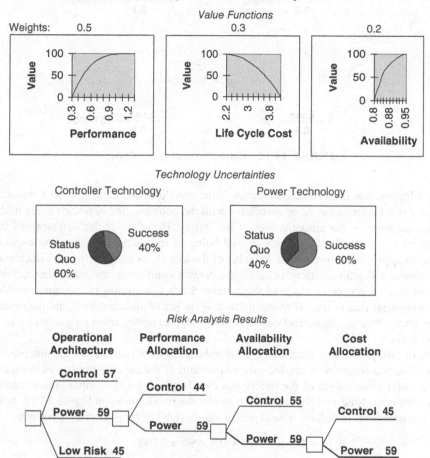

FIGURE 14.11 Summary of requirements allocation case study.

U.S. stock or what is considered a high-risk international stock. The international stock is often the lower-risk alternative because its performance is either negatively correlated or uncorrelated with the performance of your portfolio while the performance of the low-risk U.S. stock is highly correlated with your current portfolio.

Risk transference involves options that transfer risk to others, an example being the purchase of insurance. The insurance purchaser is willing to pay a fixed price and have the insurance company take the risk of a major loss.

Risk management involves the use of hedging strategies; a hedging strategy is the maintenance of fallback options in case a riskier option fails. The failure is not catastrophic because the fallback option can be used. This is common in systems engineering when multiple contractors are asked to develop the same component; one contractor is pursuing the high-risk and high-performance approach that will be used if successful, while another contractor is pursuing a more conservative approach.

Risk analysis addresses risk explicitly when decisions are made in uncertain situations. Addressing the uncertainty faced in a decision by assigning probabilities to the uncertain outcomes,

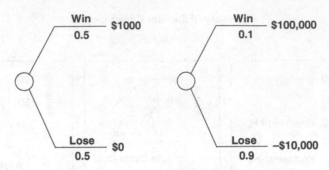

FIGURE 14.12   Comparison of two lotteries.

producing a lottery, has been discussed above. If the outcomes are measured on a numerical scale (e.g., dollars) that captures the value associated with the outcome, the expected value of the lottery is used as a measure of the attractiveness of the lottery. However, if the outcomes of the lottery are substantial compared to the wealth or well-being of the decision-maker, the expected value may not be an appropriate measure of the value of the lottery, as judged by many decision-makers. The value associated with a lottery is called the *certain equivalent,* the value the decision maker would be willing to accept in place of the lottery. Since this notion of certain equivalence is a subjective judgment that is special to the individual (or set of stakeholders) and the context at the time of the decision, a mathematical description of risk preference must be guided by the feelings of decision-makers.

A utility or risk preference function, u, is introduced to be a function of the outcome values of the lottery. If such a function exists, the inverse function of the expected utility of the lottery is the value of the certain equivalent of the lottery that can then be used to compare the attractiveness of the lottery with other lotteries. For example, consider the two lotteries in Figure 14.12, in which the outcomes are measured in dollars. The expected values (EV) of these two lotteries are:

$$EV(1) = 0.5 \times \$1000 + 0.5 \times \$0 = \$500$$

$$EV(2) = 0.1 \times \$100,000 + 0.9 \times -\$10,000 = \$1000$$

These expected values indicate that lottery 2 is preferred to lottery 1; $EV(2) > EV(1)$. Yet many people, who cannot afford a loss of \$10,000, would prefer the first lottery with the lower expected value. In other words, for those people, the *expected utility* of lottery 1 > the expected *utility* of lottery 2, or

$$0.5u(\$1000) + 0.5u(\$0) > 0.1u(\$100,000) + 0.9u(-\$10,000) \tag{14.12}$$

Mathematically, if the inverse function of $u(.)$ exists, then Eq. (14.12) can be restated as

$$u^{-1}[0.5u(500) + 0.5u(0)] > u^{-1}[0.1u(100,000) + 0.9u(-10,000)] \tag{14.13}$$

The question is: "Will such a function generally explain the decision maker's risk preference judgments over all possible lotteries?" The two expressions on either side of the inequality in Eq. (14.13) are called the certain equivalents of the two lotteries.

The *risk premium,* x_p, of a lottery is defined to be the difference between the expected value of the lottery and the certain equivalent, $\tilde{x}$

$$x_p = \bar{x} - \tilde{x} \tag{14.14}$$

For risk-averse decision makers the certain equivalent will always be less than the expected value and the risk premium will be positive.

14.5.4.1 Assessing a Risk Preference Function. Discussion of a risk preference function for a specific decision assumes that the outcomes of the decision have been characterized by a value function that collapses all dimensions of value onto one dimension, commonly called the *numeraire*. A money equivalent is the most common numeraire, but others are also possible. The risk preference function is then a function over the value numeraire.

There are two types of questions involving a certain equivalent and a two-outcome lottery that one can ask a decision-maker during a risk assessment session. These two question types are shown in Figure 14.13. The first question type assumes the probabilities of the lottery are known, and the decision maker is asked to provide one of the outcome values, typically the value of the certain equivalent. However, one could fix the certain equivalent and ask for the value of either the best outcome or the worst outcome. The second question assumes that all of the outcome values are known, including the certain equivalent, and the decision maker is asked to supply the probability value.

Unfortunately, research has shown that people do not provide coherent answers to these two types of queries. That is, in general, the answers to the second question type are going to suggest much greater risk aversion than answers by the same individual to the first question type. A not uncommon response to the first query, which has an expected value of $50, is $35, yielding a risk premium of $15. Now, if $35 is the certain equivalent in the second query, an individual might respond that the question mark for the probability of $100 in the second lottery would be 0.6. The risk premium for this second lottery is $25 (the expected value of $60 minus the certainty equivalent of $35).

The first question type is asking directly for the response that will be substituted into various analyses. Therefore, it is somewhat more appropriate to ask this question. However, very few decision-makers have thought seriously about these issues in general, and even fewer have thought about them with respect to a specific decision situation. The assessment process is therefore a learning experience for the decision maker. The responses to the early questions should be treated as a warm-up process.

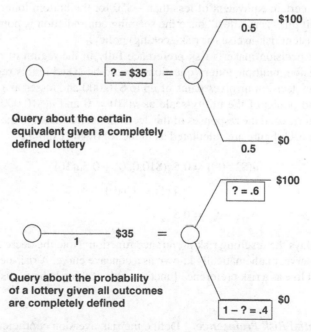

FIGURE 14.13 Simple risk preference assessment queries.

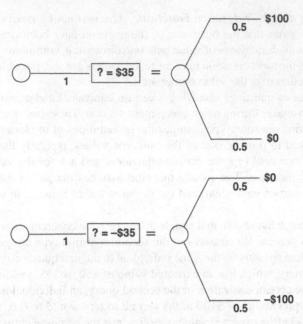

FIGURE 14.14   Illustration of the zero effect.

A second caution for the risk assessment process is that there is a very substantial zero effect. That is, people exhibit risk-averse behavior for gains but risk-seeking behavior for losses. Figure 14.14 shows responses for a certainty equivalent that demonstrates this behavior. The risk premium is $15 for the top lottery and −$15 for the bottom lottery. The risk-averse person in the top lottery would have a certain equivalent of less than −$50 for the bottom lottery. Generally, people do not want to exhibit this "zero effect" once the seeming contradiction is pointed out to them and will switch to a consistent risk-averse (or risk-seeking) policy.

To investigate the decision maker's risk preference fully in the region of outcomes associated with the current decision, multiple lottery questions should be asked in this region. For illustrative purposes, suppose the decision involves gains of up to $10,000 and losses as great as $10,000. We arbitrarily set the end points of the utility-scale as $u(\$0) = 0$ and $u(\$10,000) = 1$. Figure 14.15 provides six such lotteries and the responses of the decision maker shown in the boxes. Note that the utilities shown under each figure are calculated as in the following example:

$$u(\$2,500) = 0.5u(\$10,000) + 0.5u(\$0)$$
$$= 0.5\,(1) + 0.5\,(0)$$
$$= 0.5$$

Figure 14.16 displays the resulting risk preference function. Note the decreasing rate of increase associated with this curve, mathematically known as a concave curve. A risk-neutral decision maker would have a straight line as a risk preference function; risk-seeking behavior is typified by a convex curve.

14.5.4.2   *Exponential Risk Preference.*   Define the risk aversion coefficient $\gamma = -u''(x)/u'(x)$. If γ is a constant, it can be shown by simple integration that the risk preference function must take

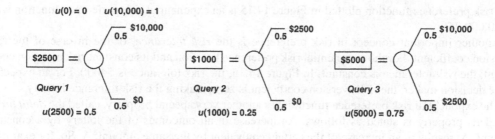

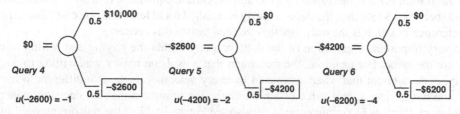

FIGURE 14.15 Assessment queries for risk preference function.

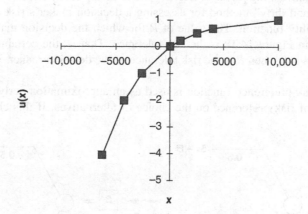

FIGURE 14.16 Assessed risk preference points.

the form.

$$u(x) = \begin{cases} k_1 x + k_2, & \text{if } \gamma = 0 \\ k_1 e^{-\gamma x} + k_2, & \text{if } \gamma \neq 0 \end{cases} \tag{14.15}$$

A common way to write such a risk preference function is

$$u(x) = \frac{1 - e^{-\gamma x}}{1 - e^{-\gamma x_{\max}}} \tag{14.16}$$

where $x_{\max}$ is the largest value that x is expected to take. Thus, for any valued outcome x, the utility of x can be calculated using the exponential utility function. Note that this format produces

$$u(x_{\max}) = 1.0$$

$$u(0) = 0$$

The risk preference function plotted in Figure 14.15 is an exponential risk preference function with $\gamma = 0.000,25$.

Another important concept in risk preference is the *risk tolerance,* or the inverse of the risk aversion coefficient. For the exponential risk preference function and its constant risk aversion coefficient, the risk tolerance is constant. In Figure 14.15, the risk tolerance is $4000. For an expected value decision maker, the risk aversion coefficient is zero, making the risk tolerance infinity.

The exponential risk preference function has another very special property, called the *delta property* This property is stated as follows: An increase in all outcomes of the lottery by a constant amount, Δ, results in an increase of the certain equivalent by the same amount, A. So, for example, in the first example above, suppose that the certain equivalent for 50 : 50 gamble of $100 and $0 was $35. Now, if each prize is increased by $100 and the certain equivalent of a 50 : 50 gamble on $200 and $100 becomes $135, then the delta property is satisfied for at least this one case. The exponential risk preference function is the only function that can satisfy this property.

One very important implication of the delta property is that the buying and selling prices of a lottery are the same. For example, the maximum that a decision maker was willing to pay, B. for a lottery, is the amount that when subtracted by every outcome, made us indifferent to having the lottery and not having it, or a value of $0. Similarly, the minimum that the decision-maker would sell the lottery for, S, is its certain equivalent; also see Figure 14.17. If the risk preference function is exponential, it can be proven that $B = S$ through the use of the delta property. For other risk preference functions, the buying and selling prices of a lottery are not necessarily equal.

There is a "quick and dirty" method for assessing a decision maker's risk aversion coefficient for an exponential utility function. The value of R for which the decision maker is indifferent to accepting the lottery in Figure 14.18 is the risk tolerance. That is, the certainty equivalent of the lottery in Figure 14.18 is 0 when R is the risk tolerance of the decision maker. It can be shown that $\gamma = 1/R$.

The exponential risk preference function is used as an approximation early in risk analyses to determine the effect of risk preference on the choice of alternatives. If this choice is sensitive in

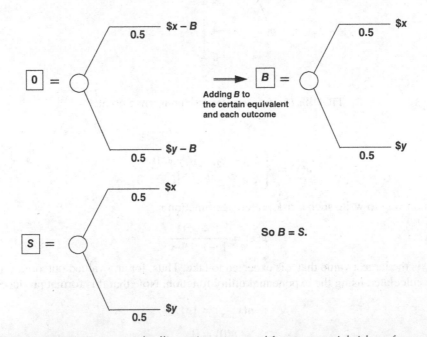

FIGURE 14.17 Buying and selling prices are equal for exponential risk preference.

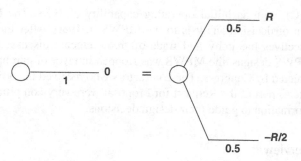

FIGURE 14.18   Risk aversion coefficient lottery.

the appropriate region of the decision maker's risk tolerance, then a more detailed analysis of the decision maker's risk preference is appropriate.

SIDEBAR 14.4: ECONOMIC MODELS

Hazelrigg [1996] provides strong motivation to use decision analysis tools in systems engineering design decisions. In his treatment, he addresses the results of Arrow's impossibility theorem [Sen, 1970] for achieving group consensus on preferences and recommends the use of the demand function from economics for defining consumer preferences for alternate design alternatives. The issue of gaining stakeholder consensus on trade-offs needed during design is real; thus the systems engineering team must resort to accepting the position of one stakeholder (the bill payer) as king when these disagreements cannot be resolve. This was the method used in the application presented in this section.

The notion of a demand function for a military system is not helpful. However, for a commercial system, the multiattribute value function can be considered to be a first-order, Taylor series approximation of the demand function. Hazelrigg [1996] does not go into detail about how to obtain the demand function; the suggestion made in this book is to elicit stakeholders' preferences and use the bill payer as king or queen to resolve disagreements.

14.6   SAMPLE APPLICATION

This application demonstrates how decision analysis can be used in the requirements development process of systems engineering. The requirements development process consists of the development of an operational concept, identification of the external systems that interact with the system and the context in which the system operates, an objectives hierarchy for the system's performance, and the requirements. These requirements are divided into requirements categories of input/output, system-wide and technology, trade-off, and test. The focus of this application is the use of multi-attribute value analysis as the approach for defining the trade-off requirements that comprise the value model to be used by the stakeholder in evaluating the available alternatives. Implicit in this approach is an objectives hierarchy for defining the value space of the stakeholder (see Sidebar 14.4). Also included is the mathematical structure for the trade-off requirements.

Throughout this discussion, a system called the Mobile Protected Weapons System (MPWS) is used to describe the development of the system engineering and decision analysis concepts. The MPWS was to be a helicopter-transportable, direct-fire support weapons system for the U.S.

Marine Corps (USMC), with an initial operating capability of 1988. The basis of the example was a real application of decision analysis to the MPWS in 1980. After the evaluation structure embodied in the objectives hierarchy and trade-off requirements discussed below was used to evaluate proposed MPWS designs, the MPWS was stopped in favor of purchasing similar vehicles "off-the-shelf," as directed by Congress. The contractors who received the objectives hierarchy and trade-off requirements as part of the Request for Proposal were very complimentary of the USMC for providing this information to guide their design decisions.

14.6.1   MPWS Overview

An intuitive need for a highly mobile, helicopter-transportable weapons system that can provide the landing force assault fire support as well as an antiarmor capability first became apparent to the USMC in the early 1970s. There were several contributing factors:

- Naval gunfire support assets, so important during an amphibious assault, were steadily decreasing.
- Navy combatant ships with suitable guns for shore bombardment were being retired without replacements or being replaced with ships less capable of providing gunfire support to amphibious forces.
- The retirement from the Fleet Marine Force (FMF) of the ONTOS, a light, mobile, antitank weapon system carrying six 106-millimeter (mm) recoilless rifles.
- The retirement of the crew-served individual 106-mm recoilless rifle.
- The deletion of the 3.5-inch rocket launcher from the Marine Corps inventory.
- At a time when naval gunfire and direct-fire weapons were decreasing, the Soviet and Soviet-aligned forces increased their capability with a wide array of armored weapons systems, including tanks, armored personnel carriers, and lightly armored weapons platforms.

In accordance with acquisition procedures contained in Circular A-109 of the U.S. Office of Management and Budget, Mission Area Analysis (MAA) was continuous, and a Mission Element Needs Statement was developed stating that:

- Amphibious forces possess capabilities that are uniquely featured by their responsiveness to the maritime aspects of the national strategy. Amphibious warfare requires the full spectrum of capabilities, from naval combat effectiveness offshore and in the air to the close combat mission ashore. The close combat capability provides the mobility, shock action, and portions of the firepower necessary to enable landing forces to successfully attack and destroy enemy personnel and materiel, breach their defenses, link up surface-borne with helicopter-borne forces, defeat infantry and mechanized counterattacks, and exploit success in combat ashore.
- Capabilities currently possessed by the landing force provide limited mobility and direct-fire combat power to enable assault units to rapidly close with and destroy enemy forces. Mobility and direct-fire support capabilities required to enhance current capabilities are:
  a. Helicopter transportability of weapons systems by heavy-lift helicopter
  b. Vehicle and crew survivability through armor protection from nearby artillery airbursts and medium-caliber direct-fire weapons firing at medium range
  c. Rapid cross-country mobility, agility, and endurance without significant degradation of on-road capability and capable of competing with the expected mobility of the threat

d. An on-board weapons suite with a long-range, high-kill probability capability against armored, light armored, materiel, and personnel targets characteristic of the threat
e. The ability to engage and defeat the target spectrum in all weather conditions
f. Nuclear, Biological, and Chemical (NBC) detection and protection

The Marine Corps requirements defined an affordable weapons system that was to be highly mobile, helicopter-transportable, compatible with amphibious operations, and able to provide direct-fire support during landing force operations. The weapons system must provide protection from suppressive fires and be capable of engaging and defeating armored, personnel, and materiel targets.

14.6.2 Operational Concept for MPWS

In defining the mission needs for the MPWS, three employment scenarios were considered. These scenarios represent the spectrum of scenarios that drive the design of MPWS. The relative importance of each parameter in the design process changes as a function of scenario.

Scenario 1: Offensive Role (assault support with the infantry) MPWS would be used with the infantry in offensive operations. A red/blue force ratio of 1 : 4 and a northern NATO environment are established as the base for the determination of relative capability requirements in this scenario.

Scenario 2: Defensive Role (blocking position) MPWS would be employed with helicopter-borne forces to establish blocking positions. Friendly tanks are not available. The mission calls for delaying the enemy and channelizing his avenues of approach. It is assumed that enemy forces are mechanized to include T62, T64, and T72 tanks, BMP, BTR, assault guns, SP artillery, and attack helicopters. MPWS will be operating at altitudes higher than sea level. A red/blue force ratio of 4 : 1 in a Middle East environment is established as the base in this scenario.

Scenario 3: Subsequent Operations MPWS would be employed with a combined arms task force and would no longer be in an amphibious assault role. Blue forces are task-organized, and there would most likely be low-mid-intensity nonnuclear conflict. Red/blue force ratio of 1 : 4 and a Middle East/Third World environment are the requirements determination base.

14.6.3 External Systems Diagram

The external systems of the MPWS during its operational and maintenance phase would be the operators (driver, gunner, and passengers), maintainers, targets (light armored vehicles, tanks, personnel, and helicopters), a heavy-lift helicopter that would have to transport the MPWS.

Figure 14.19 is an external systems diagram showing the inputs to and outputs from MPWS for the various external systems. This diagram was completed using the IDEFO Integrated Definition for Function Modeling process modeling (see Chapter 3). Four external systems are shown in Figure 14.19: the MPWS operators, the MPWS targets, the heavy-lift helicopter that will carry the MPWS from point to point, and the MPWS maintenance personnel. The interaction between the MPWS and its operators is shown by the three arrows: two leaving the operators' function and one leaving the MPWS function. Terrain forces are shown as part of the context, entering the MPWS function as input from outside the set of external systems. The primary benefits of this analytical construct are to bound the MPWS system very specifically by showing where MPWS ends and

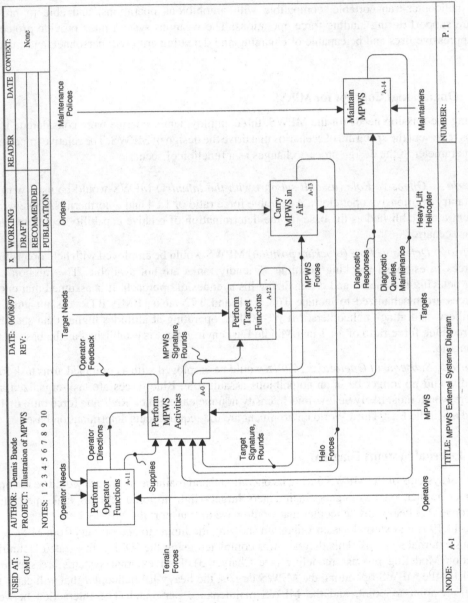

FIGURE 14.19 External systems diagram for MPWS.

other systems begin, and to specify the inputs to and outputs of MPWS so that requirements can be defined to make these inputs and outputs possible.

Figure 14.20 portrays an objectives hierarchy similar to the one developed by a team of USMC experts and the decision analysts working on the project. The three operational scenarios are the first decomposition of the hierarchy because the principal objectives of the USMC for the MPWS had different relative importance depending upon the scenario. The top-level objectives, or measures of effectiveness (MOEs), were firepower, mobility, availability, and survivability. Firepower was broken into measures of performances (MOPs): lethality, servicing rate, stowed kills (a combination of the number of stowed rounds and the lethality of those rounds), and target acquisition. Lethality is composed of the various types of targets, followed by the ranges at which those targets would be engaged. Target acquisition is composed of identification and recognition in good weather as well as the bad weather capability. Mobility is broken into capabilities related to cross-country, long-distance airlift, road, and water. Survivability is measured by means proxies for agility, protection, and signature.

14.6.4 Requirements

The focus of this application is the set of requirements called trade-off requirements, algorithms for comparing any two alternate designs on the aggregation of cost and performance objectives. As discussed in Chapter 6, these algorithms are divided into (1) performance trade-offs, (2) cost trade-offs, and (3) cost-performance trade-offs.

In the development of requirements for MPWS, substantial attention was devoted to the trade-off requirements for performance. The structure that describes the mission-related objectives on which these performance trade-offs were defined is the objectives hierarchy shown in Figure 14.20. The trade-off requirements consist of a utility or value curve for each bottom-level objective and a set of weights at each branch in the tree.

14.6.4.1 Utility Curves. Figure 14.20 portrays the many operational effectiveness variable performance parameters whose utility for improvement was quantified for guidance by the USMC committee. Inherent in these value or utility curves for the many performance parameters is the notion that design trade-offs are acceptable within the 0-to-100 range of utility; that is, MPWS performance in some areas can be sacrificed to the point of zero marginal utility, but no further, in order to achieve performance gains in other areas. The zero-utility point on each performance parameter does not mean that a system with this capability has no utility to the Marine Corps. Rather, it means that this level of performance is the minimum acceptable to the Marine Corps across its range of missions. So, for example, to be helicopter-transportable the MPWS must not weigh any more than 16 tons at 3000 feet on a 91.5 °F day. The utility curve for helicopter transportability is shown in Figure 14.21. Increased performance for each parameter has value to the Marine Corps, as shown by the shape of the utility curves.

The shapes of these utility curves are the same for all the above scenarios. However, the relative values of improvements in one parameter compared to improvements in another parameter do not vary across the three scenarios. These relative values of performance parameter improvements are described in Section 14.6.4.2.

14.6.4.2 Weights. Improvements in performance determined from the curves for each parameter are not equally important in the overall analysis of an MPWS. Therefore, a weighting procedure is applied to define the relative value of improving from the 0 to the 100 level of utility on one performance objective compared to another. The meaning of the weights can be described as follows: the

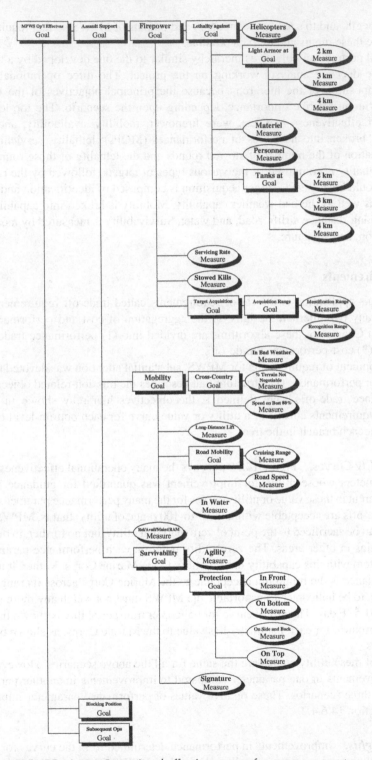

FIGURE 14.20   Operational effectiveness performance parameters.

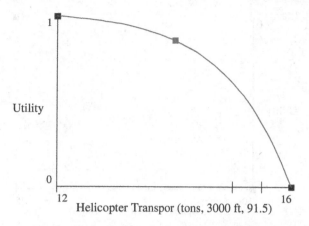

FIGURE 14.21 Utility curve for helicopter transportability, measured in tons.

weight given to parameter A reflects how much more valuable it is to improve from a score of 0 to 100 in parameter A as compared to the improvement in parameter B from 0 to 100. Note weights are not a generic measure of value but are dependent upon the swings from 0 to 100 on the associated utility curves.

For MPWS, weights played a large role in distinguishing among scenarios. While the shapes of utility curves remain constant across scenarios, their relative importance changed significantly. For example, an improvement in utility for helicopter transportability was very important in the blocking position role since the MPWS might have to be lifted into position. This same improvement was far less important in the subsequent operations role since the force would be traveling over land. Therefore, the weight that helicopter transportability has, relative to other operational effectiveness factors, was greater in the former role than in the latter.

14.6.5 Use of Utility Curves and Weights

Value (or utility) curves and weights can be used as follows: the abscissa (x axis) of each curve is a measurable attribute that provides input to the curve. The ordinate (y axis) is a measure of relative value or utility ranging from 0 to 100. As an example, value or utility curves for V80 and Percent No-Go are shown in Figures 14.22 and 14.23. Note that an improvement in V80 from 10 to 15 mph is valued as highly as a gain from 15 to 25 mph. Both improvements would net 50 utility points. Using these curves, a candidate propulsion system yielding a V80 speed of 15 mph would receive 50 utility points while one with a V80 speed of 20 mph would receive 80 points; a candidate with 6% No-Go scores 85 while one with 16% scores 35.

These value or utility scores would not be very meaningful for comparing systems without a relative measure of importance between attributes. Thus, a weighting procedure is applied to the scores to allow evaluation based upon a combination of parameters. Again, consider the value or utility curves illustrated in Figures 14.22 and 14.23: Suppose propulsion system 1 yields a V80 speed of 15 mph and Percent No-Go of 6%, while propulsion system 2 has values of 20 mph and 16%. System 1 scores would be 50 and 85, while system 2 scores would be 80 and 35. If both V80 and Percent No-Go were equally important, the weighted scores for both systems would be:

$$\text{System 1:} \ 1/2\,(50) + 1/2\,(85) = 67.5$$

$$\text{System 2:} \ 1/2\,(80) + 1/2\,(35) = 57.5$$

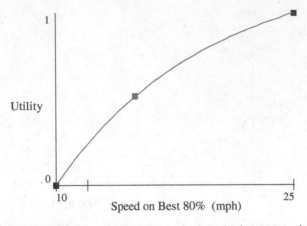

FIGURE 14.22 Utility curve for V80, speed on the best 80% of terrain.

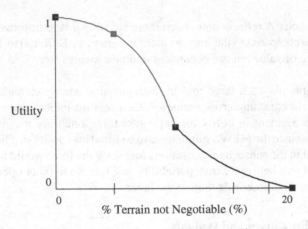

FIGURE 14.23 Utility curve for % no go, % of terrain that is not negotiable.

This would indicate that propulsion system 1 was superior on these factors. However, if V80 was considered to be two times as important as Percent No-Go, the weighted scores would be:

$$\text{System 1: } 2/3\,(50) + 1/3\,(85) = 61.7$$

$$\text{System 2: } 2/3\,(80) + 1/3\,(35) = 65$$

In this case, propulsion system 2 would be better.

It should be clear that the relative weights of the objectives play a major role in the design and evaluation processes.

14.6.6 Conclusions

As discussed in Chapter 6, the requirements development process is a systematic one that considers how the system is to be used, how the system is going to interact with other systems and the general environment, and the user objectives and priorities. Since user objectives and priorities are inherently subjective, the ultimate requirements for the system have to be subjective, reflecting trade-offs of the users. This is not to say that substantial analysis is not critical to the development of good requirements. In the case of the MPWS, the USMC used a great deal of analysis about alternate sites around the world in which it might be involved in conflict and the capabilities of the CH-53E helicopter to develop the utility curve for helicopter transportability and its relative weight to other performance objectives. By using many analysis techniques and a broad base of experts, logical and explicit statements of requirements were developed based upon informed consensus. The appropriate, detailed requirements inputs to the process can be obtained at lower organizational levels using appropriate experts and analyses, yet the more difficult, high-level requirement questions can be addressed at the highest levels of the organization.

14.7 SUMMARY

This chapter has introduced the complexities associated with decision-making in general and addressed the difficulty of decision-making in the engineering of a system. With respect to engineering a system, the definition of clear and meaningful alternatives for the design and integration of a system involves the use of sophisticated processes and modeling techniques, as described in the first 12 chapters of this book. The development of the value structure for selecting design and integration alternatives was discussed in Chapter 6 and involves complex trade-offs across stakeholders and stages of the system's life cycle. Finally, there is significant uncertainty regarding the relative effectiveness and cost of competing technologies as well as the future needs of the stakeholders.

The axioms of decision analysis, as presented in this chapter, provide a sound basis for a coherent, rational decision-making process that incorporates meaningful approaches for addressing value trade-offs and uncertainty. Multiattribute value analysis, a product of the axioms of decision analysis, uses value functions and weights to quantify the trade-offs across objectives. These value functions and weights require that the stakeholders answer questions that have meaningful interpretations to them in terms of the decision being made; the quantification of values is not an ad hoc set of numbers producing an index of goodness.

Dealing with uncertainty is a difficult problem; decision analysis relies upon probability theory to capture the uncertainty faced by the decision maker. In the engineering of a system, the uncertainty is not often described by existing data and is interpretable as the long-run frequency of a set of known events. Instead, the uncertainty deals with processes that change with time and for which no (or at most a few) known events have occurred. Instead of ignoring the uncertainty faced in the engineering of a system, decision analysis permits the engineers to capture the expert judgment of the engineers, stakeholders, and other experts and use this information to provide insights about the design choices with the best information available at the time. Recent advances in decision analysis provide graph-theoretic models for representing probabilistic dependence (relevance diagrams) and decision problems (influence diagrams).

Once uncertainty is modeled explicitly, the risk preference of the decision maker has to be addressed as part of decision analysis. The concepts of risk aversion, neutrality, and preference are defined mathematically and illustrated as part of the decision-analysis process. Using the decision maker's risk preference requires computing the certainty equivalent as the inverse of the utility function.

Clearly, it is inappropriate to use the sophisticated tools of decision analysis for every decision that is part of the engineering of a system. Many times, engineers have described the benefit of thinking about the decision in the terms of decision analysis. At other times, developing the value model and using a quick scoring and weighting evaluation provides insight into which alternatives are serious and which should be ignored. For complex and contentious decisions, the full power of decision analysis can provide an explicit and rational process for defining and discussing the alternatives to reach a conclusion consistent with the values of the stakeholders and the uncertainty as defined by relevant experts.

PROBLEMS

14.1 In defining reliability of a system, we talk about the probability of a failure. Failure here is an event or distinction, but not one that passes the clarity test. As a result, systems engineers work very hard to focus on the distinction, mission failure, where a mission failure is a failure that precludes the user from completing her/his mission. This definition still does not pass the clarity test because we have not defined the mission, a definition that is system and context-dependent.

For the elevator system where you work or go to school,

a. Define mission in a way that meets the clarity test.

b. Define as many failures as possible and show which would be classified as a mission failure. Be sure to keep the clarity test in mind when defining these failures.

c. Discuss whether it is sufficient to discuss failures one at a time or whether it is necessary to examine possible combinations of failures to define fully all *possible* mission failures.

14.2 Garbled Communications, Ltd. is designing a new system for special-purpose use that only requires three signals to be sent and received. The derived requirements below list the probability that signal s_i is received given that signal s_j is sent:

p (s_j received I s_i sent, &)	Receive s_1	Receive s_2	Receive s_3
Send s_1	0.80	0.10	0.10
Send s_2	0.05	0.90	0.05
Send s_3	0.02	0.08	0.90

For the operational concept each signal is equally likely to be sent. The stakeholders' requirement for this scenario is that each signal should have a 0.85 probability of being sent given that it was received. Is this requirement met if these derived requirements can be satisfied? Note the symbol "&" on the right-hand side stands for all prior information.

14.3 Garbled Communications, Ltd. has begun producing its new communications system and has built three assembly lines, L1, L2, and L3. L1 is the most productive, accounting for 40% of the production; L2 is the least productive, accounting for 25%. L3 accounts for the rest. Test data show that L1 has a 2% chance of producing a lemon, L2 has a 4% chance, and L3 has a 3% chance. What is the probability that a lemon picked at random will come from each of the assembly lines?

14.4 Write the joint probability distribution that is consistent with the relevance diagram shown below.

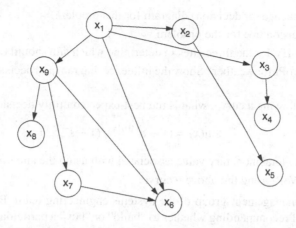

14.5 Create a relevant diagram that is consistent with the following joint probability distribution:

$$p(x_1, x_2, x_3, x_4, x_5, x_6, x_7, x_8 \mid \&) = p(x_8 \mid x_7, x_5, \&)\, p(x_7 \mid x_6, x_5, x_4, \&)\, p(x_6 \mid x_3, \&)$$

$$\times\, p(x_5 \mid x_2, x_1, \&)\, p(x_4 \mid x_3, x_1, \&)\, p(x_3 \mid \&)\, p(x_2 \mid \&)\, p(x_1 \mid \&)$$

14.6 You have been tasked with providing a recommendation for a test site at which an acceptance test will be conducted. There are three possible test sites (A, B, and C). Site A is the preferred site during good weather. Site C is the least preferred. Unfortunately, there is a long-range weather forecast for three months from now when the test needs to be conducted. The weather forecasters described the possibilities for weather as "good," "fair," and "poor." These possibilities have been defined very carefully, and their forecast for the time period of the test is: 0.3 for good, 0.6 for fair, and 0.1 for poor.

You have tried to find a way to reserve site A for a long enough period of time that the weather will certainly be good. However, site A is used by many people, and management has determined that the project cannot afford to rent site A for this extended time period. The cost at which the sites can be reserved for the time period in question is: $1000 for site A, $700 for site B, and $400 for site C.

Usage of each of these sites has varying positives and negatives for being able to analyze the results and recommend that the system be accepted. You have queried your colleagues to determine how much they would be willing to pay to change a specific site in the different weather conditions to the preferred site A and weather conditions. These relative dollar values do not include the cost of renting the site for the needed time period. The relative dollar value equivalents for sites and weather conditions are shown below:

	Weather Is Good	Weather Is Fair	Weather Is Poor
Site A	$1000	$200	$0
Site B	$950	$300	$200
Site C	$500	$450	$300

That is, site A in good weather is worth $1000 more dollars in terms of test performance than it is in poor weather. Similarly, site A in good weather is worth $500 more than site C in good weather.

a. Draw the influence (or decision) diagram for this problem.

b. Draw the decision tree for the problem.

c. Compute the EV for the three sites to determine which site should be recommended.

d. What is the VoPI for weather? Show the influence diagram and decision tree for computing the VoPI.

e. Using the following u curve, what is the best-expected utility decision?

$$u(x) = (1 - e^{-0.01x})/(1 - e^{-0.01})$$

where x is the total monetary value associated with using the site in question.

f. What is the VoPI using the above u curve.

14.7 As part of the management group of the systems engineering team, Bill D. Orby has been given the task of recommending whether to "build" or "buy" a particular component. Bill has called several manufacturers of this component and found the best "buy" alternative will cost $200,000 for the quantity needed. The performance of this component that is available from outside is categorized as moderate; this categorization includes many performance parameters and is rather coarse, but Bill hopes sufficient for an initial analysis.

Next, Bill spent significant time talking to several design engineers within his company who would be given the task of building this component, and several others who have built similar components in the past. There is uncertainty concerning both the cost and ultimate performance of this component if it is built by Bill's organization. Bill has modeled the uncertainty about total cost for developing and building the total quantity of the component as follows:

Build Cost	Probability
$100,000	0.2
$200,000	0.6
$400,000	0.2

The performance of the built component expected by the engineers with whom Bill spoke is substantially greater than the performance to be provided by the bought component. Bill has devised three performance categories to describe the uncertainty surrounding the built component: low, moderate, and high. The assessed probabilities of these performance outcomes, which are independent of the cost uncertainty, are

Build Performance	Probability
Low	0.2
Moderate	0.3
High	0.5

The last issue that must be addressed is the combination of costs and performance, including the difference between spending money outside the organization for the component versus spending the money inside the organization. You have found that management can think of an "equivalent purchase price" for the nine possible combinations of outcomes associated with building the component. The following table provides this equivalent purchase price. [Note that (1) negative numbers are equivalent to receiving money and (2) the cost of building the component has been included in the values in the table.]

Table of Equivalent Outside Purchase Price as a Function of "Built Performance" and "Built Cost"

Built Performance	Built Cost inside the organization		
	$100,000	$200,000	$400,000
High	−300,000	−200,000	0
Moderate	0	100,000	300,000
Low	100,000	200,000	400,000

Note that management prefers to build the component inside because the $200,000 build cost with moderate performance is equivalent to spending $100,000 outside. Assume that management's value function on "Outside Purchase Price" is a linear function with coefficient of −1.

a. Draw an influence diagram for this problem.
b. What is the best expected value decision?
c. What is the expected VoPI for built performance? for built cost? and for the combination of built performance and built cost? Show the influence diagram for each of these perfect information calculations.

14.8 Consider Problem 14.6. The first paragraph holds except we will drop the fair-weather condition. The probability of good weather is 0.3; the probability of poor weather is 0.7.

We are now going to enhance this model to address the need to test our system under a specified test condition. The weather affects the ability of each site to provide the necessary elements (e.g., terrain, visibility) that define the test condition. Our test experts visit each site and return with probabilities that each site can do a "good" versus "poor" job of reproducing the needed test condition. Assume that we have definitions of good and poor that meet the clairvoyant's test. (Note we could have defined more than two categories if we felt we needed to achieve more accuracy.)

Site	Weather	p (test condition good| site <row entry>, Weather <row entry>, &)	p (test condition poor| site <row entry>, Weather <row entry>, &)
A	Good	1.0	0.0
A	Poor	0.5	0.5
B	Good	0.9	0.1
B	Poor	0.5	0.5
C	Good	0.7	0.3
C	Poor	0.2	0.8

The test engineers have determined that they would be willing to pay $10,000 to move from a test site providing a poor version of the test condition to a test site providing a good version of the test condition.

Which site should we choose? Remember the rental cost of each site. What is the VoPI on the weather?

14.9   Now we are going to take Problem 14.8 and increase the modeling complexity by defining three different test conditions that must be reproduced by the test site. We call these test conditions *X*, *Y*, and *Z*. We first generate descriptions of "good" and "poor" for each test condition. Then we ask the wizard to help us elicit the values for having good versus poor representations of the three test conditions. We respond that having a poor representation of each test condition is worth no money to us. Test condition X is the most important for obtaining a good representation and we would pay $10,000. Similarly, we would pay $5000 to obtain a good representation of Y and $1667 to obtain a good representation of Z.

If we were using multiattribute value theory, what would our swing weights be for these three test conditions?

Site	Weather	p(test condition X is good| site <row entry>, Weather <row entry>, &)	p(test condition Y is good| site <row entry>, Weather <row entry>, &)	p(test condition Z is good| site <row entry>, Weather <row entry>, &)
A	Good	1.0	1.0	1.0
A	Poor	0.5	0.5	0.5
B	Good	0.9	0.9	0.6
B	Poor	0.5	0.5	0.5
C	Good	0.7	0.5	0.7
C	Poor	0.2	0.2	0.2

Which site should we choose? Remember the rental cost of each site.
What is the VoPI on the weather?

14.10   Returning to Problem 14.6, there is another way in which we could have expanded the analysis from this point. In fact, the systems engineers and stakeholders have to determine whether the system is acceptable after these tests are over and the test results are in; that is, they have to make a decision. In addition, going into the test, they are not sure whether the system has acceptable performance for the stakeholders. If the system does, and it is accepted, then there should be relatively few and inexpensive fixes needed relative to the case where the system's performance is unacceptable, but the decision is made to accept the system.

So we have two decision nodes: which test site to choose and whether to accept the system for use by the stakeholders.

The weather has two states and associated probabilities as in Problem 14.8.

The ability of the three sites to reproduce good versus poor test conditions in the weather conditions is as it was in Problem 14.8.

Now we must introduce our prior probabilities on the acceptability of the system's performance. Suppose we start with only two possibilities (acceptable and unacceptable) with probabilities of 0.8 and 0.2.

We must also introduce our uncertainty that the test will say the system is "acceptable." This uncertainty is dependent on the system's actual performance and our ability to reproduce the test condition. The table below describes this probability distribution.

Actual System Performance	Ability to Reproduce the Test Condition	p (test says accept|system is <1st col. entry>, test condition is <2nd col. entry>, &)
Acceptable	Good	0.95
Acceptable	Poor	0.60
Unacceptable	Good	0.10
Unacceptable	Poor	0.25

The quality engineers are called in to help us determine what the relative value of accepting a system is given it is or is not acceptable, over the lifetime of the system. These engineers conduct an analysis over the 10-year lifetime of our system and present the net present value (NPV) to our organization for the following conditions:

Actual System Performance	Decision to Accept or Not	Justification for Last Column	NPV over System Life Time
Acceptable	Accept	Best profit	$100,000
Acceptable	Do not accept	Make some unneeded fixes	$80,000
Unacceptable	Accept	Have many repairs under warranty, damage reputation	−$10,000
Unacceptable	Do not accept	Make needed fixes, delay hurts sales	$20,000

Which site should we choose? Remember the rental cost of each site.
What is the expected VoPI on the weather?

Chapter 15

The Science and Analysis of Systems

15.1 INTRODUCTION

This chapter addresses the scientific underpinnings of systems engineering and the criticality of systems analytics to achieving successful outcomes, with emphasis on the pervasiveness of uncertainty throughout. The science of systems covers general system theory, systems science, natural systems, cybernetics, and systems thinking. System analytics covers the quantitative characterization of systems, system dynamics, constraint theory, and approximate methods. Systems analytics covers nonlinear as well as linear systems, incomplete as well as complete information, and probabilistic as well as deterministic conditions. One purpose of this chapter is to demonstrate that there is a solid basis for systems engineering as an engineering discipline in its own right. The second purpose is to demonstrate the criticality of analytics in systems engineering to achieve successful outcomes. The performance, quality, and suitability requirements introduced in Chapter 6 are validated by the system analytics, just as the derived functional requirements are validated from the functional architecture in Chapters 6 and 7 from the functional model, and the operational architecture allocating functions to physical components and prescribing behavior are described in Chapter 10. Table 6.3 states that requirements are individually verifiable, i.e., a finite, cost-effective process can be defined to check that the requirement has been attained and the set of requirements must be attainable, i.e., solutions exist within performance, cost, and schedule constraints.

The focus of this chapter is on the practical application of science and analytics in the engineering design of systems. General systems theory is covered in Section 15.2, followed by systems science (Section 15.3), natural systems (Section 15.4), cybernetics (Section 15.5), and systems thinking (Section 15.6). The quantitative characterization of systems is covered in Section 15.7 with several examples to illustrate the application: the elevator case study used throughout the text, the soda machine introduced in Chapter 12, and an aircraft. Systems dynamics is discussed in Section 15.8, constraint theory in Section 15.9, and Fermi problems and guesstimation in Section 15.10. Artificial (or augmented) intelligence and machine learning are discussed in Section 15.11. The interested reader is encouraged to pursue one or more systems analysis topics via references for these topics.

The Engineering Design of Systems: Models and Methods, Fourth Edition. Dennis M. Buede and William D. Miller
© 2024 John Wiley & Sons, Inc. Published 2024 by John Wiley & Sons, Inc.
Companion website: www.wiley.com/go/engineeringdesignofsystems4e

15.2 GENERAL SYSTEM THEORY

Ludwig von Bertanlanffy [1968] noted that the underlying concept of systems emerged from philosophy, quoting Aristotle: "The whole is more than the sum of the parts." Von Bertanlanffy further lays out the foundations of general system theory as a basic science, its development, and applications in human-designed (engineered), biological, and social systems. He finds the successful realization, operation, and support of systems necessitates the *systems approach*, integrating multiple heterogeneous disciplines, thereby driving the need for *scientific generalists* to perform the systems function, integrating the parts into the whole greater than the sum.

Central concepts in general systems theory are feedback, control, stability, and probability. Chapter 7: Functional Architecture Development introduces the concept of feedback and control. In particular, review Figure 7.5. The purpose of feedback is to drive the system to a desired goal and ensure the stability of the system. Chapter 14: Decision Analysis for Design Trades introduces probability and is used in the science and analysis of systems to accommodate the real-life phenomena of uncertainty.

Systems are modeled mathematically in different ways. A commonly used approach is a set of simultaneous, perhaps nonlinear, perhaps partial, difference/differential equations where there is some measure of n elements x_i, $i = 1$–n and functions f_i over time t.

$$dx_i/dt = f_i(x_1, x_2, \ldots, x_n) \tag{15.1}$$

Applications found in engineered, biological, and social systems include *growth*, *competition* between parts, *behavior*, and *finality*. Wymore [1967] provides examples of systems and elements of systems described by sets of equations. A relatively simple example is the Wright Flyer that decomposed the problem of flight into the functions of lift, propulsion, and control whose behavior is described individually and collectively by a set of equations.

A growth equation for a system of a single variable x can be as simple as

$$dx/dt = f(x) \tag{15.2}$$

Expanding $f(x)$ as a Taylor series yields

$$dx/dt = a_1 x + a_2 x^2 + a_3 x^3 + \ldots \tag{15.3}$$

Making an engineering decision that the dominant term is the first term in the Taylor series expansion results in an approximation

$$dx/dt \sim a_1 x \tag{15.4}$$

that has the solution of exponential growth or decay depending on whether a_1 is a positive or negative number

$$x(t) \sim x_0 e^{a_1 t} \tag{15.5}$$

For a_1 a positive number, exponential growth means that $x(t)$ grows without bound as time t increases. For a_1 a negative number, exponential decay means that $x(t)$ goes to 0 as time t increases.

Making an engineering decision that the dominant terms are the first two terms in the Taylor series expansion results in an approximation

$$dx/dt \sim a_1 x + a_2 x^2 \tag{15.6}$$

with the solution of the well-known logistic curve that has finite limiting values.

$$x(t) \sim x_0 e^{a_1 t}/(1 - x_0 e^{a_2 t}) \tag{15.7}$$

An example of competition in an exponential growth or decay system of two players is described by the set of equations

$$dx_1/dt = a_1 x_1 \tag{15.8a}$$
$$dx_2/dt = a_2 x_2 \tag{15.8b}$$

having solutions

$$x_1(t) = b_1 e^{a_1 t} \tag{15.9a}$$
$$x_2(t) = b_2 e^{a_2 t} \tag{15.9b}$$

It can be proved that $x_1(t)$ and $x_2(t)$ are interrelated with one measure having a solution expressed as a power function of another solution

$$x_1(t) = \beta x_2(t)^\alpha \tag{15.10}$$

where

$$\alpha = a_1/a_2 \text{ and } \beta = b_1/b_2^\alpha \tag{15.11}$$

The competitive system can be extended from the simple exponential growth system of two players to more complicated cases such as parabolic growth and the logistic function.

15.3   SYSTEMS SCIENCE

Von Bertanlanffy correlates general system theory with applied science under the general name of systems science, as described by Russell Ackoff [1960]. John Warfield [2006] and Derek Hitchins [2007] characterize systems science as the science of complex systems and the science of wholes. The emphasis is on how systems behave as a whole without having to explain the behaviors of the individual parts of the system. Behavior is the response of the system to some stimulus, including that of the system's environment. Hitchins characterizes behavior mathematically in terms of physical conservation laws, queuing phenomena, and chaotic phenomena. Of special interest are emergent behaviors that are not predicted based on the integration of the behaviors of the individual parts of the system.

Not having to perform reductionism to explain system behavior is an advantage in the engineering design of systems early in their concept formulation and *roughing out* before committing to development. High-level characterizations to estimate behavior are useful to establish feasibility early on, thereby reducing technical risk.

15.4   NATURAL SYSTEMS

Natural systems constitute living and non-living systems found in nature. Von Bertanlanffy in General Systems Theory extends the discussion on competition introduced in Section 15.2. In biology, Eq. (15.10) is known as the *allometric* equation [von Bertanlanffy, 1968]. The size of an organ in an animal or human is a function of another animal or human characteristic, e.g., basal metabolism is a function of body weight. Additional examples of proportions in animals and humans are mathematically described in Bender [1978]. The balance of proportions in animals and humans has its analogs in engineered systems, e.g., the balance of weight, power for propulsion, carrying capacity, fuel, range, and acceleration for transportation systems.

Structures found in nature are reverse-engineered and applied to the engineering design of systems. One example is the design of Japan's bullet train. The nose of the bullet train mimics the beak of the Kingfisher bird to mitigate severe vibrations and pressure changes on passengers when the trains enter tunnels and mitigate loud *tunnel booms* as the trains exit tunnels [INCOSE, 2013].

15.5 CYBERNETICS

Norbert Wiener [1961] coined cybernetics, from the Greek *steersman*, to explain the regulation, and self-regulation, using compensators and closed-loop feedback servomechanisms to control and stabilize biological, social, and engineered systems in the face of uncertainties characterized by varieties of allowed inputs as well as disturbances and noise. Cybernetics integrates the applied mathematics of control theory and information theory to achieve the desired regulation, even for complex systems with complex uncertainties. Information theory, created by Claude Shannon [1948], characterizes the variety of allowed inputs to the system as well as the disturbances and noises using the measure of entropy, in bits.

Cybernetic systems are mathematically described as having time-varying inputs $x(t)$ and outputs $y(t)$ related by transfer function $g(t, x(t))$ characterized by integral-differential equations. The time-varying inputs, transfer function, and outputs are quite complicated in the time domain and are represented as differential equations. Fortunately, such systems can be more easily analyzed as algebraic equations in the frequency domain using Laplace transforms, for example, to find solutions and to assess and assure the stability and controllability of systems. The Laplace transform is a particular form of a generalized linear integral function.

The Laplace transform $L\{f(t)\}$ for a function $f(t)$ with s the complex variable defined as $s = \sigma + j\omega$ is

$$L\{f(t)\} = \int_0^\infty f(t)e^{-st}dt = F(s) \tag{15.12}$$

The inverse Laplace transform of $F(s)$ for $j = \sqrt{-1}$ and c a complex constant is the linear integral function

$$L^{-1}\{F(s)\} = \frac{1}{2\pi j} \int_{c-j\infty}^{c+j\infty} F(s)e^{st}ds = f(t) \tag{15.13}$$

The open-loop system in the frequency domain has response

$$Y(s) = G(s)X(s) \tag{15.14}$$

The inputs can be further characterized by reference inputs $r(t)$ and disturbances $d(t)$

$$x(t) = r(t) + d(t) \tag{15.15}$$

By augmenting the system with compensator, also called controller or regulator, having transformation $C(s)$ in the frequency domain and feedback servomechanism having transformation $H(s)$, we have the regulated system in Figure 15.1

$$Y(s) = \left[\frac{C(s)G(s)}{1 + H(s)C(s)G(s)} \right] X(s) \tag{15.16}$$

with the closed-loop transfer function

$$\frac{Ys}{X(s)} = \frac{C(s)G(s)}{1 + H(s)C(s)G(s)} \tag{15.17}$$

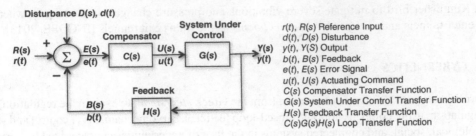

FIGURE 15.1 Generic feedback control system block diagram.

where compensator and feedback servomechanism are engineered to achieve the desired regulation. An example is the simple analog thermostat regulating heating and cooling. The system control drives the heating and cooling to a user-specified set point of desired temperature.

The disturbances $d(t)$ are unintended inputs to the system attributable to the environment or to intentional or unintentional interferences, e.g., background noise or deliberate jamming in a radio communications system. Disturbances can be characterized in terms of the entropy measured as information bits E

$$E = \sum_{p_i} \log_2(p_i) \tag{15.18}$$

where p_i is the probability of an event and,

$$\sum p_i = 1 \tag{15.19}$$

and

$$\log_2(p_i) = \log_2 10 \log_{10}(p_i) = 3.32 \log_{10}(p_i) \tag{15.20}$$

Extending the control of an aircraft to compensate for random disturbances, e.g., wind gusts, one can engineer the performance of the control subsystem based on the distribution of disturbances. Paraskevopoulos [2002] describes an automatic flight control system, approximating the aircraft engine and aircraft dynamics as second-order systems and the aircraft actuator as a first-order system for reference input commands $r(t)$, disturbances $d(t)$ and output $y(t)$ can be shown for feedback transfer function $H(s) = 1$ to have the closed-loop transfer function

$$\frac{Y(s)}{R(s)} = \frac{G_1(s)G_2(s)G_a(s)\left(1 + \frac{D(s)}{G_1(s)G_2(s)}\right)}{[1 + G_1(s)G_2(s)G_a(s)]} \tag{15.21}$$

where the Laplace transforms for the approximate model engine, actuator, and aircraft dynamics are

$$G_1(s) = \frac{1}{s(s + 6)} \tag{15.22}$$

$$G_2(s) = \frac{1}{s + 8} \tag{15.23}$$

$$G_a(s) = \frac{K(s + 4)}{s^2 + 2s + 2} \tag{15.24}$$

Parameter $K > 0$ changes during flight conditions, e.g., takeoff, landing, steady flight, etc. Paraskevopoulos shows that the closed-loop system in Figure 15.2 is stable, assuming no disturbances, $D(s) = 0$.

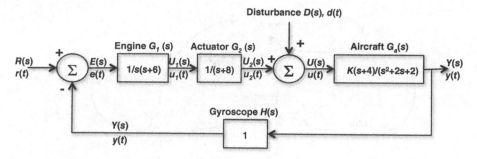

FIGURE 15.2 Automatic flight control system simplified block diagram.

The time-domain input/output response of the closed-loop system can be determined from the inverse Laplace transform of the transforms of the engine, actuator, and aircraft transfer functions, reference input, and disturbances

$$y(t) = L^{-1} \left\{ \frac{G_1(s)G_2(s)G_a(s)\left(1 + \frac{D(s)}{G_1(s)G_2(s)}\right)}{[1 + G_1(s)G_2(s)G_a(s)]} R(s) \right\} \tag{15.25}$$

Again, the transfer function $H(s)$ for the feedback loop does not show in the inverse Laplace transform since $H(s) = 1$ in this example.

Time-domain methods analyze transient and steady-state responses, including rise times, overshoots, settling times, and steady-state errors, and are often formed as state equations [Kuo and Golnaraghi, 2003].

Another approach for the time-domain analysis is to start with the time-domain functions of the engine, actuator, aircraft, and gyroscope rather than their Laplace transforms. The time-domain functions of the engine $g_1(t)$, actuator $g_2(t)$, and aircraft $g_a(t)$ are

$$g_1(t) = \frac{1 - e^{-6t}}{6} \tag{15.26}$$

$$g_2(t) = e^{-8t} \tag{15.27}$$

$$g_a(t) = Ke^{-t}(3\sin(t) + \cos(t)) \tag{15.28}$$

The outputs of the functions are then found from convoluting the functions and their inputs for $e(t) = r(t) - y(t)$ and $u(t) = u_2(t) + d(t)$

$$u_1(t) = \int_0^t e(\tau)g_1(t - \tau)d\tau \tag{15.29}$$

$$u_2(t) = \int_0^t u_1(\tau)g_2(t - \tau)d\tau \tag{15.30}$$

$$y(t) = \int_0^t (u(\tau) + d(\tau))g_a(t - \tau)d\tau \tag{15.31}$$

15.6 SYSTEMS THINKING

Systems thinking [Kauffman, 1980; Senge, 1990; Checkland, 1993; Boardman and Sauser, 2008] provides a holistic look at the *whole* of systems, interrelationships with outside forces (Fig. 2.1), and the properties of systems, especially emergence and stability. Senge enumerates a set of archetypes

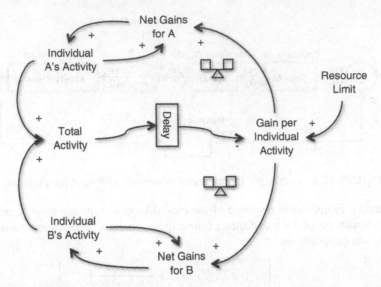

FIGURE 15.3   Systems thinking archetype of tragedy of the commons.

structured as causal loop diagrams to characterize system behaviors: *balancing process with delay, limits to growth, shifting the burden, eroding goals, escalation, success to the successful, tragedy of the commons, fixes that fail, and growth and underinvestment.* Figure 15.3 is an example causal loop diagram for the *tragedy of the commons* archetype, where multiple activities place demands on a common resource. When unregulated, the resource becomes depleted over time. This particular archetype is relevant to resource availability during system development and system operation. Causal loop diagrams, static diagrams that describe dynamic behavior over time, convey a qualitative sense of behavior in terms of the inherent stability or instability of a system. Causal loop diagrams can be the basis for quantitative modeling to assess both spatial and temporal performance, e.g., predicting the time interval to depletion of resources in a *tragedy of the commons* archetype.

Checkland details a systems taxonomy and distinguishes hard systems thinking as an engineering contribution contrasted with human-centered *soft* systems thinking. Boardman and Sauser integrate their predecessors' contributions and introduce the *systemigram* as a conceptual model to show the relationships and influences in systems and systems of systems. The current practice of systemigram creation is inherently qualitative and informal, with no built-in checking of the correctness and integrity of the model. Systemigrams can be converted into executable Petri net models to perform behavioral and what-if analysis [Sagoo and Boardman, 1998]. Figure 15.4 is a systemigram of the elevator system context and can be compared to the information in the IDEF0 diagram shown in Figure 6.8. The systemigram denotes circles/ovals labeled as *things*, unlike the functions in IDEF0 diagrams, whereas the labels on the arrows can be verbs, nouns, or both. A good practice for systemigrams would be to label the arrows as either verbs or nouns, but not both, to avoid confusion in translating to an executable model. The systemigram in Figure 15.4 specifically highlights the initial operational concept scenario from Table 6.4 for an elevator moving passengers between floors. The relations between *Passengers* and the *Elevator System* invoked in the scenario are sequentially numbered from 1 to 4.

15.7 QUANTITATIVE CHARACTERIZATION OF SYSTEMS

This section illustrates the quantitative characterization of systems using the elevator case study used throughout the text, the soda machine introduced in Chapter 12, and an aircraft. Although

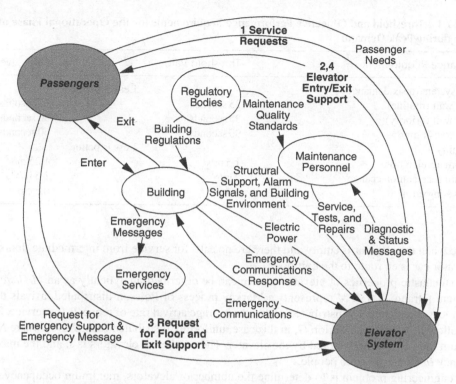

FIGURE 15.4 Systemigram of an elevator system context.

all are complex systems operating in a context of uncertainty, the treatment here is simplified to emphasize key concepts using approximate models to assess the feasibility of meeting performance requirements.

15.7.1 Elevator

Elevators in the context of vertically moving people and cargo exhibit stochastic, complex behaviors. The topic of control systems is of interest in terms of the algorithms to dispatch elevator cars driven by stochastic demands for service and is left to other treatments. Rather, the focus here is on the feasibility of meeting the stated performance requirements.

For the elevator case study, Figure 2.4 shows the fundamental objectives hierarchy for the operational phase of elevator. Performance requirements include threshold and objective values for peak demand *time in system* (*average waiting times* for *routine* and *priority* service, as well as *the average transit time*) and *ride quality* (*maximum acceleration, maximum acceleration change,* and *floor leveling error*). See Table 15.1. The performance requirements in Figure 2.4 also include *availability* (*operational MTBF* and *operational MTTR*), which is not addressed here.

The performance requirements in Table 15.1 are interdependent. Average wait times and transit times are dependent on the accelerations and something not shown in Table 15.1, the maximum velocity of the elevator. Velocity, then, is a derived performance requirement. Note that the constraints on acceleration and change in acceleration bound the maximum number of building floors. Also, average wait and transit times are critically dependent on the number of elevator cars in service and stochastic demands for service. Stochastic demands increase average wait and transit times compared to the low probability during peak demand periods that only a single call for service will

TABLE 15.1   Threshold and Objective Performance Requirements for the Operational Phase of Elevator during Peak Demand

Performance Requirement	Threshold Value	Objective Value
Time in system (peak demand)	Less is better	
Average wait (routine)	35 seconds	27 seconds
Average wait (priority)	35 seconds	30 seconds
Average transit time	90 seconds	60 seconds
Ride quality	Less is better	
Maximum acceleration	1.5 m/s^2	1.25 m/s^2
Maximum acceleration change	2 m/s^3	1.5 m/s^3
Floor leveling error	0.7 cm	0.3 cm

be made when the elevator is empty and there are no calls for service from intermediate floors while the elevator car is en route to the exit floor.

The stochastic properties of elevator service can be described analytically as an $M/G/s/c$ queuing system [Braun, 2003]. M represents a Markov process of Poisson-distributed arrivals that are independent and identically distributed with λ the average arrival rate of requests for service. Transit times follow a general distribution G, in this case, uniformly distributed from *1 to N* where N is the maximum number of floors served by the elevator. The number of elevators is s, and the maximum occupancy of an elevator is c people.

The engineering problem is to determine the number of elevators, maximum occupancy of elevator cars, and uninterrupted transit times between any two floors to assure the feasibility of the performance requirements early in design and to then manage changes to the values to assure feasibility throughout development and qualification. Table 15.2 shows the external inputs, external outputs, and engineered values for the elevator system. The arrival of people seeking service on any floor and the floors from *1* to *N* selected by these people are nondeterministic external inputs to the system. The number of elevators s, the occupancy c of an elevator, and the transit time between levels subject to the ride quality acceleration and floor leveling error requirements are engineered values of the system. The external outputs of the system are the individual waiting times and individual transit times, characterized as average wait time and average transit time. From this analytic model, the number of elevators s, elevator maximum occupancy c, average wait time, and average transit time can be determined from the nondeterministic external inputs to ensure the feasibility of performance requirements in Table 15.1.

There are two simple approaches for operating the elevator between uninterrupted entry and exit floors. One approach is to accelerate the elevator from a stop to a constant velocity and then decelerate the elevator to the next stop, which could be an exit stop, another entry stop, or both. The second

TABLE 15.2   External Inputs, External Outputs, and Engineered Values for the Operational Phase of Elevator

External Inputs	External Outputs	Engineered Values
Individuals seeking service (Poisson distribution λ)	Individual wait times (Average wait time)	Number elevators s Max elevator occupancy c
Floor selection (Uniform distribution 1 to N)	Individual transit times (Average transit times)	Transit time between floors (1 floor to $N - 1$ floor)

approach is to accelerate the elevator to the mid-point between stops and then decelerate beginning at the mid-point. The first approach allows for the possibility of the next stop being scheduled by a call for service at an intermediate floor, and less so for the second approach. For simplification, the first approach is selected with one exception discussed below. The constant velocity approach is modeled piece-wise. The set of equations to determine transit time for acceleration a, constant velocity v and distance id between stops, i the number of floors and d the height of a floor, ignoring inertia is approximately

$$t = t_{\text{accelerate}} + t_{\text{constant}} + t_{\text{decelerate}} \tag{15.32}$$

$$id = d_{\text{accelerate}} + d_{\text{constant}} + d_{\text{decelerate}} \tag{15.33}$$

$$t_{\text{accelerate}} = \frac{v - v_0}{\alpha} = \frac{v}{\alpha} \quad \text{for } v_0 = 0 \text{ and } d_{\text{accelerate}} = \frac{1}{2}at^2_{\text{accelerate}} \tag{15.34}$$

$$t_{\text{constant}} = \frac{d_{\text{constant}}}{v} \quad \text{and } d_{\text{constant}} = id - \frac{1}{2}at^2_{\text{accelerate}} - \frac{1}{2}a \tag{15.35}$$

$$t_{\text{decelerate}} = \frac{v - v_0}{a} = \frac{v}{a} \quad \text{for } v_0 = 0 \text{ and } d_{\text{decelerate}} = \frac{1}{2}at^2_{\text{decelerate}} \tag{15.36}$$

An exception to the determination of time t occurs when the time to accelerate to constant velocity v is greater than the time to transit to a stop, typically when the exit floor is one floor above or below the entry floor. In this case, the second approach is used to accelerate the elevator to the mid-point between stops and then decelerate beginning at the mid-point. In this case, time t is

$$t = t_{\text{accelerate}} + t_{\text{decelerate}} \tag{15.37}$$

$$d = d_{\text{accelerate}} + d_{\text{decelerate}} \tag{15.38}$$

$$d_{\text{accelerate}} = \frac{1}{2}d = \frac{1}{2}at^2_{\text{accelerate}} \tag{15.39}$$

$$d_{\text{decelerate}} = \frac{1}{2}d = \frac{1}{2}at^2_{\text{decelerate}} \tag{15.40}$$

$$t_{\text{accelerate}} = \left(\frac{d}{a}\right)^{0.5} \tag{15.41}$$

$$t_{\text{decelerate}} = \left(\frac{d}{a}\right)^{0.5} \tag{15.42}$$

From Table 15.1, Threshold and Objective Performance Requirements for the Operational Phase of Elevator, the maximum velocity is derivable from the requirements for the maximum acceleration and the maximum change in acceleration. Note when the elevator is stationary, the acceleration a_0 is 0. Thus, the maximum change in acceleration is the difference between the maximum acceleration a and a_0 divided by the time to accelerate $t_{\text{accelerate}}$:

$$\Delta a = \frac{a - a_0}{\frac{v}{a}} = \frac{a^2}{v} \quad \text{for } a_0 = 0 \tag{15.43}$$

Thus,

$$v = a^2/\Delta a \tag{15.44}$$

The derived values for maximum velocity of the elevator from the thresholds and objectives for maximum acceleration and maximum change in acceleration are given in Table 15.3.

The derived maximum velocity is over the range of 0.781–1.5 m/s.

TABLE 15.3   Derived Values for Maximum Velocity of the Elevator in m/s

	Maximum Change in Acceleration	
Maximum Acceleration	Threshold 2 m/s^3	Objective Value 1.5 m/s^3
Threshold 1.5 m/s^2	1.125 m/s	1.5 m/s
Objective 1.25 m/s^2	0.781 m/s	1.042 m/s

The average transit time $E(T)$ is determined from the assumed uniform distribution of requests for floors from 1 to N and the transit time between any two floors:

$$E(T) = \left(\frac{1}{N-1}\right) \sum t_i \quad \text{for } i = 1 \text{ to } N - 1 \tag{15.45}$$

The maximum number of floors that can be supported with an average transit time of *90 seconds* (threshold) and *60 seconds* (objective ranges) for the constraints on acceleration, maximum change in acceleration, and derived maximum velocity can be determined. Using *4* m as the height of a floor in a building built for business use, the maximum numbers of building floors constrained by the maximum acceleration, maximum change in acceleration, derived maximum velocity, opening/closing time of elevator doors, and time for passengers to enter/exit the elevator car are shown in Tables 15.4 and 15.5. The threshold performance requirements for maximum acceleration, maximum change in acceleration, and average transit time can be met for a building of 33 floors.

The peak period average waiting time $E(W)$ is approximated from the interval time $E(t_{0i})$ for an elevator to arrive at the lobby level and queue size q to arrive at the lobby level Barney [2003] by the equations

$$E(W) \sim 0.4E(t_{0i}) \quad \text{elevator car occupancy} < 0.5 \tag{15.46}$$

$$E(W) \sim \left[0.4 + \frac{1.8q}{c} - 0.77^2\right] E(t_{0i}) \quad \text{elevator car occupancy} \geq 0.5 \tag{15.47}$$

TABLE 15.4   Maximum Number of Building Floors to Meet Performance Constraints for Threshold Average Transit Time $E(T)$ 90 Seconds

	Maximum Change in Acceleration	
Maximum Acceleration	Threshold 2 m/s^3	Objective Value 1.5 m/s^3
Threshold 1.5 m/s^2	50 floors	66 floors
Objective 1.25 m/s^2	34 floors	46 floors

TABLE 15.5   Maximum Number of Building Floors to Meet Performance Constraints for Objective Average Transit Time $E(T)$ 60 Seconds

	Maximum Change in Acceleration	
Maximum Acceleration	Threshold 2 m/s^3	Objective Value 1.5 m/s^3
Threshold 1.5 m/s^2	33 floors	44 floors
Objective 1.25 m/s^2	23 floors	30 floors

where the interval time is a function of traffic demand, number of elevator cars s, elevator car maximum occupancy c.

The average wait time $E(W)$ is sufficiently complex to warrant a Monte Carlo simulation to engineer the appropriate number of elevator cars s, elevator car capacity c, and transit times between any two floors subject to acceleration and floor leveling constraints.

SIDEBAR 15.1: ELEVATOR TRAFFIC ANALYSIS AND SIZING

The elevator industry, working with the building construction industry and facilities operations managers, uses standardized tables and graphs with built-in assumptions and rules of thumb for sizing the elevators in a building to determine the number of elevator cars s, maximum occupancy c *people*, and car speed v m/s. The initial givens are the building population P, the number of floors N, and the floor height h. The sizing is determined to meet throughput and waiting requirements for peak periods of usage. These peak periods are different for office buildings, hotels, residence buildings, and schools. For example, office building peak periods are driven by morning arrivals on the building entrance, or lobby, floor.

For an office building, the heuristically derived throughput requirement is in the range of 10–20% of the building population in a 5-minute period and assumes that floor selection from the second to the Nth floor above the lobby level is uniformly distributed for the morning peak period. A typical floor height is approximately *4* m, excepting the lobby level, which is assumed to be twice the height of the floors above.

Also, for an office building, the average waiting time requirement for an elevator drives the specification of the (average) interval time $E(t_{0i})$ between elevators opening on the lobby level. Average interval time is assumed to be 0.6 of the average waiting time.

The fill factor for the elevator in the morning peak period is assumed to be *0.8*. For example, a standardized elevator car with a rated capacity of *2500 lbs.* or *17 people* on average then, would carry *13 people*.

The average trip time is determined from a graph with number of floors above the lobby level and curves with standardized elevator motor speeds s *fpm*. The average round trip time to the lobby level is determined from a similar graph with number of floors above the lobby level and curves with standardized elevator motor speeds s *fpm*. The times on these graphs have built-in assumptions on elevator door opening/closing times, passenger entry/exit times, and acceleration/constant velocity/deceleration times. The average trip time is used to verify the average transit time across the ensemble of floors, and the average round trip time verifies the interval time.

SIDEBAR 15.2: FLAW OF AVERAGES

The psychological irritation of humans in waiting for service increases in proportion to the square of waiting and task completion times. The tails of distributions are preferable to means or averages for setting performance requirements reflecting desired user experiences. As an example, half of the occurrences are above the mean value for symmetric distributions, e.g., uniform and *Gaussian* or *Normal* distributions. Many distributions that occur in nature and in human-engineered systems have asymmetric distributions with longer tails to the right of the mean, resulting in more than half the occurrences being greater than the mean. The use of averages of the parameter set to recast systems with uncertainties present as simpler deterministic systems leads to actual system performances worse than the expected value of the solution. Savage [2009] has a compelling and entertaining argument for using the tails of distributions and not averages for setting performance requirements.

15.7.2 Soda Machine

For the soda machine, Figure 12.4, performance requirements include threshold and objective values for *mean time between supplies*, *probability correct change*, and *probability correct soda*. Figure 12.4 also includes *availability* as a measure of quality. See Table 15.6.

The stochastic properties of the soda machine can be described analytically as an *M/G/1* queuing system [Cooper, 1991]. *M* again represents a Markov process of Poisson-distributed arrivals that are independent and identically distributed with λ the average arrival rate of requests for soda. The demands for individual soda flavors have Poisson-distributed arrivals that are independent and identically distributed with λ_i the average rate of requests for *i*th flavor of soda.

$$\lambda = \sum \lambda_i \quad \text{for } i = 1, \dots, m \tag{15.48}$$

Service times follow a general distribution *G*, identically distributed random variables independent of the arrival process and each other, with an average service time τ. The soda machine is a single server. Customers arrive when there are one or more customers ahead in order of appearance and wait until the soda machine is available to them. The offered load ρ_i to the system for soda flavor *i* and the overall offered load ρ for sodas to the system are

$$\rho_i = \lambda_i \tau_i \tag{15.49}$$

$$\rho = \lambda \tau \tag{15.50}$$

where $\tau = \tau_i$, i.e., there is no difference in service times for the different flavors of soda.

From queuing theory, for the system to be stable, $\rho < 1$. The average waiting time $E(W)$ for a *M/G/1* system is known to have a closed-form solution

$$E(W) = \left(\frac{\rho}{1-\rho}\right)\left(\frac{\tau}{2} + \frac{\sigma^2}{2\tau}\right) \tag{15.51}$$

where σ^2 is the variance of the service time distribution.

TABLE 15.6 Threshold and Objective Performance Requirements for the Operational Phase of the Soda Machine

Performance Requirement	Threshold Value	Objective Value
Time in system	More is better	
Mean time between supplies	12 hours	36 hours
Probability of correctness	More is better	
Probability correct change	0.999	0.999 99
Probability correct soda	0.99	0.999 9
Quality	More is better	
Availability	0.9	0.99

TABLE 15.7 External Inputs, External Outputs, and Engineered Values for the Operational Phase of the Soda Machine

External Inputs	External Outputs	Engineered Values
Queue at soda machine time t (Poisson distribution λ)	Individual wait time (Average wait time)	Number sodas type i
Select soda flavor i time $t + \delta_1$ (Poisson distribution λ_i)	Transaction time (Average service time) Soda type i	
Deposit money time $t + \delta_1 + \delta_2$ (General distribution currency and coins)	Change in coins (Coin distribution)	Initial amount of change to be dispensed

Table 15.7 shows the external inputs, external outputs and engineered values for the soda machine. The external input to the soda machine is a customer making the selection of soda and depositing money. The external outputs of the soda machine are changed if the customer enters more than the correct amount and the dispensed soda(s). The engineered values of the system are the number of different sodas stored upon resupply and the initial amount of change available to be dispensed.

The number of sodas of each flavor is n_i, and the total number of sodas upon resupply are

$$n = \sum n_i \quad \text{for } i = 1, \dots, m \text{ and } n < N \qquad (15.52)$$

where N is the maximum soda capacity of the machine.

There is a decision if the soda machine should be resupplied before any soda flavor is totally dispensed or rather before the machine is empty of all soda flavors. For this analysis, it is assumed that waiting times for service are negligible. In the first case, the average time $E(T_i)$ to dispense all sodas of any flavor is

$$E(T_i) = \frac{n_i}{\lambda_i} \quad \text{for } i = 1, \dots, m \qquad (15.53)$$

Then, the minimum average time to dispense the soda flavor most in demand is $\text{Min}[E(T_i)]$, which, per the requirement in Table 15.6, must be between the threshold and objective values. For the decision to resupply the soda machine before the soda machine is empty of all soda flavors is

$$E(T) = \frac{n}{\lambda} \qquad (15.54)$$

which, per the requirement in Table 15.6, must be between the threshold and objective values. A consequence of the decision to resupply before all sodas are dispensed is that customers may forgo purchasing a soda of another flavor, resulting in lost revenues.

There is a separate stochastic problem to determine the amount and types of coins placed in the soda machine for change upon resupply. The external inputs are the amounts of paper currency and coins of different denominations that customers enter into the machine, and that can be used to provide change for subsequent customers.

15.7.3 Aircraft

For an aircraft, one can imagine a set of performance requirements for payload, speed, altitude, range, endurance, and energy maneuverability. The Breguet range equation, Ruijgrok [2009], is an approximation for an aircraft in level flight at a given altitude and constant speed relating aircraft range R m to velocity V m/s, fuel efficiency c, initial weight W_i, final weight W_f, lift L, and drag D, where ln is the natural logarithm:

$$R = \left(\frac{V}{c}\right)\left(\frac{L}{D}\right)\ln\left(\frac{W_i}{W_f}\right) \tag{15.55}$$

Equations (15.55) and (15.56) are the formulas for jet aircraft whose propulsion is stated in terms of thrust. The approximation is for the cruise portion of a flight only and does not include takeoff, ascent to cruise altitude, descent for landing, and landing. Nor does it include evasive and aggressive maneuvering for combat aircraft. The final weight should include payload as well as reserve fuel for required safety margins. Fuel efficiency c is defined to be the ratio of fuel flow to thrust, e.g., fuel flow g/s divided by the propulsion thrust g. The air density for a given altitude only affects the ratio (V/c) in determining the range. Note also that the ratios L/D and W_i/W_f are dimensionless.

Similarly, the Breguet endurance equation to determine the maximum cruise time T_{max}s is

$$T_{max} = \left(\frac{1}{c}\right)\left(\frac{L}{D}\right)\ln\left(\frac{W_i}{W_f}\right) \tag{15.56}$$

The range and endurance are closed-form equations that can be the basis for quickly assessing the balance among range, endurance, velocity, fuel efficiency, lift, drag, and weight parameters in the engineering design of (jet) aircraft. Table 15.8 shows an example of range and endurance values for different velocities, fuel efficiencies, lift-to-drag ratios, and weight ratios arranged as a 4-factor, 3-level orthogonal array, requiring only 27 combinations rather than the full factorial 3^4 or 81 combinations.

SIDEBAR 15.3: C-5A CARGO AIRCRAFT *WEIGHT EMPTY* REQUIREMENT

A cautionary tale in balancing aircraft design parameters is the *weight-empty* requirement imposed by the United States Air Force and agreed to by the prime contractor for the C-5A Galaxy cargo aircraft as documented by an Air Force Institute of Technology systems engineering case study [Griffin, 2005]. The case study uses English units rather than metric units. The C-5 is a large aircraft with a maximum design weight of 920,000 pounds, maximum payload of 265,000 pounds, maximum fuel weight of 335,000 pounds, and maximum landing weight of 635,850 pounds. Cruise performance is 440 knots at 30,000 feet. Punitive cost penalties on the prime contractor for every delivered aircraft were prescribed at $10,000 per pound that exceeded the *weight empty guarantee* performance requirement of 302,494 pounds, subsequently raised to 318,469 pounds. The *weight-empty guarantee* drove every aspect of aircraft design, with severe unintended consequences because it was unattainable. Every delivered aircraft did exceed the *weight empty guarantee,* and the prime contractor incurred cost penalties, even after the prime contractor aggressively reduced weights, including a significant redesign of the wing. Within a relatively short time period, the aircraft in operation developed structural fatigue cracks. The initial mitigations were to reduce payload weights, restrict maneuvers, reduce flying hours, and prohibit takeoffs and landings on unimproved and short runways. Eventually, the Air Force paid the contractor to strengthen the wings to improve operational service life, without imposing additional cost penalties, as the redesigned wings added to the already overage weights.

TABLE 15.8 Range and Endurance Example for an Aircraft

Velocity (m/s)	Propulsion Efficiency (/s)	L/D Ratio	Weight Ratio	Range (1000s km)	Endurance (hours)
120	0.000 1	10	1.5	5	11
120	0.000 1	15	1.7	10	22
120	0.000 1	20	2	17	39
120	0.000,125	10	1.7	5	12
120	0.000,125	15	2	10	23
120	0.000,125	20	1.5	8	18
120	0.000,15	10	2	6	13
120	0.000,15	15	1.5	5	11
120	0.000,15	20	1.7	8	20
180	0.000,1	10	1.7	10	15
180	0.000,1	15	2	19	29
180	0.000,1	20	1.5	15	23
180	0.000,125	10	2	10	15
180	0.000,125	15	1.5	9	14
180	0.000,125	20	1.7	15	24
180	0.000,15	10	1.5	5	8
180	0.000,15	15	1.7	10	15
180	0.000,15	20	2	17	26
240	0.000,1	10	2	17	19
240	0.000,1	15	1.5	15	17
240	0.000,1	20	1.7	25	29
240	0.000,125	10	1.5	8	9
240	0.000,125	15	1.7	15	18
240	0.000,125	20	2	27	31
240	0.000,15	10	1.7	8	10
240	0.000,15	15	2	17	19
240	0.000,15	20	1.5	13	15

15.8 SYSTEM DYNAMICS

System dynamics, originally described as industrial dynamics, is credited to Jay Forrester [1961], beginning with his early work in the mid-1950s [1961]. Forrester's stock-flow-feedback structure modeling of General Electric appliance-manufacturing plants revealed that the observed three-year employment cycle of hiring and layoffs was attributable to the internal structure of the firm and not to the external forces of the business cycle in that the internal structure amplified rather than dampened the external forces of the business cycle. A key lesson in Forrester's work is that answers and solutions to observed phenomena may be non-intuitive without analysis. Stocks define the states of the system, and the variables defining the changes in states are the flows. The stock-flow-feedback metaphor models nth-order difference/differential equations that describe the behavior of a system. Nouns represent stocks, whereas verbs represent flows. Stocks send out signals representing information about the state of the system to the rest of the system. Stocks have the following characteristics: memory, ability to change the time shape of flows, decouple flows, and create delays. Figure 15.5 shows a simplified systems dynamics model for the soda machine performance.

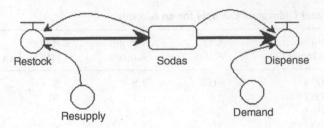

FIGURE 15.5 Soda machine system dynamics model.

15.9 CONSTRAINT THEORY

Constraint theory is the work of George Friedman [2005] and extended by Phan Phan [2011] to determine the well-posedness of the set of often disparate, interrelated multidimensional models, often from different organizations within the enterprise, to describe, and perhaps optimize, systems. (Constraint theory is different from the theory of constraints applied in business operations management problems as defined by Goldratt and Cox (1984).) Models can be based on physics, objectives, behaviors, preferences, costs, and combinations thereof. Constraint theory applies set theory (Chapter 4) and graph theory (Chapter 5) to determine whether a system's integrated mathematical model is internally consistent and whether requests for information out of the model are allowable. In the latter case, the approach can detect if a system model is over-constrained or under-constrained. Variable relationships of models are visualized with bipartite graphs that provide topological structures of the models. The graphs are mapped to constraint matrices, which can then be computationally solved in much less time than brute-force computations of all possible combinations to determine consistency and identify clashes. The rules cited by Friedman to determine the consistency of a mathematical model, i.e., constraint theory toolkit, are enumerated in Table 15.9.

To illustrate, the well-posedness of the elevator system objectives hierarchy in Figure 2.4 and Table 15.1 is determined. The hierarchy in Figure 2.4 is recast as a set of equations for the overall objective Y as a weighted function of the value function of cost C, and performance P. Note that all value functions $v\{.\}$ are on the interval $[0,1]$ so Y, P, S, Q, and A are on the interval $[0,1]$ because the weights in each of the equations sum to 1.

$$Y = 0.1v\{C\} + 0.9P \quad Y \varepsilon [0,1] \tag{15.57}$$

where P is the weighted function of time in the system S, ride quality Q, and availability A.

$$P = 0.35S + 0.3Q + 0.35A \quad P \varepsilon [0,1] \tag{15.58}$$

Time S is the weighted function of the value functions of routine average wait $E[W_r]$, priority average wait $E[W_p]$, and average transit time $E[T]$.

$$S = 0.3v\{E[W_r]\} + 0.35v\{E[W_p]\} + 0.35v\{E[T]\} \quad S \varepsilon [0,1] \tag{15.59}$$

Ride quality Q is the weighted function of the value functions of maximum acceleration a_{max}, maximum acceleration change a'_{max}, and floor leveling error e.

$$Q = 0.3v\{a_{max}\} + 0.5v\{a'_{max}\} + 0.2v\{e\} \quad Q \varepsilon [0,1] \tag{15.60}$$

TABLE 15.9   Rules to Determine Consistency of Mathematical Models

1. Organize the set of variables and their relations into a bipartite graph
   - There are two types of vertices: relations nodes N and variable knots K
   - Edges connect node relations to relevant variable knots where the graph represents a math model with undirected edges, and the computation has directed edges
   - Degree of vertex $d(v)$ is the number of edges which intersect the vertex.

2. Constraint matrix is a companion form to the bipartite graph that is more convenient for computational analysis

3. Constraint propagation rules across the bipartite graph
   - For relation nodes $d(n) - 1$ edges flowing in and 1 edge flowing out
   - For variable knots 1 edge flowing in and $d(k) - 1$ edges flowing out.

4. Propagate connectivity along edges of graph or entries in matrix to find connected components
   - $K + N - E = 1$ is a tree-like connected component with no intrinsic constraint
   - Circuit rank $K + N - E \neq 1$ is not tree-like, and result is number of independent circuits in component
   - For constraint potential of a submodel $p = N - K > 0$, an intrinsic constraint exists.

5. Submodel $N = K$ is a nodal square and nodal square with no internal nodal squares is a basic nodal square, i.e., a kernel of a constraint
   - Overlapping basic nodal squares with $p > 0$ indicate over-constrained, i.e., inconsistency.
   - Constraint propagates from basic nodal squares.
   - Overlapping resultant constraint domains are either redundant or inconsistent.
   - Inconsistencies must be negotiated to resolve over constraints.

6. Basic nodal squares only exist within circuit clusters.

7. Systematic search for basic nodal squares requires separating connected components, trimming external trees, eliminating internal trees, separating *kissing* circuit clusters by removing separate vertices, and computing constraint potential of remaining circuit clusters.

8. No computations are allowable for inconsistent mathematical models.

9. Extrinsic constraints, i.e., independent variables and variables held constant, are added to intrinsic constraints and propagated through the bipartite graph
   - Computation is not allowable if over-constraint occurs at any vertex
   - Search for basic nodal squares which can continue the propagation if constraint does not propagate to dependent variable
   - Computation is not allowable if there is an under-constraint, i.e., the constraint flow does not reach the dependent variable
   - Computations are allowable if there occurs neither an over-constraint or an under-constraint.

Availability A is the weighted function of the value functions of operational mean time between failure *MTBF* and operational mean time to repair *MTTR*.

$$A = 0.5v\{MTBF\} + 0.5v\{MTTR\} \quad A \in [0,1] \tag{15.61}$$

Equations (15.57) through (15.61) are modeled as a directed bipartite graph, as shown in Figure 15.6, where circles represent the variables (knots) and squares represent the relations (equations). The labels in the squares are the equation numbers. The constraint matrix representing the directed bipartite graph is shown in Figure 15.7. The number of relation nodes $N = 5$, variable knots $K = 14$ and edges $E = 18$ yields a circuit rank of $K + N - E = 1$ and hence the circuit is tree-like with no intrinsic constraint in the component. Each submodel is tree-like. The constraint potential of each submodel $N - K < 0$ so no intrinsic constraints exist. The tree-like nature of the model and

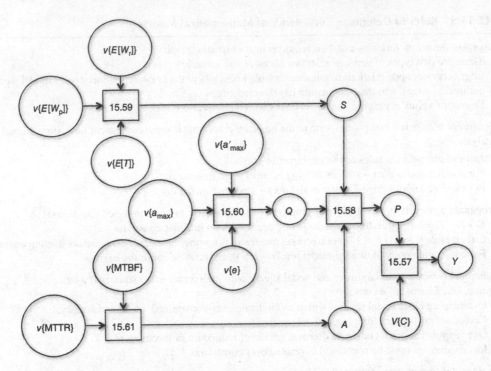

FIGURE 15.6   Directed bipartite graph of elevator system objectives hierarchy.

	$v\{E[W_l]\}$	$v\{E[W_p]\}$	$v\{E[T]\}$	$v\{a'_{max}\}$	$v\{a_{max}\}$	$v\{e\}$	$v\{MTBF\}$	$v\{MTTR\}$	S	Q	A	P	$v\{C\}$	Y
15.57												1	1	−1
15.58									1	1	1	−1		
15.59	1	1	1				−1							
15.60				1	1	1		−1						
15.61							1	1			−1			

FIGURE 15.7   Constraint matrix of elevator system objectives hierarchy.

submodel is verifiable by inspection of the original objectives' hierarchy shown in Figure 2.4. The model and no submodel constitute a basic nodal square, so there are no inconsistencies. The model is computational, as there are no inconsistencies, over-constraints, or under-constraints.

15.10   FERMI PROBLEMS AND GUESSTIMATION

A Fermi problem, named after physicist Enrico Fermi, also called guesstimation [Weinstein and Adam, 2008], is an approach to making good, approximate estimates with little or no data. An unknown, and perhaps unknowable, dependent variable is determined to be within an order of magnitude of its true value by structuring the solution as the product of its independent variables, making assumptions about their values. A famous Fermi problem is to estimate the number of piano tuners in the city of Chicago. An engineering of systems example is to estimate how much energy is needed to get a payload from Earth to the International Space Station. The unknown quantity of any independent variable is estimated by using the geometric mean, rather than the arithmetic mean, of the known or assumed upper and lower bounds of the variable. If the individual estimates of the independent variables are unbiased, then the estimate of the dependent variable tends to be within an order of magnitude of the true value, as the number of independent variables increases.

For Fermi's classic piano tuner estimate, rough order of magnitude guesses are made for each of the variables in the product equation to estimate the number of piano tuners y

$$y = (x_1)(x_2)(x_3)(x_4)(x_5) \tag{15.62}$$

for x_1 population, x_2 proportion of pianos per person, x_3 annual rate that piano is tuned, x_4 time to tune a piano, and x_5 piano tuner annual work time. Each of the variables can be refined, e.g., including travel time for the piano tuner and it is always good practice to document underlying assumptions used to estimate each of the x_i. An acceptable estimate is within the order of magnitude of the true value.

The value of making good approximate estimates, with little to no data, and little time, is a critical engineering skill in all phases of the life cycle of a system [Alber, 2012]. In the early phases of identification of need, concept definition, and preliminary design, determining the technical, cost, and schedule feasibility is especially critical to success (Fig. 1.2). In the later phases of detailed design, integration and test, production, operation, maintenance and disposal, fast approximate estimates provide a sanity check of more detailed models and data.

15.11 ARTIFICIAL (OR AUGMENTED) INTELLIGENCE AND MACHINE LEARNING

Artificial (or augmented) intelligence and machine learning (AI/ML) advances pose challenges in the engineering design of systems. AI methods can be broadly classified as rule-based and neural-network-based [Miller and Verma, 2020; Raz et al., 2023]. Rule-based methods are relatively mature, with decades of experience in application, and are well-understood. In contrast, much is unknown about the contextually driven behavioral characteristics of neural-network-based AI, commonly referred to as machine learning, deep learning, and including reinforcement learning. Neural network implementation takes the form of mapping outputs to inputs through multiple hidden layers from large data sets that are split for training and validation. The implementations may be software executed on conventional computers, graphics processing units, or specialized neural networks. Neural networks are characterized as black boxes with no information available on the decision-making constructs of the hidden layers and the underlying input-output map. Neural network performance is critically dependent on the datasets used to train the algorithms and whose actions currently cannot be guaranteed to be *fit for purpose* to meet the attributes of elegance that systems accomplish their intended purposes, be resilient to effects in real-world operation, while minimizing unintended actions, side effects, and consequences [Griffin, 2010].

Systems engineering for AI (SE4AI) methods may offer valuable insight for optimizing a large-scale system comprised of a workflow of various information fusion and AI/ML technologies. AI for systems engineering (AI4SE) may provide new tools for systems engineering that develop insights into optimizing the design and engineering of a larger system of systems. Furthermore, there may be opportunities to optimize these systems such that they not only provide an optimized design space (e.g., software maintenance costs, minimizing system failures, constructing workflows of AI/ML algorithms that perform superior as a sum-of-parts) but also are balanced human-centered design (e.g., minimize workload, increase situational awareness, manage attention) [Lawless et al., 2021a,b].

Incorporating artificial intelligence leveraging statistical machine learning into complex systems poses numerous challenges to traditional test and evaluation (T&E) methods [Freeman, 2020]. As AI handles varying decision levels, the underlying ML needs confidence to ensure testable, repeatable, and auditable decisions. Additionally, we need to understand failure modes and failure mitigation techniques. We need AI assurance–certifying ML and/or AI algorithms function as intended and are vulnerability-free, either intentionally or unintentionally designed or inserted as data/algorithm parts. T&E provides a process for AI assurance.

Incorporating AI components into complex systems exacerbates the T&E challenge by adding complexity via stochastic algorithms. T&E for AI algorithms based on statistical learning is still in its nascent phases. Both expert systems-enabled AI and statistical machine learning-enabled AI pose new challenges for T&E. Expert systems are based on explicit models covering all models, which can expand test scope. However, the newest wave in AI is machine learning algorithms; learning from experience and improving by analyzing large data quantities and identifying patterns adds a complication in changing systems outcomes based on new data inputs or the algorithms themselves changing over time.

Research is looking into explainable AI (xAI) for developing methods to help understand, explain, and interpret ML algorithms [Arrieta et al., 2020]. These xAI methods are expected to become the centerpiece of T&E for systems with ML components. Another research path is counterfactual test and evaluation (cT&E), which imagines situations that may not occur or are known not to be represented by the model and/or the training/validation data sets. Then it investigates the model response to such conditions [Roese, 1997]. Counterfactual testing creates hypothetical scenarios and examines how the system would have behaved if those scenarios had taken place, exploring the model outputs beyond what the model is trained for or exposed to under nominal and expected operational conditions.

15.12   SUMMARY

This chapter has introduced the underpinnings of the science of systems in general and the analytics in the engineering of a system. Von Bertanlanffy describes the foundations of *general system theory* as a basic science, applying to human-designed (engineered), biological, and social systems. The successful realization, operation, and support of systems necessitate the *systems approach*, integrating multiple heterogeneous disciplines, with *scientific generalists* performing the systems function, integrating the parts into the whole greater than the sum. Ackoff, Warfield, and Hitchins characterize *systems science* as the science of complex systems and the science of wholes, with emphasis on how systems behave as a whole without having to explain the behaviors of the individual parts of the system. System behavior can be explained without reductionism with the advantage that high-level characterizations to estimate behavior can establish feasibility early on in the engineering of a system, thereby reducing technical risk. *Natural systems* constitute living and non-living systems found in nature. Structures found in nature can be reverse-engineered and applied to the engineering design of systems. Wiener's *cybernetics* integrates the applied mathematics of control theory and information theory to achieve the desired regulation for complex systems with complex uncertainties. Shannon's *information theory* characterizes the variety of allowed inputs to the system as well as the disturbances and noises using the measure of entropy. *Systems thinking* as described by Senge, Checkland, Boardman, and Sauser, provides a holistic look at the *whole* of systems, interrelationships with outside forces, and the properties of systems, especially emergence and stability.

The *quantitative characterization of systems* is illustrated using the elevator case study used throughout the text, the soda machine introduced in Chapter 12, and an aircraft. All three are complex systems operating in a context of uncertainty. The treatment is simplified to emphasize key concepts using approximate models to assess the feasibility of meeting performance requirements. Forrester's *system dynamics* is a stock-flow-feedback structure modeling of systems in operation. A key is that answers and solutions to observed phenomena may be non-intuitive without analysis. Friedman's *constraint theory*, extended by Phan, determines the well-posedness of the set of often disparate, interrelated multidimensional models, often from different organizations within the enterprise, to describe systems. Models can be based on physics, objectives, behaviors, preferences, costs, and combinations thereof. Constraint theory determines the internal consistency of a system's integrated mathematical model and whether requests for information out of the model are allowable. *Fermi problems*, also called guesstimation, is an approach to making good, approximate estimates with little or no data. An unknown, and perhaps unknowable dependent variable is determined to be within an order of magnitude of its true value by structuring the solution as the product of its independent variables, making assumptions about their values. The value of making good approximate estimates, with little to no data, and little time, is a critical engineering skill in all phases of the life cycle of a system and is of value as a sanity check of more detailed models and data.

AI/ML poses V&V challenges to the engineering design of systems and is an emerging area of systems engineering research focusing on explainable AI and counterfactual testing. See the companion website for additional materials, including examples.

PROBLEMS

15.1 Von Bertanlanffy [1968] describes the competition of two players in a predator-prey dynamic. Show that the viability of the predator depends on the viability of the prey. He extends the concept of competition to that of the two players competing for the same resources. Show that competition for resources leads to the elimination of the player with the smaller growth potential.

15.2 Analyze the technical feasibility for a diesel-powered automobile with an empty weight 1200 kg, capacity of five people including driver with a total maximum passenger weight 400 kg, luggage 200 kg, acceleration from 0 to 30 m/s in 5 seconds, maximum speed 60 m/s, and range 600 km and fuel efficiency 12 km/l highway driving at 30 m/s. Assume the following: length 4.5 m, height 1.5 m, width 1.75 m, automobile drag coefficient 0.33, and rolling resistance friction coefficient 0.03.

15.3 For the automatic flight control system in Section 15.5, show the system transfer function of the controlled system for a sustained disturbance of a 50 m/s headwind with time series function $d(t) = -50t$ m and transfer function of $D(s) = -50/s^2$. Assess the stability of the controlled system.

15.4 The *limits to growth* archetype in systems thinking is a qualitative model. Derive an equivalent quantitative system dynamics model.

15.5 The elevator industry engineers capacity based on peak period average waiting time $E(W)$ approximations. Contrast peak period 95th percentile waiting times $T P(W < T) = 0.95$ with average waiting time approximations.

15.6 Replace the objectives hierarchy for the elevator system with the following objectives: minimize cost C and maximize performance P. Performance P is to be maximized by minimizing expected routine waiting time W_r, expected priority waiting time W_p, and expected transit time T. Availability is to be maximized from maximizing mean time between failure $MTBF$

and minimizing mean time to repair *MTTR*. Ride quality is not considered in the revised set of objectives.

$$Y = f(\min C, \max P)$$

$$P = f(\min S, \max A)$$

$$S = f(\min E[W_r], \min E[W_p], \min E[T])$$

$$A = f(\max MTBF, \min MTTR)$$

Determine if the set of objectives is a feasible set, i.e., perfectly constrained and not under-constrained or over-constrained.

15.7 Estimate the volume of water in liters required for six astronauts to complete a return mission to Mars.

Glossary

Acceptance Stakeholder function for agreeing that the designed system, as tested or otherwise evaluated by the stakeholders, is acceptable.

Acceptance Plan How the qualification data will be used to determine that the real system is acceptable to the stakeholders.

Agile Systems Engineering An iterative, incremental approach continually modeling, analyzing, developing, and trading options in the engineering of systems.

Allocated Architecture Complete description of the system design, including the functional architecture allocated to the physical architecture, derived input/output, technology, and system-wide, trade-off and qualification requirements for each component, an interface architecture that has been integrated as one of the components, and complete documentation of the design and major design decisions.

Apportionment Requirements flowdown approach that spreads a system-level requirement among the system's components of the system, maintaining the same units.

Approximation Approach used in engineering to find almost correct solutions that are deemed good enough for problems when the resources required to solve the problem with more fidelity are not called for; approaches include rounding quantitative values, solving for only first and perhaps second-order terms of higher nth order systems of equations and models, linearizing non-linear models, and assuming probabilistic distributions to characterize uncertainty.

Attainable Solutions exist within performance, cost, and schedule constraints.

Behavior Model Defines the control, activation, and termination of system functions that is needed to meet the performance requirements of the system.

Bipartite Graph (Digraph) Graph (digraph) whose set of nodes can be partitioned into two sets A and B such that no edge connects a node in A to another node in A and, similarly, no edge connects a node in B to another node in B.

The Engineering Design of Systems: Models and Methods, Fourth Edition. Dennis M. Buede and William D. Miller
© 2024 John Wiley & Sons, Inc. Published 2024 by John Wiley & Sons, Inc.
Companion website: www.wiley.com/go/engineeringdesignofsystems4e

Black Box Testing Outputs are determined correct or incorrect based upon inputs; the inner workings of the module are ignored. Both positive and negative testing have to be employed.

Cartesian Product of Two Sets, $A \times B$ Set of all possible ordered pairs of those two sets.

Centralized Architecture Architecture with a central location for the execution of the transformation and control functions of the system.

Client–Server Architecture Architecture that distinguishes between client processes (requestors) and server processes (task completers).

Comparable Pertaining to requirements, the relative necessity of the requirements is included.

Complete Pertaining to requirements, (1) everything the system is required to do throughout the system's life cycle is included, (2) responses to all possible (realizable) inputs throughout the system's life cycle are defined, (3) the document is defined clearly and self-contained, and (4) there are no "to be defined" (TBD) or to be refined (TBR) statements; completeness is a desired property but cannot be proven at the time of requirements development, or perhaps ever.

Component Subset of the physical realization (and the physical architecture) of the system to which a subset of the system's functions has been (will be) allocated. A component could be the integration of hardware and software, a specific piece of hardware, a specific segment of the system's software, a group of people, facilities, or a combination of all of these.

Composability of Systems A systems design approach is made to increase agility and accelerate application development by reusing existing assets and reassembling them in unique ways to satisfy specific user requirements. Composability is the opposite of TTDSE. The principles of composability are modularity, autonomy, and discoverability.

Conceptual Validity Correspondence between the stakeholders' needs and the operational concept.

Concise Pertaining to requirements, no unnecessary information is included in the requirement.

Configuration Items Lowest level components in the physical architecture.

Consistent Pertaining to requirements (1) internal – no two subsets of requirements conflict and (2) external – no subset of requirements conflicts with external documents from which the requirements are traced.

Constraint Theory Credited to George Friedman, determines the well-posedness of the set of often disparate, interrelated multidimensional models, often from different organizations within the enterprise, to describe, and perhaps optimize, systems, not to be confused with the Theory of Constraints.

Context of a System Set of entities that can impact the system but cannot be impacted by the system.

Correct Pertaining to requirements, what the system is in fact required to do.

Cost Requirement Requirement addressing the payment of money during the appropriate life cycle phase for the system in question to be useful.

Cyber-Physical Systems (CPS) Engineered systems that are built from, and depend upon, the seamless integration of computational algorithms and physical components [NSF, 2023].

Cybernetics Coined by Norbert Wiener, from the Greek *steersman*, to explain the regulation, and self-regulation, using compensators and closed-loop feedback servomechanisms to control and stabilize biological, social, and engineered systems in the face of uncertainties characterized by varieties of allowed inputs as well as disturbances and noise.

Data Model   Defines the relationships among the inputs and outputs of a system.

Deadlock   Undesired state of the system in which activity ceases, and throughput is nonexistent. Deadlock can occur for two reasons: contention over resources and waiting for communication.

Decentralized Architecture   Architecture with multiple, specific locations at which the same or similar transformational or control functions are performed.

Decision   Irrevocable allocation of resources to affect some chosen change or the continuance of the status quo.

Definitive Model   Addresses the question of how an entity should be defined.

Descriptive Model   Attempts to predict answers to questions for which the truth may or may not be obtained in the future.

Design   Preliminary activity that has the purpose of satisfying the needs of the stakeholders begins in the mind of the lead engineer but must be transformed into models employing visual formats in a highly skilled manner for success to be achieved.

Design Independent   Pertaining the requirements, each requirement does not specify a particular solution or a portion of a particular solution.

Design Validity   Congruence between the *Originating Requirements Document* (ORD) and the derived requirements.

Digital Engineering   An integrated digital approach that uses authoritative sources of systems data and models as a continuum across disciplines to support life cycle activities from concept through disposal.

Digital Thread   (1) The use of digital tools and representations for design, evaluation, and life cycle management. (2) A data-driven architecture that links together information generated from across the product life cycle and is envisioned to be the primary or authoritative data and communication platform for a company's products at any time. (3) More narrowly, the lowest level design and specification for a digital representation of a physical item. The digital thread is a critical capability in model-based systems engineering (MBSE) and the foundation for a digital twin. (4) The traceability of the digital twin back to the requirements, parts, and control systems that make up the physical asset.

Digital Twin   (1) A related yet distinct concept to digital engineering. The digital twin is a high-fidelity model of the system that can be used to emulate the actual system [SEBoK, 2022]. (2) An integrated multiphysics, multiscale, probabilistic simulation of an as-built system, enabled by a Digital Thread, that uses the best available models, sensor information, and input data to mirror and predict activities/performance over the life of its corresponding physical twin [U.S. DoD, 2018].

Directed Graph or Digraph   Pair of sets, $V(G)$ and $E(G)$. $V(G)$ $\{n_1, n_2, \ldots, n_N\}$ is the set of vertices or nodes. $V(G)$ is a finite, nonempty set. $E(G) = e_{ij}$ is a subset of $V \times V$ or *ordered* pairs of nodes. e_{ij} is said to be from n_i to n_j. $E(G)$ may be empty.

Distributed Architecture   Architecture in which there are two or more autonomous processors connected by a communications interface and running a distributed operating system.

Early Validation   Determination that the right problem is being defined at the current level of abstraction, given the validity of the problem definition at a higher level of abstraction.

Engineering   Discipline for transforming scientific concepts into cost-effective products using analysis and judgment.

Engineering of a System   Engineering discipline that develops, matches, and trades off requirements, functions, and alternate system resources to achieve a cost-effective, life cycle-balanced product based upon the needs of the stakeholders.

Entity-Relationship Diagrams   Model the data structure or relationships between data entities.

Equivalence   Simple requirements flowdown approach that causes the component requirement to be the same as the system requirement.

Error   Subset of the system state, which may lead to a failure. The system can monitor its own state, so errors are observable in principle.

External Interface Requirements   Limitations placed upon the receipt of inputs and transmission of outputs by the interfaces of the external systems.

External Systems Diagram   View of the model of the interaction of the system with other (external and enabling) systems in the relevant contexts, thus providing a definition of the system's boundary in terms of the system's inputs and outputs.

Failure   Deviation in behavior between the system and its requirements. Since the system does not maintain a copy of its requirements, a failure is not observable by the system.

Fault   Defect in the system that can cause an error. Faults can be permanent (e.g., a failure of a system component that requires replacement) or temporary due to either an internal malfunction or external transient.

Feedback and Control   Comparison of the actual characteristics of an output with the desired characteristics of that output for the purpose of adjusting the process of transforming inputs into that output.

Fermi Problems and Guesstimation   Approach attributed to Enrico Fermi to making good, approximate estimates with little or no data. An unknown, and perhaps unknowable, dependent variable is determined to within an order of magnitude of its true value by structuring the solution as the product of its independent variables, making assumptions about their values; the unknown quantity of any independent variable is estimated by using the geometric mean, rather than the arithmetic mean, of the known or assumed upper and lower bounds of the variable; if the individual estimates of the independent variables are unbiased, then the estimate of the dependent variable tends to be within an order of magnitude of the true value, as the number of independent variables increase.

Figure of Merit (FOM)   Describes a specific system property or attribute for a given environment and context; an FOM is measured within the system, also called a measure of performance.

Function (Mathematical)   Binary relation from A to B such that every element of A is mapped to one and only one element of B.

Function (Engineering)   Process that transforms inputs into outputs.

Functionality   Set of functions required to produce a particular output. *Simple functionality* is an *ordered* sequence of functional processes that operates on a single input to produce a specific output. Note there may be many inputs required to produce the output in question, but this simple functionality is only related to one of the inputs. *Complete functionality* is a complete set of coordinated processes that operate on all of the necessary inputs for producing a specific output.

Functional Architecture   (1) Logical architecture that defines what the system must do, a decomposition of the system's top-level function. This very limited definition of functional architecture is the most common and is represented as a directed tree. (2) Logical model that captures the transformation of inputs into outputs using control information. This definition adds the flow of inputs and outputs throughout the functional decomposition. (3) Logical model of a functional decomposition plus the flow of inputs and outputs, to which input/output requirements have been traced to specific functions and items (inputs, outputs, and controls).

Functional Requirements The two to seven functions are the first-level decomposition of the system's function.

Fundamental Objective Aggregation of the essential set of objectives that summarizes the current decision context and is yet relevant to the evaluation of the options under consideration.

Fundamental Objectives Hierarchy Subdivision of the fundamental objective into value objectives that more meaningfully define the fundamental objective, thereby forming a value structure.

General Systems Theory Defined by Ludwig von Bertanlanffy as a basic science, its development, and applications in human-designed (engineered), biological, and social systems; successful realization, operation, and support of systems necessitates the *systems approach*, integrating multiple heterogeneous disciplines, thereby driving the need for *scientific generalists* to perform the systems function, integrating the parts into the whole greater than the sum.

Graph, *G* A pair of sets, $V(G)$ and $E(G)$. $V(G) = \{n_i, n_2, \ldots, n_N\}$ is the set of vertices or nodes. $E(G) = \{e_{ij}\} \subseteq V(G) \times V(G)$ is a relation that defines the set of edges that are unordered, not necessarily distinct pairs of nodes. $V(G)$ is a finite, nonempty set. $E(G)$ may be empty and is a subset of the Cartesian product of $V(G)$ with itself.

Hardware Redundancy Use of extra hardware to enable the detection of errors as well as to provide additional operational hardware components after errors have occurred. *Passive hardware redundancy* masks or hides the occurrence of errors rather than detecting them; recovery is achieved by having extra hardware available when needed. *Active hardware redundancy* attempts to detect errors, confine damage, recover from the errors, and isolate and report the fault.

ICOMs The inputs, controls, outputs, and mechanisms of a function in IDEF0.

IDEF0 IDEF acronym comes from the U.S. Air Force's Integrated Computer-Aided Manufacturing (ICAM) program that began in the 1970s. IDEF is a complex acronym that stands for ICAM Definition. The number, *0*, is appended because this modeling technique was the first of many techniques developed as part of this program. In 1993, the U.S. Department of Commerce (National Institute of Standards and Technology [NIST]) issued Federal Information Processing Standard (FIPS) Publication 183 [1993a] that defines the IDEF0 language and renames the acronym Integrated Definition for Function Modeling. In 2008, FIPS PUB 183 was withdrawn but is still online at https://nvlpubs.nist.gov/nistpubs/Legacy/FIPS/fipspub183.pdf. In 2012, ISO/IEC/IEEE 31320-1:2012 identified the basic components of Integration Definition 0 (IDEF0) syntax (the drawn, visual elements of the language and how they may be used together) and IDEF0 semantics (what it means when the visual elements are used together in specific, allowable ways), specifies the rules that govern the use of these modeling components, and describes the types of diagrams used in an IDEF0 model. ISO/IEC/IEEE 31320-1:2012 identifies and discusses the model pages with which each diagram in an IDEF0 model is associated and discusses specific features found in an IDEF0 diagram. It describes IDEF0 reference expressions and IDEF0 diagram feature references. It also presents an abstract formalization of the IDEF0 language.

Information Redundancy Addition of extra bits of information to enable error detection and correction using special codes.

Information Theory Credited to Claude Shannon, determines fundamental limits on information and the capacity of information-carrying communications channels and characterizes the variety of allowed inputs to a system as well as the disturbances and noises using the measure of entropy.

Input/Output Requirements Requirements about sets of acceptable inputs and outputs, trajectories of inputs to and outputs from the system, interface constraints imposed by the external systems, and eligibility functions that match system inputs with system outputs for the life cycle phase of interest. This category is partitioned into four subsets: (1) inputs, (2) outputs, (3) external interface constraints, and (4) functional requirements. Invalid inputs and undesired outputs should also be characterized in order to detect and mitigate their effects.

Input/Output Trace Has a timeline associated with each major actor (our system and other systems) in the scenario. The systems involved are listed across the top of the diagram, with the timelines running vertically down the page under each of the systems. Time moves from top to bottom in an input/output trace; the system of concern is highlighted with a bold label and heavier line. Interactions involving the movement of data and horizontal arcs from the originating system to the receiving system designate energy or matter among systems. A label is shown just above each arc to describe the data or item being conveyed. Double-headed arcs are permissible to represent dialog in a compact manner. Having two or more arcs in quick succession is also common to illustrate that the same item is being transmitted from one system to multiple systems or that multiple systems are potentially transmitting the same item to one system.

Input Requirements State what inputs the system must receive and any performance or constraint aspects of each.

Integration Process of assembling the system from its components, which must be assembled from their configuration items (CIs).

Interface Connection for hooking to another system (an external interface) or for hooking one system component to another (an internal interface). The interface of a system contains both a logical element and a physical element (or link) that are responsible for carrying items (materiel, electromechanical energy, or information) from one component or system to another.

Items Inputs that are received by the system, the outputs that are sent by the system to other systems, and the inputs that are generated internally to the system and sent to other parts of the system to assist in the transformation process for which the system is responsible. Items may be materiel, electromechanical energy, or information.

Laplace Transform Particular form of a generalized linear integral function to convert a dynamic deterministic or stochastic function or variable in the time domain to the equivalent in the complex frequency domain; the transformation simplifies what would be complex convolutional integral solutions in the time domain to a much simpler algebraic solution in the complex frequency domain; controllability and stability of systems can also be determined; the inverse Laplace transform converts the solution in the complex frequency domain back to the time domain.

Life Cycle Begins with the gleam in the eyes of the users or stakeholders, is followed by the definition of the stakeholders' needs by the systems engineers, includes developmental design and integration, goes through production and operational use, usually involves refinement, and finishes with the retirement and disposal of the system.

Livelock Undesired state of the system in which resources are being routed in cycles (oscillating) while waiting for the proper allocation of resources to enable the completion of necessary activities; unfortunately, this proper allocation of resources is never achieved, and the system cycles continuously, never reaching the desired outputs.

Manufacturing Using resources to perform operations on materials to produce products.

Measure of Effectiveness (MOE)   Variable that describes how well a system carries out a task or set of tasks within a specific context; a MOE is measured outside the system for a defined environment and state of the context variables.

Measure of Performance (MOP)   Variable that describes a specific system property or attribute for a given environment and context; a MOP is measured within the system.

Mental Model   Abstraction of thought.

Mission Requirements   Requirements that relate to objectives of the stakeholders that are defined in the context of the supersystem, not the system itself.

Mode of a System   Distinct operating capability of the system during which some or all of the system's functions may be performed to a full or limited degree.

Model   Any incomplete representation of reality, an abstraction. The *essence* of a *model* is the question or set of questions that the model can reliably answer for us.

Model-Based Systems Engineering (MBSE)   The formalized application of modeling to support system requirements, design, analysis, verification, and validation activities begins in the conceptual design phase and continues throughout development and later life cycle phases [SEBOK, 2021].

Modifiable   Pertaining to requirements, changes can be made easily, consistently (free of redundancy), and completely.

Morphological Box   Matrix in which the columns (or rows) represent the components in the generic physical architecture. The boxes in each column (or row) then represent alternate choices for fulfilling that generic component.

Multiattribute Value Analysis   Quantitative method for aggregating a stakeholder's preferences over conflicting objectives to find the alternative with the highest value when all objectives are considered.

Natural Systems   Constitute living and non-living systems found in nature; balance of proportions in animals and humans has its analogs in engineered systems, e.g., balance of size, weight, power, speed, acceleration, and endurance.

Normative Model   Model that addresses how individuals or organizational entities ought to think about a problem and guide decision-making.

Object-Oriented Design   A bottom-up process that begins by defining a set of objects that need to be part of the system to achieve the system-level functionality desired. Objects are thought to be basic building blocks that can perform functions (methods) and contain information. Key properties of OO design are inheritance and information hiding.

Objectives Hierarchy   Hierarchy of objectives that are important to the system's stakeholders in a value sense; that is, the stakeholders would (should) be willing to pay to obtain increased performance (or decreased cost) in any one of these objectives. The definition of the natural subsets of the fundamental objective into a collection of performance requirements.

Observance Requirement   Requirement stating how the estimates (qualification data) for each input/output and system-wide requirement will be obtained. Typically, one of the four major qualification methods (test, analysis and simulation, inspection, or demonstration) is assigned to each input/output and system-wide requirement.

Open Architecture   Architecture in which the hardware and software interfaces are sufficiently well-defined that additional resources can be added to the system with little or no adjustment.

Operational Concept Vision for what the system is (in general terms), a statement of mission requirements, and a description of how the system will be used. The shared vision is based on the perspective of the system's stakeholders on how the system will be developed, produced, deployed, trained, operated, maintained, refined, and retired to overcome some operational problems and achieve the stakeholders' operational needs and objectives. The mission requirements are stated in terms of measures of effectiveness. The operational concept includes a collection of scenarios, one or more for each group of stakeholders in each relevant phase of the system's life cycle.

Operational Validity Matching of the capabilities of the designed system to the operational concept naturally occurs late in the integration phase after the designed system has been verified.

Output Requirements Requirements that state what outputs the system must produce and any performance aspects.

Overlap in the Functional Architecture Redundancy in functionality that is not needed to achieve additional performance.

Partition on a Set A Collection P of disjoint subsets of A whose union is A.

Performance Analysis Analysis for the purpose of discovering the range of performance that can be expected from a specific design or a set of designs that are quite similar.

Performance Requirement Requirement defined on some index that establishes a range of acceptable performance from a minimum acceptable threshold to a design goal.

Physical Architecture Resources for every function identified in the functional architecture. The *generic physical architecture* is a description of the partitioned elements of the physical architecture without any specification of the performance characteristics of the physical resources that comprise each element (e.g., central processing unit). An *instantiated physical architecture* is a generic physical architecture to which complete definitions of the performance characteristics of the resources have been added.

Physical Model Representation of an entity in three-dimensional space and can be divided into full-scale mock-up, subscale mock-up, breadboard, and electronic mock-up.

Power Set of Set A Set of all sets that are subsets of A.

Process Model Model that defines the functional decomposition of the system function and the flow of inputs and outputs for those functions.

Prototype Physical model of the system that ignores certain aspects of the system, glosses over other aspects and is fairly representative of a third segment of aspects of the system. The prototype can range from a subscale model of the system to a paper display (storyboard) of the user interface of the system.

Qualification Process of verifying and validating the system design and then obtaining the stakeholders' acceptance of the design.

Qualification Methods Inspection, analysis and simulation, instrumented test, and demonstration.

Qualification Requirements Requirements that address the need to qualify the system as being designed right, the right system, and an acceptable system. There are four primary elements: (1) observance: to state which qualification data for each input/output and system-wide requirement will be obtained by (i) demonstration, (ii) analysis and simulation, (iii) inspection, or (iv) instrumented test; (2) verification plan: to state how the qualification data will be used to determine that the real system conforms to the design that was developed; (3) validation plan: to state how the qualification data will be used to determine that the real system complies with the originating performance, cost and trade-off requirements; and (4) acceptability: to state how the qualification data will be used to determine that the real system is acceptable to the stakeholders.

Qualitative Model Model that provides symbolic, textual, or graphic answers. Symbolic models are based on logic or set theory. Textual models are based on verbal descriptions. Graphical models use either elements of mathematical graph theory or simply artistic graphics to represent a hierarchical structure, the flow of items or data through a system's functions, or the dynamic interaction of the system's components.

Quantitative Model Model that provides answers that are numerical; these models can be either analytic, simulation, or judgmental models.

Queuing (Theory) The behavior and characterization of waiting lines, subject to stochastic inputs on demands for service, the numbers of servers, and the possibly stochastic duration or rate of service.

Regression Testing Retesting a portion of the system after a change has been made to ensure that new problems were not introduced.

Relation (Binary) Relation that relates elements of A to elements of B and is a subset, R, of $A \times B$.

Relation (Unary) on a Set A Relation that relates elements of A to itself and is a subset, R, of $A \times A$.

Requirements Flowdown Derivation of requirements from one level of the operational architecture for a lower level of the architecture. Includes three approaches: apportionment, equivalence, and synthesis.

Requirements Statements That define the needs and objectives of stakeholders.

Requirements Validity Correspondence between the operational concept and the originating requirements.

Risk Combination of the probability of an event occurring and the significance of the consequence of the event occurring.

Risk Analysis Analysis is done early in the development process to examine the ability of the divergent concepts to perform up to the needed level of performance across a wide range of operational scenarios. At this time, there remains substantial uncertainty about the stakeholders' needs, the state of technology under consideration, and the details of the operational architecture.

Risk Avoidance Selection of the low-risk alternative; unfortunately, what seems to be low risk intuitively is high risk in some cases.

Risk Management Use of hedging strategies; a hedging strategy is the maintenance of fallback options in case a riskier option fails.

Risk Transference Transfer of risk to others; an example being the purchase of insurance.

Scenario Defines how the system will respond to inputs from other systems in order to produce a desired output. Included in each scenario are the relevant inputs to and outputs from the system and the other systems that are responsible for those inputs and outputs. The scenario should not describe how the system is processing inputs to produce outputs; rather, the scenario focuses on the exchange of inputs and outputs by the system with other systems.

Schedule Requirement Requirement addressing a timing issue for the relevant system for the phase of life cycle in question.

Semantics Study of relationships between signs and symbols and what they represent.

Set   A collection of well-defined objects, called elements or members.

Shortfall in the Functional Architecture   Absence of a functionality that is required to produce a desired output from one or more inputs.

Software Redundancy   Use of multiple versions of the same software functionality to provide multiple operational software components in the event of a software failure.

Specification   Collection of requirements that completely define the constraints and performance requirements for a specific physical entity that is part of the system.

Stakeholder   Owner and/or bill payer, developer, producer or manufacturer, tester, deployer, trainer, operator, user, victim, maintainer, sustainer, product improver, and decommissioner. Each stakeholder has a significantly different perspective of the system and the system's requirements.

Stakeholders' Requirements   Statements by the stakeholders about the system's capabilities that define the constraints and performance parameters within which the system is to be designed. These stakeholders' requirements focus on the boundary of the system in the context of these mission requirements, are written in the stakeholders' language, are produced in conjunction with the stakeholders of the system and are based upon the operational needs of these stakeholders.

Stakeholders' Requirements Document (StkhldrsRD)   Document that contains the stakeholders' requirements. Sometimes called the Originating Requirements Document (ORD) or Operational Requirements Document.

Starvation   Undesired state of the system that occurs when a function needs a particular resource for execution, but the resource is always allocated to other functions due to a poorly designed resource assignment algorithm.

State of the System   Static snapshot of the set of metrics or variables needed to describe fully the system's capabilities to perform the system's functions.

Suitability Requirements   Requirements that address quality concerns of a system and are system-wide in scope. Examples are availability and safety.

Surge or Race   Undesired state of the system that occurs in relatively uncontrolled systems when components are competing with each other to perform a task.

Syntax   Way in which words are put together to form phrases and sentences.

Synthesis   Requirements flowdown approach for those situations in which the system-level requirement is comprised of complex contributions from the components, causing the component requirements that are flowed down from the system to be based upon some analytic model. The derived requirements for each component will have significantly different units than the system-level requirement.

Systems Modeling Language (SysML)   A general-purpose modeling language published by the Object Management Group (OMG) for systems engineering applications. It supports the specification, analysis, design, verification, and validation of a broad range of systems and systems-of-systems. SysML is an open-source specification project and includes an open-source license for distribution and use. SysML v1 is defined as an extension of a subset of the Unified Modeling Language (UML) using UML's profile mechanism designed to support systems engineering. SysML v2 introduces a metamodel based on formal semantics and not constrained by UML while preserving most UML modeling capabilities. SysML v2 has flexible graphical, tabular, and textual view and viewpoint specifications and execution, as well as a standardized application programming interface (API) for the system model to be interoperable with other tools.

System Set of components (subsystems, segments) acting together to achieve a set of common objectives via the accomplishment of a set of tasks.

System Context Set of entities that can impact the system but cannot be impacted by the system.

System Dynamics Credited to Jay Forrester, originally described as industrial dynamics, is a stock-flow-feedback structure that models nth-order difference/differential equations that describe the behavior of a dynamic, time-varying system subject to uncertainties and disturbances.

System Requirements Document (SysRD) Document that contains the system requirements.

System (Human-Designed) Specially defined set of segments (hardware, software, physical entities, humans, and facilities) acting as planned, via a set of interfaces, which are designed to connect the components, to achieve a common mission or fundamental objective (i.e., a set of specially defined objectives), subject to a set of constraints, through the accomplishment of a predetermined set of functions.

System of Systems Set of systems or system elements that interact to provide a unique capability that none of the constituent systems can accomplish on its own. A system of systems (SoS) is characterized by managerial and operational independence of its constituent systems. A SoS exhibits emergent capabilities beyond the mere aggregation of the constituent systems, just as a system has its own emergent capabilities. SoS types are acknowledged, collaborative, directed, and virtual.

System Task or Function Set of functions that must be performed to achieve a specific objective.

System Requirements Translation (or derivation) of the originating requirements into engineering terminology.

Systems (External of a System) Set of entities that interact with the system via the system's external interfaces.

Systems Science Science of complex systems and the science of wholes; emphasis is on how systems behave as a whole without having to explain the behaviors of the individual parts of the system; behavior is the response of the system to some stimulus, including that of the system's environment.

Systems Thinking Holistic looks at the *whole* of systems, interrelationships with outside forces (systems, external systems, and context), and the properties of systems, especially emergence and stability.

Technology and System-wide Requirements Constraints and performance index thresholds are placed upon the physical resources of the system. This category can be partitioned into four subsets: (1) technology, (2) suitability and quality issues, (3) cost for the relevant system (e.g., development cost, operational cost), and (4) schedule for the relevant life cycle phase (e.g., development time period, operational life of the system).

Technology Requirement Constrains the engineering creativity and should result from the other requirements if they are justifiable. These requirements are usually justified on the basis of interoperability or compatibility with an existing product line, which ultimately should be reflected in cost savings.

Time Redundancy Use of extra processing when time is available to perform the same computation multiple times with a single hardware and software combination and then compare the results.

Trade Study Analysis that focuses on finding ways to improve the system's performance on some highly important objective while maintaining the system's capability in other objectives.

Tree Graph, *G,* with no loops in which there is a unique, simple (no loops), nondirected path (or semipath in the case of a digraph) between each pair of nodes. A *rooted tree* is a tree in which there is a designated "root" node. In a graph, the root node must have a degree of 1. In a directed tree, the root node must have no parents, or an in degree of 0. A *directed tree* is a rooted tree in which there is a (directed) path from the root to every other node.

Traceable Pertaining to requirements, each derived requirement must be traceable to an originating requirement via some unique name or number.

Traced Pertaining to requirements, each requirement is traced to some document or statement of the stakeholders.

Trade-off Requirements Algorithms for comparing any two alternate designs on the aggregation of cost and performance objectives. These algorithms can be divided into (1) performance trade-offs, (2) cost trade-offs, and (3) cost–performance trade-offs.

Traditional Tod-Down Systems Engineering (TTDSE) A process for systems engineering that begins with a thorough analysis of what the problem is that needs to be solved. Several potential competing concepts for implementing the system of interest are defined, with the most favorable concept selected for implementation. An operational concept and system-level requirements are defined for the solution concept. A layered iterative process creates an architecture, deriving requirements and refining the needed test system and associated data collection requirements. The bottom layer addresses the configuration items (CIs) that the discipline engineers will design. Once the CIs have been designed and delivered for integration, the verification, validation, and acceptance testing process begins. Each layer of the decomposition process is verified against the associated derived requirements. During the process, requirements may be adjusted, or the architecture and design of the system may be modified as needed. At the system level, validation against the concept of operations and acceptance testing (as defined by the stakeholders) is conducted. Many different organizations can be tasked to design one or several pieces and all of the pieces can be integrated easily and effectively to achieve the desired system.

Unambiguous Pertaining to requirements, every requirement has only one interpretation.

Understandable Pertaining to requirements, interpretation of each requirement is clear.

Unique pertaining to requirements, those that are not overlapping or redundant with other requirements.

Usability Includes ease of learning (learnability), ease of use (efficiency), ease of remembering (memorability), error rate, and subjective pleasing (satisfaction).

Usability Testing Obtaining samples of users and eliciting the reactions of these users about their needs and desires as they interact with prototypes.

Validation Process of determining that the systems engineering process has produced the *right system*, based upon the needs expressed by the stakeholder.

Validation Plan How the qualification data will be used to determine that the real system complies with the originating requirements.

Verifiable Finite, cost-effective process has been defined to check that the requirement has been attained.

Verification Matching of *Configuration items* (CIs), components, subsystems, and the system to their corresponding requirements to ensure that each has been *built right*.

Verification Plan   How the qualification data will be used to determine that the real system conforms to the design that was developed.

White Box Testing   Inner workings of the module are examined as part of the testing to ensure proper functioning. Usually used at the CI level of testing, this method becomes impractical at the system level.

References

Ackoff, R. L. (1960). Systems, organizations, and interdisciplinary research. *Gen. Syst.*, 5, 1–8.

Agha, G. (1985). *Actors: A Model of Concurrent Computation in Distributed Systems*. MIT Press.

Akao, Y. (ed.) (1990). *Quality Function Deployment: Integrating Customer Requirements into Product Design*. Productivity Press, Cambridge, MA.

Alber, I. E. (2012). *Aerospace Engineering on the Back of an Envelope*. Springer, New York.

Alexander, C. (1964). *Notes on the Synthesis of Form*. Harvard University Press, Cambridge, MA

Alford, M. W. (1977). A requirements engineering methodology for real-time processing requirements. *IEEE Trans. Software Eng.*, 3(1), 60–69.

Alford, M. (1985). A graph model based approach to specifications. In *Distributed Systems: Methods and Tools for Specification*. M. Paul and H. J. Siegert (eds.), Berlin: Springer-Verlag, 131–202.

Allen, R. H. (1962). *Morphological Creativity*. Prentice-Hall, Englewood Cliffs, NJ.

Ambler, S. W. (1997). *Building Object Applications that Work*. Cambridge University Press.

Ambler, S. W. (2004). *The Object Primer: Agile Model Driven Development with UML 2.0*. Cambridge University Press.

Anderson, T. and Lee, P. A. (1981). *Fault Tolerance Principles and Practice*. Prentice-Hall, Englewood Cliffs, NJ.

Arciszewski, T. (1988). ARIZ77: An innovative method. *J. Des. Methods Theor.* 22(2), 796–820.

Arrieta, A. B., Díaz-Rodríguez, N., Del Ser, J., Bennetot, A., Tabik, S., Barbado, A., Garcia, S., Gil-Lopez, S., Molina, D., Benjamins, R., Chatila, R., and Herrera, F. (2020). Explainable artificial intelligence (XAI): Concepts, taxonomies, opportunities and challenges toward responsible AI. *Inf. Fusion*,58, 82–115.

Baier, C. and Katoen, J.-P. (2008). *Principles of Model Checking*. MIT Press.

Barney, G. C. (2003). *The Elevator Traffic Handbook: Theory and Practice*. Taylor & Francis, New York.

Barron, F. H. and Barrett, B. E. (1996). Decision quality using ranked attribute weights. *Manage. Sci.*,42(11), 1515–1523.

Baylin, E. N. (1990). *Functional Modeling of Systems*. Gordon & Breach, New York.

Beizer, B. (1990). *Software Testing Techniques*. Van Nostrand-Reinhold, New York.

Bender, E. A. (1978). *An Introduction to Mathematical Modeling*. Dover, Mineola, NY.

von Bertanlanffy, L. (1968). *General System Theory: Foundations, Developments, and Applications*. Braziller, New York.

The Engineering Design of Systems: Models and Methods, Fourth Edition. Dennis M. Buede and William D. Miller
© 2024 John Wiley & Sons, Inc. Published 2024 by John Wiley & Sons, Inc.
Companion website: www.wiley.com/go/engineeringdesignofsystems4e

Berube, M. S. (1991). *The American Heritage Dictionary.* Houghton Mifflin, Boston, MA.

Bias, R. G. and Mayhew, D. J. (eds.) (1994). *Cost-Justifying Usability.* Academic Press, Boston, MA.

Birnbaum, J. (1989). New qualms about the DC-10. *Time,* August 7, p. 20.

Blanchard, B. S. and Fabrycky, W. J. (1998). *Systems Engineering and Analysis.* Prentice-Hall, Upper Saddle River, NJ.

Blum, B. I. (1992). *Software Engineering: A Holistic View.* Oxford University Press, New York.

Boar, B. H. (1984). *Application Prototyping: A Requirements Strategy for the 80's.* Wiley-Interscience, New York.

Boardman, J. and Sauser, B. (2008). *Systems Thinking: Coping with 21st Century Problems.* CRC Press, Boca Raton, FL.

Bock, C. (2006). SysML and UML 2 support for activity modeling. *Syst. Eng.,* 9(2), 160–186.

Boehm, B. W. (1976). Software engineering. *IEEE Trans. Comput.,* C-25, 1226–1241.

Boehm, B. W. (1981). *Software Engineering Economics.* Prentice-Hall, Englewood Cliffs, NJ.

Boehm, B. W. (1986). A spiral model of software development and enhancement. *ACM SIGSOFT Softw. Eng. Notes,* 11(4), 14–24.

Boehm, B. W. (1988). A spiral model of software development and enhancement. *IEEE Comput.,* 21(5), 61–72.

Boehm, B. W. and Papaccio, P. N. (1988). Understanding and controlling software costs. *IEEE Trans. Software Eng.,* 14(10), 1462–1477.

Bohm, C. and Jacopini, G. (1966). Flow diagrams, turing machines, and languages with only two formation rules. *Commun. ACM,* 9(5), 366–371.

Braasch, M. S. (1990). A signal model for UPS. *Navigation,* 37(4), 363–377.

Braun, R. (2003). Need a Lift? An Elevator Queueing Problem. *US Workshop on Mathematical Problems in Industry.* 19th MPI, Worcester, MA.

Brooks, C. G., Grimwood, J. M., and Swenson, L. S., Jr. (1979). *Chariots, for Apollo: A History of Manned Lunar Spacecraft.* NASA, Washington, DC.

Brown, C. M. L. (1988). *Human-Computer Interface Design Guidelines.* Ablex, Norwood, NJ.

Brown, W. H., Malveau, R. C., McCormick, III, H. W., and Mowbray, T. J. (1998). *AntiPatterns: Refactoring Software, Architectures, and Projects in Crisis.* Wiley.

Browning, T. (2001). Applying the design structure matrix to system decomposition and integration problems: A review and new directions, *IEEE Trans. Eng. Manage.,* 48(3), 292–306.

Buede, D. M. (1997). Developing originating requirements: Defining the design problem. *IEEE Trans. Aerosp. Electron. Syst.,* 33(2), 596–609.

Buede, D. M. (1998). The air hag system: What went wrong with the systems engineering. *Syst. Eng.,* 1(1), 90–94.

Buede, D. M. (1999). Functional analysis. In *Handbook of Systems Engineering and Management.* A. P. Sage and W. B. Rouse (eds.), New York: Wiley, 997–1036.

Buede, D. M. and Bresnick, T. A. (2007). Applications of decision analysis to the military systems acquisition process. In *Advances in Decision Analysis: from Foundations to Applications.* W. Edwards, R. F. Miles, and D. von Winterfeldt (eds.), Cambridge University Press.

Buede, D. M. and Choisser, R. W. (1992). Providing an analytic structure for key system design choices. *J. Multi-Criter. Decis. Anal.,* 1, 17–27.

Buede, D. M. and Maxwell, D. T. (1995). Rank disagreement: A comparison of multi-criteria methodologies. *J. Multi-Criter. Decis. Anal.,* 4(1), 1–21.

Chambers, G. J. and Manos, K. L. (1992). Requirements: Their origin, format and control. In *Systems Engineering for the 21st Century.* A. F. Monision and J. M. Wirth (eds.), 2nd Annual International Symposium of NCOSE, 83–90. NCOSE.

Chapanis, A. (1996). *Human Factors in Systems Engineering.* Wiley, New York.

Chapman, W. L., Bahill, A. T., and Wymore, A. W. (1992). *Engineering Modeling and Design.* CRC Press, Boca Raton, FL.

Charbonneau, S. M. (1996). Generation of originating requirements: Use of functional decomposition and state transition diagrams. M.S. Thesis, George Mason University, Fairfax, VA.

Checkland, P. (1993). *Systems Thinking, Systems Practice.* Wiley, Chichester, West Sussex, England.

Childers, S. R. and Long, J. E. (1994). A concurrent methodology for the system engineering design process. In *Systems Engineering: A Competitive Edge in a Changing World*. J. T. Whalen, D. McKinney, and S. Shreve (eds.), 4th Annual International Symposium of NCOSE, 243–248. INCOSE.

Chrissis, M. B., Konrad, M., and Shraun, S. (2011). *CMMI for Development: Guidelines for Process Integration and Product Improvement*. Addison-Wesley, Reading, MA, 3rd Edition.

Chu, W. W. and Tan, L. M.-T. (1987). Task allocation and precedence relations for distributed real-time systems. *IEEE Trans. Comput.*, C-36(6), 667–679.

Chusho, T. (1987). Test data selection and quality estimation based on the concept of essential branches for path testing. *IEEE Trans. Software Eng.*, 13(5), 509–517.

Cigital, (2003). Case study: Finding defects earlier yields enormous savings. Available at www.cigital.com.

Clausing, D. (1994). *Total Quality Development*. ASME Press, New York.

Clemen, R. T. and Reilly, T. (2001). *Making Hard Decisions*. Duxbury Thomson Learning, Pacific Grove, CA.

Cockburn, A. (1997a). Structuring use cases with goals, Part 1. *J. Object-Oriented Program.*, 10(5), 45–51.

Cockburn, A. (1997b). Structuring use cases with goals, Part 2. *J. Object-Oriented Program.*, 10(6), 56–62.

Cockburn, A. (2001). *Writing Effective Use Cases*. Addison-Wesley, New York.

Connell, J. L. and Shafer, L. (1989). *Structured Rapid Prototyping*. Prentice-Hall, Englewood Cliffs, NJ.

Cook, S. C. (2000). What the lessons from large, complex, technical projects tell us about the art of systems engineering. *INCOSE Symposium*, Minneapolis, MN.

Cooper, R. B. (1991). *Introduction to Queueing Theory*. North Holland, New York.

Coulouris, G., Dollimore, J., and Kindberg, T. (1994). *Distributed Systems Concepts and Design*. Addison-Wesley, Wokingham, UK.

Cox, M. E., O'Neal, P., and Pendley, W. L. (1994). LTPAR analysis: Dollar measurement of a usability indicator for software products. In *Cost-Justifying Usability*. R. G. Bias and D. J. Mayhew (eds.), Boston, MA: Academic Press 145–158.

Craik, K. J. W. (1943). *The Nature of Explanation*. Cambridge University Press, Cambridge, UK.

Crowe, D. Smith, H., Haberli, G., Cohen, R. M., and Lykins, H. (1996). Adaptation of a software requirements engineering method to the system level for software-intensive systems. In *Systems Engineering: Practices & Tools*. M. J. Ross and E. E. Barker (eds.), 6th Annual International Symposium of INCOSE, 665–672. Raytheon Electronics.

Dahmann, J. S. (2015). Systems of Systems Characterization and Types. The MITRE Corporation, McLean, VA. NATO/OTAN EN-SCI-276-01.pdf

Daly, E. (1977). Management of software development. *IEEE Transactions of Software Engineering*, 3(3), 229–242.

Dam, S. (2006). *DoD Architecture Framework: A Guide to Applying System Engineering to Develop Integrated, Executable Architectures*. BookSurge Publishing.

Dam, S. (2014). *DoD Architecture Framework 2.0: A Guide to Applying System Engineering to Develop Integrated, Executable Architectures*. SPEC Innovations, Manassas, VA.

Dam, S. H. (2019). *Real MBSE: Model-Based Systems Engineering (MBSE) using LML and Innoslate®*. SPEC Innovations.

Daniels, J., Werner, P. W., and Bahill, A. T. (2001). Quantitative methods for tradeoff analyses, *Syst. Eng.*, 4(3), 190–212.

Davis, A. M. (1990). A comparison of techniques for the specification of external system behavior. In *System and Software Requirements Engineering*. R. H. Thayer and M. Dorfman (eds.), Los Alamitos, CA: IEEE Computer Society Press, 200–217.

Davis, A.M. (1993). *Software Requirements: Objects, Functions, and States*. Prentice Hall, Englewoods Cliffs, NJ.

Davis, A. M. (2005). *Just Enough Requirements Management: Where Software Development Meets Marketing*. Dorset House, New York.

Davis, A. M., Bersoff, E. H., and Corner, E. R. (1990). A strategy for comparing alternative software development life cycles. In *System and Software Requirements Engineering*. R. H. Thayer and M. Dorfman (eds.), Los Alamitos, CA: IEEE Computer Society Press, 496–504.

De Finetti, B. (1974). *Theory of Probability, A Critical Introductory Treatment*, Vol. 1. Wiley, Chichester, UK.

De Marco, T. (1979). *Concise Notes on Software Engineering*. Yourdon Press, New York.

Defense Systems Management College. (1989). *Risk management: Concepts and guidance.* Defense Systems Management College, Ft. Belvoir, VA.

DeFoe, J. C. (ed.) (1993). An identification of pragmatic principles. INCOSE Report, January 21. INCOSE.

Delligatti, L. (2013). *SysML Distilled: A Brief Guide to the Systems Modeling Language.* Addison-Wesley.

DeMarco, T. and Lister, T. (2003). Risk management during requirements. *IEEE Softw.*, 20(5), 99–101.

Denning, P. J., Dennis, J. B., and Qualitz, J. E. (1978). *Machines, Languages, and Computation.* Prentice-Hall, Englewood Cliffs, NJ.

Dickinson, B. W. (1991). *Systems: Analysis, Design, and Computation.* Prentice-Hall, Englewood Cliffs, NJ.

Dietrich, B. L. (1991). A taxonomy of discrete manufacturing systems. *Oper. Res.*, 39(6), 886–902.

Director of Systems Engineering, Office of the Director, Defense Research and Engineering, Department of Defense. (2010). *Systems Engineering Guide for Systems of Systems: Summary.* Washington, DC.

Dixon, A. F. (1926). Development of Communication Systems. *Bell Lab. Rec.*, 2(2), 67–68.

DoD (Department of Defense). (2018). Digital Engineering Working Group. June 2018. Retrieved 11 December 2022.

Dori, D. (2016). *Model-Based Systems Engineering with OPM and SysML.* Springer-Verlag, New York.

Dorny, C. N. (1993). *Understanding Dynamic Systems: Approaches to Modeling, Analysis and Design.* Prentice-Hall, Englewood Cliffs, NJ.

Douglass, B. P. (2002). *Real-Time Design Patterns: Robust Scalable Architecture for Real-Time Systems.* Addison-Wesley, Reading, MA.

Driscoll, P. J. (2007). System life cycle. In *Decision Making in Systems Engineering and Management.* G. S. Parnell, P. J. Driscoll, and D. L. Henderson (eds.), New York: Wiley.

Duato, J., Yalamanchili, S., and Ni, L. (1997). *Interconnection Networks: An Engineering Approach.* IEEE Computer Society Press, Los Alamitos, CA.

Duffy, M. A. and Buede, D. M. (1996). Structured programmatic decision support. Unpublished Technical Report.

Dyer, J. S. (1990). Remarks on the analytic hierarchy process. *Manage. Sci.*, 36, 249–258.

Edwards, W. (1977). How to use multiattribute utility measurement for social decision making. *IEEE Transactions on Systems, Man, and Cybernetics*, 7, 326–340.

Edwards, W. and Barron, F. H. (1994). SMARTS and SMARTER: Improved simple methods for multiattribute utility measurement. *Organizational Behavior and Human Performance*, 60, 306–325.

Elam, J. and Mead, M. (1990). Can software influence creativity? *Inf. Syst. Res.*, 1, 1–22.

Engstrom, E. W. (1957). Systems engineering: A growing concept. *Electr. Eng.*, 76, 113–116.

Eppinger, S. D. (1997). A planning method for integration of large-scale engineering systems. *Proceedings of the International Conference on Engineering Design IDED-97*, Tampere, Finland.

Ericsson, A. and Erixon, G. (1999). *Controlling Design Variants: Modular Product Platforms.* Society of Manufacturing Engineers.

Eriksson, H-E. and Penker, M. (2000). *Business Modeling with UML:Business Patterns at Work.* Wiley, NY.

Fagan, M. (1974). Design and code inspections and process control in the development of programs. IBM Rep. IBM-SDD-TR-21-572.

Fagen, M. D. (ed.) (1978). *A History of Engineering and Science in the Bell System: National Service in War and Peace (1925–1975).* Bell Telephone Laboratories, Inc.New York.

Faulk, S., Brackett, J., Ward, P., and Kirby, J., Jr. (1992). The CoRE method for real-time requirements. *IEEE Software*, 9(5), 22–33.

Federal Information Processing Standards (FIPS) Pub. No. 183. (1993a). *Integration Definition for Function Modeling (IDEFO).* U.S. Dept. of Commerce, Washington, DC.

Federal Information Processing Standards (FIPS) Pub. No. 184. (1993b). *Integration Definition for Information Modeling (IDEFIX).* U.S. Dept. of Commerce, Washington, DC.

Ferrarini, L. and Maroni, M. (1997). A control algorithm for deadlock-free scheduling of manufacturing systems. *1997 IEEE International Conference on Systems, Man and Cybernetics.* Orlando, FL, pp. 3762–3767.

Fienberg, R. T. (1990). The space telescope: Picking up the pieces. *Sky & Telescope*, 80(4), 352–358.

Findley, P. B. (1926). The systems development department. *Bell Lab. Rec.*, 2(2), 69–73.

Fitts, P. M. (ed.) (1951). *Human Engineering for an Effective Air-Navigation and Traffic-Control System*. Ohio State University Research Foundation, Columbus, OH.

Flynn, M. J. (1972). Some computer organizations and their effectiveness. *IEEE Trans. Comput.*, C-21(9), 948–960.

Forrester, J. W. (1961). *Industrial Dynamics*. MIT Press, Cambridge, MA.

Forsberg, K. and Mooz, H. (1992). The relationship of systems engineering to the project cycle. *Eng. Manage. J.*, 4(3),36–43.

Forsberg, K. and Mooz, H. (1995). Application of the 'Vee' to incremental and evolutionary development. In *Systems Engineering in the Global Market Place*. C. Kirkpatrick and C. Wilke (eds.) 5th Annual International Symposium of INCOSE, 801–808.

Forsberg, K. and Mooz, H. (1996). Risk and opportunity management. In *Systems Engineering: Practices and Tools*. M. J. Ross and B. M. McCay (eds.), Vol. 2, 6th Annual International Symposium of INCOSE, 24–36.

Fowler, M. (1960). *Analysis Patterns: Reusable Object Models*. Addison-Wesley, Reading, MA.

Frankel, E. G. (1988). *Systems Reliability and Risk Analysis*. Kluwer Academic Press, Dordrecht, The Netherlands.

Franklin, G. F., Powell, J. D., and Emarni-Naeini, A. (1994). *Feedback Control of Dynamic Systems*. Addison-Wesley, Reading, MA.

Frantz, W. F. (1993). Requirements: A practical, tested approach for breakthrough systems. In *Systems Engineering in the Workplace*. J. E. McAuley and W. H. McCumber (eds.), 3rd Annual International Symposium of NCOSE, 801–810.

Freeman, L. (2020). Test and evaluation for artificial intelligence. *INSIGHT*, 23(1), 27–30.

French, S. (1986). *Decision Theory: An Introduction to the Mathematics of Rationality*. Wiley, Chichester, UK.

Fricke, E. and Schulz, A. P. (2005). Design for changeability (DfC): Principles to enable changes in systems throughout their entire lifecycle. *Syst. Eng.*, 8(4), 342–359.

Friedenthal, S. and Moore, A. (2014). *A Practical Guide to SysML: The Systems Modeling Language*. Morgan Kaufmann.

Friedman, G. J. (2005). *Constraint Theory: Multidimensional Mathematical Model Management*. Springer, New York.

Friend, J. and Hickling, A. (1987). *Planning Under Pressure: The Strategic Choice Process*. Pergamon, Oxford, UK.

Gawer, A. and Cusumano, M. A. (2002). *Platform Leadership: How Intel, Microsoft, and Cisco Drive Industry Innovation*. Harvard Business School Press.

Gentner, D. and Stevens, A. L. (eds.) (1983). *Mental Models*. Erlbaum, Hillsdale, NJ.

Ghahramani, S. (1996). *Fundamentals of Probability*. Prentice-Hall, Upper Saddle River, NJ.

Gilman, G. W. (1953). Systems engineering in bell telephone laboratories. *Bell Lab. Rec.*, 31(1), 1–8.

Glegg, G. L. (1981). *The Development of Design*. Cambridge University Press, Cambridge, UK.

Gobinath, P. and Gupta, R. (1990). Applying compiler techniques to scheduling in real-time systems. *1990 IEEE Real-Time Systems Symposium*, pp. 247–256.

Goldratt, E. M. and Cox, J. (1984). *The Goal: A Process of Ongoing Improvement*. North River Press, Great Barrington, MA.

Gomaa, H. (1993). *Software Design Methods for Concurrent and Real-Time Systems*. Addison-Wesley, Reading, MA.

Goodaire, E. G. and Parmenter, M. M. (1998). *Discrete Mathematics with Graph Theory*. Prentice-Hall, Upper Saddle River, NJ.

Goode, H. H. and Machol, R. E. (1957). *System Engineering — An Introduction to the Design of Large-Scale Systems*. McGraw-Hill, New York.

Gotel, O. C. and Finkelstein, A. C. W. (1994). An analysis of the requirements traceability problem. *Proceedings of the 1st International Conference on Requirements Engineering,* Colorado Springs, CO, 94–101.

Grady, J. O. (1993). *System Requirements Analysis*. McGraw-Hill, New York.

Grady, J. O. (1997). *System Validation and Verification*. CRC Press, Boca Raton, FL.

Griffin, J. M. (2005). *C-5A Systems Engineering Case Study*. Center for Systems Engineering, Air Force Institute of Technology, Wright-Patterson Air Force Base, OH.

Griffin, M. D. (2010). How do we fix system engineering? *61st International Astronautical Congress*, Prague, Czech Republic.

Griffith, P. B. (1994). Different philosophies/different methods: RDD and IDEF. In *Systems Engineering: A Competitive Edge in a Changing World*. J. T. Whalen, D. McKinney, and S. Shreve (eds.), 4th Annual International Symposium of NCOSE, 489–495.

Guindon, R. (1990). Designing the design process: Exploiting opportunistic thoughts. *Human–Computer Interaction*, 5, 305–344.

Haefele, J. W. (1962). *Creativity and Innovation*. Van Nostrand-Reinhold, New York.

Hall, A. (1962). *A Methodology for Systems Engineering*. Van Nostrand, Princeton, NJ.

van den Hamer, P. and Lepoeter, K. (1996). Managing design data: The five dimensions of CAD frameworks, configuration management, and product data management. *Proc. IEEE*, 84(1), 42–56.

Harary, F. (1972). *Graph Theory*. Addison-Wesley, Reading, MA.

Harary, F., Norman, R. Z., and Cartwright, D. (1965). *Structural Models: An Introduction to the Theory of Directed Graphs*. Wiley, New York.

Harel D. (1987). Statecharts: A visual formalism for complex systems. *Sci. Comput. Program.*, 8, 231–273.

Harker, P. T. and Vargas, L. G. (1990). Reply to 'Remarks on the analytic hierarchy process' by J. S. Dyer. *Management Science*, 36, 269–273.

Harwell, R., Aslaksen, E., Hooks, I., Mengot, R., and Ptack, K. (1993). What is a requirement? In *Systems Engineering in the Workplace*, J. E. McAuley and W. H. McCumber (eds.), 3rd Annual International Symposium of NCOSE, pp. 17–24.

Haskins, B., Stecklein, J., Brandon, D., Moroney, G., Lovell, R., and Dabney, J. (2004). Error cost escalation through the project life cycle. *Proceedings of the INCOSE Symposium 2004*.

Hatley, D. J. and Pirbhai, I. A. (1988). *Strategies for Real-Time System Specification*. Dorset House, New York.

Hazelrigg, G. A. (1996). *Systems Engineering: An Approach to Information-Based Design*. Prentice-Hall, Upper Saddle River, NJ.

Hewitt, C. and de Jong, P. (1983). Analyzing the roles of descriptions and actions in open systems. *Proceedings of the National Conference on Artificial Intelligence*.

Hitchins, D. K. (2007). *Systems Engineering: A 21st Century Systems Methodology*. Wiley, Chichester, West Sussex, England.

Hoffman, D. 2001 *Software Engineering Education: Difficult Decisions in Turbulent Times*. In Proc. Canadian Conference on Engineering Education.

Hogarth, R. M. (1980). *Judgment and Choice: The Psychology of Decision*. Wiley, Chichester, UK.

Holmberg, K. and Folkeson, A. (eds.) (1991). *Operational Reliability and Systematic Maintenance*. Elsevier, London.

Honour, E. C. (2006). A practical program of research to measure SE ROI. *Proceedings of the Systems Engineering/Test and Evaluation Conference*, Melbourne, Australia.

Honour, E. (2013). Systems engineering return on investment. Ph.D. Dissertation submitted to University of South Australia.

Hooks, I. (1994). Writing good requirements. In *Systems Engineering: A Competitive Edge in a Changing World*, J. T. Whalen, D. McKinney, and S. Shreve (eds.), 4th Annual International Symposium of NCOSE, pp. 197–203.

Hooks, I. and Farry, K. (2001). *Customer*-Centered Products: Creating Successful Products Through Smart Requirements Management. American Management Association, New York.

Hoppe, M., Levardy, V., Vollerthun, S., and Wenzel, S. (2003). Interfacing a verification, validation, and testing process model with product development methods. *Proceedings of the 13th International INCOSE Symposium*, Crystal City, VA.

Howard, R. A. (1968). The foundations of decision analysis. *IEEE Transactions on Systems, Science, and Cybernetics*, 4, 211–219.

Howard, R. A. (1992). In praise of the old time religion. In *Utility Theories: Measurements and Applications*, W. Edwards (ed.), Kluwer Academic Publishers, Boston, MA, 27–55.

Howard, R. A. (1993). Professional decision analysis. Unpublished manuscript.

Hunger, J. W. (1995). *Engineering the System Solution*. Prentice-Hall, Englewood Cliffs, NJ.

INCOSE. (2007). *Systems Engineering Vision 2020*. INCOSE-TP-2004-004-02. International Council on Systems Engineering, San Diego, CA.

INCOSE 2019, https://www.incose.org/about-systems-engineering/system-and-se-definitions/systems-engineering-definition.

INCOSE. (2022a). *Systems Engineering Principles*. International Council on Systems Engineering, San Diego, CA.

INCOSE. (2022b). *Systems Engineering Vision 2035*. International Council on Systems Engineering, San Diego, CA.

INCOSE (International Council on Systems Engineering). (1999). http://www.incose.org/whatis-html.

INCOSE Natural Systems Working Group. (2013). Influencing Bullet Train System Engineering. https://docs.google.com/viewer?a=v&pid=sites&srcid=ZGVmYXVsdGRvbWFpbnxpbmNv2Vuc3dnfGd4OjdlYzI5NWY2MWM1ZjFlNTk (accessed 7 September 2014).

International Standards Organization (ISO). (2015). Automation Systems and Integration: Object Process Methodology (OPM). ISO/PAS 19450:2015.

International Standards Organization (ISO). (2019). Systems and software engineering — Taxonomy of systems of systems. ISO/IEC/IEEE 21841:2019(en).

Jackson, S. (2007). System resilience: Capabilities, culture and infrastructure. *Proceedings of the 17th International INCOSE Symposium*, San Diego, CA, June, 2007.

Jacky, J. (1990). Risks in medical electronics. *Commun. ACM*, 33(12),138.

Jacobson, I. (1995). *The Object Advantage: Business Process Reengineering with Object Technology*. Addison-Wesley, Workingham, UK.

Jacobson, I., Christerson, M., Jansson, P., and Overgaard, G. (1992). *Object-Oriented Software Engineering, A Use Case Driven Approach*. Addison-Wesley, Reading, MA.

Jagacinski, R. J. and Miller, R. A. (1978). Describing the human operator's internal model of a dynamic system. *Hum. Factors*, 20, 425–433.

Jalote. P. (1994). *Fault Tolerance in Distributed Systems*. Prentice-Hall, Englewood Cliffs, NJ.

Jelassi, M. and Foroughi, A. (1989). Negotiation support systems: An overview of design issues and existing software. *Decision Support Systems*, 5,167–181.

Johnson, B. W. (1989). *Design and Analysis of Fault Tolerant Digital Systems*. Addison-Wesley, Reading, MA.

Johnson, S. B. (2002). *The Secret of Apollo: Systems Management in American and European Space Programs*. Johns Hopkins University Press, Baltimore, MD.

Johnson-Laird, P. (1983). *Mental Models*. Harvard University Press, Cambridge, MA.

Jones, M. (1997). What really happened on Mars Rover *Pathfinder*. Email message, December, 11.

Jones, D. R. and Schkade, D. A. (1995). Choosing and translating between problem representations. *Organizational Behavior and Human Decision Processes*, 61(2), 214–223.

Karangelen, N. E. and Hoang, N. T. (1994). Partitioning complex system design into five views. In *Systems Engineering: A Competitive Edge in a Changing World*, J. T. Whalen, D. McKinney, and S. Shreve (eds.), 4th Annual International Symposium of NCOSE, 675–681.

Kauffman, Jr., D. L. (1980). *Systems One: An Introduction to Systems Thinking*. S. A. Carlton, Minneapolis, MN.

Kee, C., Parkinson, B. W., and Axlerad, P. (1991). Wide area differential GPS. *Navigation*, 38(2), 123–144.

Keeney, R. L. (1992). *Value-Focused Thinking*. Harvard University Press, Boston, MA.

Keeney, R. L. and Raiffa, H. (1976). *Decisions with Multiple Objectives: Preferences and Value Tradeoffs*. Wiley, New York.

Keller, L. and Ho, J. (1988). Decision problem structuring: Generating options. *IEEE Transactions on Systems, Man, and Cybernetics*, 15, 715–728.

Kelly, M. J. (1950). The bell telephone laboratories – An example of an institute of creative technology. *Proceedings of the Royal Society, London B*, 137, 419–433.

Kidder, J. T. (1981). *The Soul of a New Machine*. Little Brown, Boston, MA.

Kirkwood, C. W. (1997). *Strategic Decision Making*. Duxbury Press, Belmont, CA.

Kirkwood, C. W. and Corner, J. L. (1993). The effectiveness of partial information about attribute weights for ranking alternatives in multiattribute decision making. *Organizational Behavior and Human Performance*, 54, 456–476.

Kleindorfer, P. R., Kunreuther, H. C., and Schoemaker, P. J. H. (1993). *Decision Sciences: An Integration Perspective*. Cambridge University Press, Cambridge, UK.

Klir, G. J. (1985). *Architecture of Systems Problem Solving*. Plenum Press, New York.

Kopp, C. (1994). *The Sidewinder Story: The Evolution of the AIM-9 Missile*. Australian Aviation.

Kossiakoff, A. and Sweet, W. N. (2003). *Systems Engineering Principles and Practice*, Wiley, New York.

Kuo, B. C. and Golnaraghi F. (2003). *Automatic Control Systems*. Wiley, Hoboken, NJ.

Kwinn, Jr., M. J. and Parnell, G. S. (2007). Decision making. In *Decision Making in Systems Engineering and Management*, G. S. Parnell, P. J. Driscoll, and D. L. Henderson (eds.), Wiley, New York.

Lake, J. (1992). Systems engineering re-energized: Impacts of the revised DoD acquisition process. *Eng. Manage. J.*, 4(3),8–14.

Lano, R. I. (1990a). A structured approach for operational concept formulation. In *System and Software Requirements Engineering*,R. H. Thayer and M. Dorfman (eds.), IEEE Computer Society Press, Los Alamitos, CA, 48–59.

Lano, R. J. (1990b). The N^2 chart. In *System and Software Requirements Engineering*,R. H. Thayer and M. Dorfman (eds.), IEEE Computer Society Press, Los Alamitos, CA, 244–271.

Larsen, R. F. and Buede, D. M. (2002). Theoretical framework for the continuous early validation (CEaVa) method, *Syst. Eng.*, 5(3), 223–241.

Lawless, W. F., Llinas, J., Sofge, D. A., and Mittu, R. (eds.) (2021a). *Engineering Artificially Intelligent Systems: A Systems Engineering Approach to Realizing Synergistic Capabilities*, Vol. 13000. Lecture Notes in Computer Science. Cham: Springer International Publishing.

Lawless, W. F., Mittu, R., Sofge, D. A., Shortell, T., and McDermott, T. A. (eds.) (2021b). *Systems Engineering and Artificial Intelligence*. Cham: Springer International Publishing.

Lee, D. and Yannakakis, M. (1996). Principles and methods of testing finite state machines — A survey. *Proc. IEEE*, 84(8), 1090–1123.

Levardy, V., Hoppe, M., and Honour, E. (2004). Verification, validation, and testing strategy and planning procedure. *Proceedings of the 14th International INCOSE Symposium*, Toulouse, France.

Levi, S. and Agrawala, A. K. (1994). *Fault Tolerant System Design*. McGraw-Hill, New York.

Levis, A. (1993). *National Missile Defense (NMD) Command and Control Methodology Development*. Contract Data Requirements List A005 report for US Army Contract MDA 903-88-0019, Delivery Order 0042. George Mason University, Center of Excellence in Command, Control, Communications, and Intelligence, Fairfax, VA.

Levis, A. H. and Wagenhals, L. W. (2000). C4ISR architectures: I. Developing a process for C4ISR architecture design. *Syst. Eng.*, 3(4), 225–247.

Levis, A. H., Moray, H., and Flu, B. (1994). Task decomposition and allocation problems and discrete event systems. *Automatica*, 30(2), 203–216.

Lifecycle Modeling Organization (LM0). (2022). Lifecycle Modeling Language (LML) Specification. http://www.lifecyclemodeling.org/spec/current.

Lindley, D. (1994). Foundations. In *Subjective Probability*,G. Wright and P. Ayton (eds.), Wiley, Chichester, UK, 3–15.

Lions, J. L. (1996). Ariane 5: Flight 501 failure. Report by the Inquiry Board, Paris.

Lovell, J. and Kluger, J. (1994). *Apollo 13 (previously titled Lost Moon)*. Pocket Books, New York.

MacKinnon, D., McCrum, W., and Sheppard, D. (1990). *An Introduction to Open Systems Interconnection*. Computer Science Press, New York.

Magee, C. L. and de Weck, O. L. (2004). Complex system classification. *Proceedings of the 14th Annual International Symposium of INCOSE*.

Magnuson, E. (1989). Brace! Brace! Brace! *Time*, July 31, pp. 12–15.

Maier, M. W. (1998). Architecting principles for systems of systems. *Syst. Eng.*, 1(4), 267–284.

Manna, Z. and Waldinger, R. (1978). The logic of computer programming. *IEEE Trans. Software Eng.*, 4, 199–220.

Mar, B. W. (1994). Requirements for development of software requirements. In *Systems Engineering: A Competitive Edge in a Changing World*,J. T. Whalen, D. McKinney, and S. Shreve (eds.), 4th Annual International Symposium of INCOSE, 39–44.

Marca, D. A. and McGowan, C. L. (1988). *SADT: Structured Analysis and Design Technique*. McGraw-Hill, New York.

Marshall, C., Nelson, C., and Gardiner, M. M. (1987). Design guidelines. In *Applying Cognitive Psychology to User-baerface Design*,M. M. Gardiner and B. Christie (eds.), Wiley, Chichester, UK, 221–278.

Martin, J. N. (2004). The seven samurai of systems engineering: Dealing with the complexity of the 7 interrelated systems. *Proceedings of the 14th International INCOSE Symposium*.

Maxwell, J. C. (1868). On governors. *Proceedings of the Royal Society of London* 16. (Reprinted in (1964) *Selected Papers on Mathematical Trends in Control Theory*. Dover, New York.

Mayhew, D. J. (1992). *Principles and Guidelines in Software User Interface Design*. Prentice-Hall, Englewood Cliffs, NJ.

Mayr, O. (1970). *The Origins of Feedback and Control* [translated from *Zur Fruhgeschichte der technischen Regelungen]*. MIT Press, Cambridge, MA.

McGibbon, T., (2003). Return on investment from software process improvement. Available at www.dacs.dtic .mil.

McLean, W. B. (1960). Management and the creative scientist. *California Management Review*, 3(1), 9–11.

McLean, W. B. (1962). The sidewinder missile program. In *Science, Technology, and Management, The Proceedings of the National Advanced-Technology Management Conference*, F. Kast and J. Rosenzweig (eds.), McGraw-Hill, New York.

McMenamin, S. M. and Palmer, J. F. (1984). *Essential Systems Analysis*. Prentice-Hall, Englewood Cliffs, NJ.

Meisenzahl, J., de la Cruz, M., and Vollerthun, A. (2006). Establishing a verification and validation process in automotive development: Increasing product quality while reducing costs. *Proceedings of the 16th International INCOSE Symposium*, Orlando, FL.

Merkhofer, M. W. (1987). Quantifying judgmental uncertainty: Methodology, experiences, and insights. *IEEE Transactions on Systems, Man, and Cybernetics*, 17, 741–752.

Meyer, M. H. and Lehnerd, A. P. (1997). *The Power of Product Platforms*. The Free Press, New York.

Military Standard. (1974). MIL-STD 499A. Systems Engineering.

Military Standard. (1993a). MIL-STD 499B (draft). Systems Engineering.

Military Standard. (1993b). MIL-STD 881B. Work Breakdown Structure.

Military Standard. (2022). MIL-STD 881F. Work Breakdown Structure.

Miller, J. G. (1978). *Living Systems*. McGraw-Hill, New York.

Miller, W. and D. Verma. (2020). From the editor-in-chief and systems engineering research center director. *INSIGHT*,23(1), 6–7.

Milliken, W. F. and Milliken, D. L. (1995). *Race Car Vehicle Dynamics*. SAE International, Warrendale, PA.

Milner, R. (1999). *Communicating and Mobile Systems: The π-calculus*. Cambridge University Press, Cambridge, UK.

Mindell, D. A. (2002). *Between Human and Machine: Feedback, Control, and Computing Before Cybernetics*. Johns Hopkins University Press, Baltimore, MD and London.

Mott, J. L., Kandel, A., and Baker, T. P. (1986). *Discrete Mathematics for Computer Scientists and Mathematicians*. Prentice-Hall, Englewood Cliffs, NJ.

Mowbray, T. J. and Malveau, R. C. (1997). *CORBA Design Patterns*. Wiley.

Mowbray, T. J. and Ruh, W. A. (1997). *Inside CORBA: Distributed Object Standards and Applications*. Addison-Wesley, Reading, MA.

Mowbray, T. J. and Zahavi, R. (1995). *The Essential CORBA: Systems Integration Using Distributed Objects*. Wiley, New York.

Murata, T. (1989), Petri nets: Properties, analysis and applications. *Proc. IEEE*, 77(4), 541–580.

Murray, C. and Cox, C. B. (1989). *Apollo: The Race to the Moon*. Simon & Schuster, New York.

Nagel, S. S. (1989). *Evaluation Analysis with Microcomputers*. JAI Press, Greenwich, CT.

Newell, A. (1969). Heuristic programming: Ill structured problems. In *Progress in Operations Research*,J. Aronofsky (ed.), Wiley, New York, 362–414.

Nielsen, J. (1993). *Usability Engineering*. AP Professional, Boston, MA.

Nii, H. P. (1986). Blackboard systems: Blackboard applications systems, blackboard systems from a knowledge engineering perspective. *AI Magazine*, 7(3),82–106.

NSF. (2023). Cyber-physical Systems.

Object Management Group (OMG). (2017a). Unified Modeling Language (UML). https://www.uml.org/.

Object Management Group (OMG). (2017b). Systems Modeling Language (SysML). https://sysml.org/.

ODASD (SE) Office of the Deputy Assistant Secretary of Defense (Systems Engineering). (2017). *DAU Glossary: Digital Engineering*. Defense Acquisition University.

Oliver, D. W., Kelliher, T. P., and Keegan, J. G., Jr. (1997). *Engineering Complex Systems with Models and Objects*. McGraw-Hill, New York.

Orfali, R., Harkey, D., and Edwards, J. (1997). *Instant CORBA*. Wiley, New York.

Ottaway, D. B. (1996). A safety device with a fatal flaw. *Washington Post,* October 27, pp. Al, A8–A9.

Pages, A. and Gondran, M. (1986). *System Reliability: Evaluation and Prediction in Engineering*. Springer-Verlag, New York.

Paraskevopoulos, P. N. (2002). *Modern Control Engineering*. Marcel Dekker, Inc., New York.

Parnell, G. S., Bresnick, T. A., Tani, S. N., and Johnson, E. R. (2013). *Handbook of Decision Analysis*. Wiley, Hoboken, NJ.

Pavlina, S. (2003). Zero-defect software development, Dexterity Software, Available at www.dexterity.com.

Pennington, N. (1985). Stimulus structures and mental representations in expert comprehension of computer programs, Tech. Rep. No. 2-ONR. University of Chicago, Graduate School of Business, Chicago.

Perdu, D. M. and Levis, A. H. (1993). Requirements determination using the Cube tool methodology and Petri nets. *IEEE Transactions on Systems, Man, and Cybernetics*, 23(5), 1255–1264.

Perry, W. E. (1988). *A Structured Approach to Systems Testing*. QED Information Sciences, Wellesley, MA.

Petersen, C. C. and Brandt, J. C. (1995). *Hubble Vision: Astronomy with the Hubble Space Telescope*. Cambridge University Press, Cambridge, UK.

Petroski, H. (1994). *Design Paradigms: Case Histories of Error and Judgment in Engineering*. Cambridge University Press, New York.

Phan, P. (2011). *Expanding Constraint Theory to Determine Well-Posedness of Large Mathematical Models*. UMI Dissertation Publishing, Ann Arbor, MI.

Pohl, E. (2007). System effectiveness. In *Decision Making in Systems Engineering and Management*, G. S. Parnell, P. J. Driscoll, and D. L. Henderson (eds.), Wiley, New York.

Pohl, E. and Nachtmann, H. (2007). Life cycle costing. In *Decision Making in Systems Engineering and Management*, G. S. Parnell, P. J. Driscoll, and D. L. Henderson (eds.), Wiley, New York.

Powell, R. A. (2002). A definition of decisions in the engineering of systems. Ph.D. dissertation, Stevens Institute of Technology, Hoboken, NJ.

Prang, J. (1992). Controlling life-cycle costs through concurrent engineering. In *Addendum to the ATE & Instrumentation Conference Proceedings*, Miller-Freeman, Anaheim, CA, p. 1.

Prasad, B. (1996). *Concurrent Engineering Fundamentals: Integrated Product and Process Organization*, Vol. 1. Prentice-Hall, Upper Saddle River, NJ.

Price, H. E. (1985). The allocation of functions in systems. *Hum. Factors*, 27(1), 33–45.

Pugh, S. (1991). *Total Design— Integrating Methods for Successful Product Engineering*. Addison-Wesley, Reading, MA.

Rasmussen, J. (1979). On the structure of knowledge—A morphology of mental models in a man-machine system context, Tech. Rep. No. Riso-M-2192. Riso National Laboratory, Roskilde, Denmark.

Rational Software Corporation. (1997). *Unified Modeling Language: Notation Guide*. Rational Software Corporation, Cupertino, CA.

Raz, A. K., W. Miller, K.-C. Chang, Y. Lin, and E. Blasch. (2023). Explainable AI and counterfactuals for test and evaluation of intelligent engineered systems *Proceedings of the 32nd International INCOSE Symposium*, Honolulu, HI.

Reason, J. (1990). *Human Error*. Cambridge University Press, Cambridge, UK.

Reed, M. A. (1993). Requirements traceability on the F-22 program. In *Systems Engineering in the Workplace*, J. E. McAuley and W. H. McCumber (eds.), 3rd Annual International Symposium of NCOSE, 293–300.

Reitman, W. R. (1965). *Cognition and Thought*. Wiley, New York.

Richardson, D. J. and Clarke, L. A. (1985). Partition analysis: A method combining testing and verification. *IEEE Trans. Software Eng.*, 11(12), 1477–1490.

Rittel, H. (1972). On the planning crisis: Systems analysis for the first and second generations. *Beprifts Konotnen*, 8, 390–396.

Roberts, R. A. (1992). *An Introduction to Applied Probability*. Addison-Wesley, Reading, MA.

Roese, N. J. (1997). Counterfactual thinking. *Psychological Bulletin*, 121(1), 133–48.

Rosen, K. H. (1995). *Discrete Mathematics and Its Applications*. McGraw-Hill, New York.

Ross, A. M., Diller, N. P., Hastings, D. E., and Warmkessel, J. M.2004). Multi-attribute tradespace exploration with concurrent design as a front-end for effective space system design. *Journal of Spacecraft and Rockets*, 41(1), 20–28.

Rothman, J., 2002. "What does it cost to fix a defect?" *StickyMinds.com*. Available at http://www.stickyminds .com/stqeletter/archive/20020220nl.asp

Royce, W. W. (1970). Managing the development of large systems: Concepts and techniques. *Proceedings of the 9th International Conference on Software Engineering*, pp. 328–338. ACM, New York.

Ruijgrok, G. J. J. (2009). *Elements of Airplane Performance*. VSSD. Delft, The Netherlands.

Saaty, T. L. (1980). *The Analytical Hierarchy Process*. McGraw-Hill, New York.

Saaty, T. L. (1986). Axiomatic foundation of the analytic hierarchy process. *Manage. Sci.*, 32, 841–855.

Sabbagh, K. (1996). *21st Century Jet: The Making and Marketing of the Boeing 777*. Scribner, New York.

Sage, A. P. (1992). *Systems Engineering*. Wiley, New York.

Sagoo, J. S. and Boardman, J. T. (1998). Towards the formalisation of soft systems models using Petri net theory, *IEE Proceedings-Control Theory and Applications*, 145, 463–471.

Sailor, J. D. (1990). System engineering: An introduction. In *System and Software Requirements Engineering*,R. H. Thayer and M. Dorfman (eds.), IEEE Computer Society Press, Los Alamitos, CA, 35–47.

Samson, D. (1993). Knowledge-based test planning: Framework for a knowledge-based system to prepare a system test plan from system requirements. *J. Syst. Software*, 20, 115–124.

Sangiorgi, D. and Walker, D. (2001). *The π-calculus: A Theory of Mobile Processes*. Cambridge University Press, Cambridge, UK.

Savage, L. J. (1954). *The Foundations of Statistics*. Wiley, New York.

Savage, S. L. (2009). *The Flaw of Averages*. Wiley, Hoboken, NJ.

Scheiber, S. F. (1995). *Building a Successful Board-Test Strategy*. Butterworth-Heinemann, Boston, MA.

Schlager, K. J. (1956). Systems engineering — Key to modern development. *IRE Transactions of Professional Group Engineering Management*, 3, 64–66.

Schmekel, H. and Wingard, L. (1993). Consistency and completeness of multiple models in product development. In *Concurrent Engineering: Methodology and Applications*,P. Gu and A. Kusiak (eds.), Elsevier, Amsterdam, 31–68.

Schwartz, M. (1987). *Telecommunication Networks: Protocols, Modeling and Analysis*. Addison-Wesley, Reading, MA.

SEBoK. (2021). Model-Based Systems Engineering (MBSE) (glossary). www.sebokwiki.org. Retrieved 2021-04-06.

SEBoK. (2022). Digital Engineering. www.sebokwiki.org. Retrieved 2022-12-12.

SEBoK Editorial Board. (2023). *The Guide to the Systems Engineering Body of Knowledge (SEBoK)*, v. 2.8, R. J. Cloutier (Editor in Chief). Hoboken, NJ: The Trustees of the Stevens Institute of Technology. www .sebokwiki.org. Accessed 12 August 2023. BKCASE is managed and maintained by the Stevens Institute of Technology Systems Engineering Research Center, the International Council on Systems Engineering, and the Institute of Electrical and Electronics Engineers Systems Council.

Sen, A. K. (1970). *Collective Choice and Social Welfare*. Holden-Day, San Francisco, CA.

Senge, P. M. (1990). *The Fifth Discipline: The Art & Practice of The Learning Organization*. Doubleday, New York.

Shachter, R. D. (1986). Evaluating influence diagrams. *Oper. Res.*, 34, 871–882.

Shachter, R. D. (1990). An ordered examination of influence diagrams. *Networks*, 20, 535–563.

Shannon, C. (1948). A mathematical theory of communication. *The Bell System Technical Journal*, 27(379–423), 623–656.

Sheridan, T. B. and Verplanck, W. L. (1978). *Human and Computer Control of Undersea Teleoperators*. Report of Man-Machine Systems Lab. Dept. of Mech. Eng. MIT, Cambridge, MA.

Shin, I. and Levis, A. H. (2003). Performance prediction of networked information systems via Petri nets and queuing nets, *Syst. Eng.*, 6(1), 1–18.

Shlaer, S. and Mellor, S. (1996). *How to Build Object Models*. Yourdon Press, Upper Saddle River, NJ.

Shneiderman, B. (1992). *Designing the User Interface*. Addison-Wesley, Reading, MA.

Shuey, R. L., Spooner, D. L., and Frieder, O. (1997). *The Architecture of Distributed Computer Systems.* Addison-Wesley, Reading, MA.

Simon, H. A. (1973). The structure of ill-structured problems. *Artif. Intell.*, 4, 145–180.

Simpson, T. W., Siddique, Z., and Jiao, R. J. (eds.). (2007). *Product Platform and Product Family Design: Methods and Applications.* Springer.

Sinnott, R. W. (1990). HST's magnificent optics... What went wrong? *Sky & Telescope*, 80(4), 356–357.

Spetzler, C. S. and von Holstein, C. A. S. (1975). Probability encoding in decision analysis. *Manage. Sci.*, 22, 340–385.

Stevens, R. and Martin, J. (1995). What is requirements management? In *Systems Engineering in the Global Market Place*, Vol. 2, C. Kirkpatrick and C. Wilke (eds.), 5th Annual International Symposium of INCOSE, 11–32.

Stillwell, W. G., Seaver, D. A., and Edwards, W. (1981). A comparison of weight approximation techniques in multiattribute utility decision making. *Organizational Behavior and Human Performance*, 28, 62–77.

Suh, N. P. (1990). *The Principles of Design.* Oxford University Press, New York.

Taguchi, G. (1993). *Taguchi on Robust Technology.* ASME Press, New York.

Terninko, J., Zusman, A., and Zlotin, B. (1996). *Step-by-step TRIZ: Creating Innovative Solution Concepts.* Responsible Management, Nottingham, NH.

Thurston, D. L. and Carnahan, J. V. (1993). Intelligent evaluation of designs for manufacturing cost. In *Concurrent Engineering: Automation, Tools, and Techniques*, A. Kusiak (ed.), Wiley, New York, 437–461.

Ulvila, J. W. and Snider, W. D. (1980). Negotiation of international oil tanker standards: An application of multiattribute value theory. *Oper. Res.*, 28, 81–96.

US DoD Digital Engineering Working Group. June 2018. Retrieved 11 Dec 2022.

Van de Vegte, J. (1994). *Feedback Control Systems.* Prentice-Hall. Englewood Cliffs, NJ.

VanGundy, A. B. (1988). *Techniques of Structured Problem Solving.* Van Nostrand-Reinhold, New York.

Veldhuyzen, W. and Stassen, H. G. (1977). The internal model concept: An application to modeling human control of large ships. *Hum. Factors*, 19, 367–380.

Voges, U. and Taylor, J. R. (1985). Systematic testing. In *Verification and Validation of Real-Time Software*, W. J. Quirk (ed.), Springer-Verlag, Berlin, 115–146.

Von Neumann, J. and Morgenstern, O. (1947). *Theory of Games and Economic Behavior.* Princeton University Press, Princeton, NJ.

Walters, J. M. (1994). Systems engineering applied to strategic planning: The LASE follow-on study. In *Systems Engineering: A Competitive Edge in a Changing World*, J. T. Whalen, D. McKinney, and S. Shreve (eds.), 4th Annual International Symposium of INCOSE, 889–895.

Walton, M. and Hastings. D. (2004). Applications of uncertainty analysis to architecture selection of satellite systems. *Journal of Spacecraft and Rockets*, 41(1), 75–84.

Warfield, J. N. (1990). *A Science of Generic Design: Managing Complexity through Systems Design*, Vols. 1 and 2. Intersystems Publications, Salinas, CA.

Warfield, J. N. (2006). *An Introduction to Systems Science.* World Scientific, Hackensack, NJ.

Watson, S. R. and Buede, D. M. (1987). *Decision Synthesis: The Principles and Practice of Decision Analysis.* Cambridge University Press, Cambridge, UK.

Weinstein, L. and Adam, J. A. (2008). *Guesstimation: Solving the World's Problems on the Back of a Cocktail Napkin.* Princeton University Press, Princeton, NJ.

Wenzel, S., Bauch, T., Fricke, E., and Negele, H. (1997). Concurrent engineering and more ... A systematic approach to successful product development. In *Systems Engineering: A Necessary Science*, L. M. Hritz and E. E. Barker (eds.). 7th Annual International Symposium of INCOSE, 617–624.

West, P. D. (2007). Solution design. In *Decision Making in Systems Engineering and Management*, G. S. Parnell, P. J. Driscoll, and D. L. Henderson (eds.), Wiley, New York.

Westrum, R. (1999). *Sidewinder: Creative Missile Development at China Lake.* Naval Institute Press, Annapolis, MD.

Wiener, N. (1961). *Cybernetics: Or Control and Communication in the Animal and the Machine.* MIT Press, Cambridge, MA.

Wieringa, R. J. (1995). Combining static and dynamic modeling methods: A comparison of four methods. *The Computer Journal*, 38(1), 17–30.

Wiklund, M. E. (ed.) (1994). *Usability in Practice*. AP Professional, Boston, MA.

Wilner, D. (1997). *Vx-Files: What Really Happened on Mars*. Keynote address at IEEE Real-Time Systems Symposium, San Francisco, CA.

von Winterfeldt, D. and Edwards, W. (1986). *Decision Analysis and Behavioral Research*. Cambridge University Press, New York.

Wolverton, M. (2001). The Spacecraft That Will Not Die. *Invention and Technology*, Winter, 47–58.

Wright, G. and Ayton, P. (eds.) (1994). *Subjective Probability*. Wiley, Chichester, UK.

Wymore, A. W. (1967). *Mathematical Theory of Systems Engineering*. Wiley, New York.

Wymore, A. W. (1993). *Model-based Systems Engineering*. CRC Press, Boca Raton, FL.

Yager, R. R. (1978). Fuzzy decision-making including unequal objectives. *Fuzzy Sets Syst.*, 1, 87–95.

Yoon, K. (1980). Systems selection by multiple attribute decision making. Ph.D. Dissertation for Kansas State University, Manhattan.

Yourdon, E. (1989). *Modern Structured Analysis*. Yourdon Press, Englewood Cliffs, NJ.

Yourdon Inc. (1993). *Yourdon Systems Method*. Yourdon Press, Englewood Cliffs, NJ.

Zwicky, F. (1969). *Discovery, Invention, Research through the Morphological Approach*. Macmillan, New York.

Historical References

Affel, H. A. Jr. (1964). System engineering. *Int. Sci. Technol.*, 35, 18–26, 79–82.

Alexander, C. (1964). *Notes on the Synthesis of Form*. Harvard University Press, Cambridge, MA.

Allen, T. H. (1967). Systems engineering bibliography. *Ind. Qual. Control*, 24, 317–321.

Asimow, M. (1962). *Introduction to Design*. Prentice Hall, Englewood Cliffs, NJ.

Barlow, R. E. (1965). *Mathematical Theory of Reliability*, Wiley, New York.

Blanchard, B. S. (1967). Cost effectiveness, system effectiveness, integrated logistic support, and maintainability. *IEEE Trans. Reliab.*, R-16, 117–126.

Blanchard, B. S. and Lowery, E. E. (1969). *Maintainability Principles and Practices*. McGraw-Hill Book Co., New York.

Bogusla, R. (1965). *The New Utopians: A Study in Systems Design and Social Change*. Prentice-Hall, Englewood Cliffs, NJ.

Brooks, F. P. (1962). Architectural philosophy. In *Planning a Computer System*, W. Bucholz (ed.), New York: McGraw-Hill, 5–16.

Chapanis, A. (1961). Men, machines, and models. *Am. Psychol.*, XVI(3), 113–131.

Chestnut, H. (1965). *Systems Engineering Tools*. Wiley, New York.

Chestnut, H. (1967). *Systems Engineering Methods*. Wiley, New York.

Churchman, C. W. (1968). *The Systems Approach*. Delta.

Connelly, M. E. (1961). System design. In *Handbook of Automation, Computation and Control*, S. Ramo, E. M. Grabble, and D. E. Woolridge (eds.), Section III. New York: Wiley.

Dept. of Army. (1969). *A Guide to Systems Engineering*.

Deutsch, R. (1969). *Systems Analysis Techniques*. Prentice Hall, Englewood Cliffs, NJ.

Dienemann, P. F. (1966). *Estimating Cost Uncertainty Using Monte Carlo Techniques*. The Rand Corporation, Santa Monica CA, January, RM-4854-PR.

Dixon, J. R. (1966). *Design Engineering — Inventiveness, Analysis, and Decision-Making*. McGraw-Hill, New York.

Dommasch, D. O. and Landeman, C. W. (1962). *Principles Underlying Systems Engineering*. Pitman, New York.

Eckman, D. P. (ed.) (1961). *Systems: Research and Design* (Proceedings of The First Systems Symposium at Case Institute of Technology). Wiley, New York.

The Engineering Design of Systems: Models and Methods, Fourth Edition. Dennis M. Buede and William D. Miller
© 2024 John Wiley & Sons, Inc. Published 2024 by John Wiley & Sons, Inc.
Companion website: www.wiley.com/go/engineeringdesignofsystems4e

Ellis, D. O. and Ludwig, F. J. (1962). *Systems Philosophy*. Prentice-Hall, Englewood Cliffs, NJ.

Fields, D. S. (1966). Cost/effectiveness analysis: Its tasks and their relationships. *Oper. Res.*, 14, 515–527.

Flagle, C. D., Huggins, W. H., and Roy, R. H. (eds.) (1960). *Operations Research and Systems Engineering*, Johns Hopkins.

Forrestor, J. W. (1968). *Principles of Systems*. MIT Press, Cambridge, MA.

Frosch, R. A. (1969). A new look at systems engineering. *IEEE Spectr.*, 6, 24–28.

Gagne, R. M. (ed.) (1962). *Psychological Principles in System Development*. Holt, Rinehart and Winston, New York.

Gibson, J. E. (1968). *Introduction to Engineering Design*. Holt, Rinehart, and Winston, New York.

Glegg, G. L. (1969). *The Design of Design*. Cambridge University Press, Cambridge.

Goldman, A. and Slattery, T. (1967). *Maintainability – A Major Element of System Effectiveness*. Wiley, New York.

Goode, H. H. and Machol, R. E. (1957). *System Engineering – An Introduction to the Design of Large-Scale Systems*. McGraw-Hill, New York.

Gosling, W. (1962). *The Design of Engineering Systems*. Heywood.

Hall, A. (1962). *A Methodology for Systems Engineering*. Van Nostrand, Princeton, NJ.

Heaton, D. H. (1969). System/cost effectiveness analysis in the system engineering process. *Defense Ind. Bull.*, 34–37.

Hill, P. H. (1968). *The Science of Engineering Design*. Holt, Rinehart, and Winston, New York.

Hino, K. (1965). A standardized approach to systems engineering. *Signal*, 41–42.

Jenkins, G. M. (1969). The systems approach. *J. Syst. Eng.*, I, 3–49.

Jerger, J. L. (1960). *Systems Preliminary Design, Principles of Guided Missile Design*. Van Nostrand, Princeton, NJ.

Johnson, R. A., Kast, F. E., and Rosenzweig, J. E. (1963). *The Theory and Management of Systems*. McGraw-Hill, New York.

Jones, J. C. and Thornley, D. G. (1963). *Conference on Design Methods*. Pergamon, Oxford.

Krick, E. V. (1965). *An Introduction to Engineering and Engineering Design*. Wiley, New York.

Laut, S. (1968). Subsystem optimization effectiveness improvement by the option trade-off analysis process. *IEEE Trans. Syst. Sci. Cybern.*, SSC-4, 133–137.

Machol, R. E. (1965). *Systems Engineering Handbook*. McGraw-Hill.

Masso, A. H., and Rudd, D. F. I. (1969). The synthesis of system design: [Part] II, Heuristic structuring. *AIChE J.*, 15(1), 10–17.

Matousek, R. (1963). *Engineering Design: A Systematic Approach*. Blackie, Glasgow.

McCormick, E. J. (1957). *Human Factors Engineering*. McGraw-Hill, New York.

Meister, D. and Rabideau, G. F. (1965). *Human Factors Evaluation in System Development*. Wiley, New York.

Miles, L. D. (1961). *Techniques of Value Analysis and Engineering*. McGraw-Hill, New York.

Morgan, C. T., Cook, J. S. III, Chapanis, A., and Lund, M. W. (eds.) (1963). *Human Engineering Guide to Equipment Design*. McGraw-Hill, New York.

Nadler, G. (1967). An investigation of design methodology. *Manage. Sci.*, 13, B642–B655.

Nadler, G. (1967). *Work Systems Design: The IDEALS Concept*. Irwin, Hammond, IL.

O'Keefe, J. K. (1964). An introduction to systems analysis. *J. Ind. Eng.*, 163–167.

Rechtin, E. (1968). Systems engineering – But isn't that what I've been doing all along? *Astron. Aeron.*, 6, 70–74.

Rudwick, B. H.(1969). *Systems Analysis for Effective Planning: Principles and Cases*. Wiley, New York.

Sargent, K. N. (1966). Insight into SEEing: A discussion of the philosophy underlying systems effectiveness engineering. *IEEE Trans. Aerosp. Electron. Syst.*, AES-2, 506–510.

Savas, E. S. (1965). *Computer Control of Industrial Processes*. McGraw Hill, New York.

Shinners, S. M. (1967). *Techniques of Systems Engineering*. McGraw-Hill, New York.

Starr, M. K. (1963). *Product Design and Decision Theory*. Prentice Hall, Englewood Cliffs, NJ.

Teasdale, A. R. (1966). Methodology of modeling. *Electra Tech*, 78, 65–74.

Timson, F. S. (1968). *Measurement of Technical Performance in Weapon System Development Programs: A Subjective Probability Approach*. The Rand Corporation, Memorandum RM-5207-ARPA.

Von Bertalanffy, L. (1968). *General Systems Theory.* George Braziller Press, New York.

Walton, T. (1963). *Technical Data Requirements in Systems Engineering.* Wiley, New York.

Warfield, J. N. (1956). Systems Engineering, United States Department of Commerce PB111801.

Williams, T. J. (1961). *Systems Engineering in the Process Industries.* McGraw Hill, New York.

Wilson, W. E. (1965). *Concepts of Engineering System Design.* McGraw-Hill, New York.

Wymore, A. W. (1967). *A Mathematical Theory of Systems Engineering: The Elements*, Wiley, New York.

Index

Note: a page number in *italics* (e.g., *84*) indicates the page reference is to a figure; Similarly, a page number in **bold** (e.g., **132**) indicates the page references a table.

The Engineering Design of Systems: Models and Methods, Fourth Edition. Dennis M. Buede and William D. Miller
© 2024 John Wiley & Sons, Inc. Published 2024 by John Wiley & Sons, Inc.
Companion website: www.wiley.com/go/engineeringdesignofsystems4e